S0-CAH-396

CONTEMPORARY THEORY OF CHEMICAL ISOMERISM

UNDERSTANDING CHEMICAL REACTIVITY

Contemporary Theory of Chemical Isomerism

ZDENĚK SLANINA

Czechoslovak Academy of Sciences,
The J. Heyrovský Institute of Physical Chemistry
and Electrochemistry, Prague

D. Reidel Publishing Company

A MEMBER OF THE KLUWER ACADEMIC PUBLISHERS GROUP

Dordrecht / Boston / Lancaster / Tokyo

Library of Congress Cataloging-in-Publication Data CIP

Slanina, Zdenek, 1948–
Contemporary theory of chemical isomerism.
(Understanding Chemical Reactivity)
Translation of: Teoretické aspekty fenoménu chemické isomerie.
Bibliography: p.
Includes index.
1. Isomerism. I. Title.
QD471.S5713 1986 541.2'252 85-14629
ISBN 90-277-1707-9 (Reidel)

Published by D. Reidel Publishing Company, P. O. Box 17, 3300 AA Dordrecht, Holland in co-edition with Academia, Publishing House of the Czechoslovak Academy of Sciences, Prague, Czechoslovakia.

Sold and distributed in the U.S.A. and Canada by
Kluwer Academic Publishers, 190 Old Derby Street, Hingham, MA 02043, U.S.A.

Sold and distributed in Albania, Bulgaria, China, Czechoslovakia, Cuba, German Democratic Republic, Hungary, Mongolia, Northern Korea, Poland, Rumania, U.S.S.R., Vietnam, and Yugoslavia by
Academia, Publishing House of the Czechoslovak Academy of Sciences, Prague, Czechoslovakia

Sold and distributed in all remaining countries by
Kluwer Academic Publishers Group, P.O.Box 322, 3300 AH Dordrecht, Holland

Original title: Teoretické aspekty fenoménu chemické isomerie

Scientific Editor
Prof. Ing. Ladislav Valko, DrSc.
Reviewer
Prof. Ing. Vladimír Kvasnička, DrSc.

Printed in Czechoslovakia

CONTENTS

Preface . . . VII

1 Introduction . . . 1
References . . . 6

2 Basic Concepts in Chemical Isomerism . . . 8
2.1 Definition and Chemical Significance of the Concept of Isomerism . . . 8
2.2 Classification of Isomerism . . . 10
2.3 Stereochemical Non-rigidity of Molecules . . . 21
2.4 Summary . . . 23
References . . . 24

3 Quantum-Chemical Description of Isomerism . . . 29
3.1 The Born-Oppenheimer Approximation — The Concept of a Potential Energy Hypersurface . . . 29
3.1.1 The Concept of Molecular Structure . . . 35
3.2 Molecular Energy . . . 44
3.3 Localization and Identification of Stationary Points on the Energy Hypersurfaces . 46
3.4 Curves on the Potential Energy Hypersurfaces . . . 54
3.5 Calculation of the Potential Energy Hypersurfaces . . . 62
3.6 Vibrational Motions of the Atomic Nuclei . . . 70
3.7 Summary . . . 75
References . . . 77

4 Quantum Mechanics and Isomerism . . . 91
4.1 Criticism and Generalization of the Concept of Molecular Structure . . . 91
4.2 The Molecular Wave Function and Isomerism . . . 101
4.3 Summary . . . 108
References . . . 108

5 Algebraic Aspects of Isomerism . . . 111
5.1 Algebraic Generalization of the Concept of Isomerism . . . 112
5.1.1 The Equivalence Relation . . . 112
5.1.2 Isomeric Ensembles of Molecules . . . 113
5.2 Computer-assisted Design of Syntheses . . . 116
5.3 Enumeration of Isomeric Structures . . . 118
5.3.1 Pólya's Theorem . . . 118
5.3.2 Iterative Enumeration . . . 125

5.3.3 Asymptotic Enumeration Formulae 127
5.3.4 Enumeration for Non-rigid Molecules 127
5.3.5 The Graph-like State of Molecules 129
5.3.6 The Importance of Enumeration for the Theory of Chemical Reactivity . . . 132
5.4 Topological Reduction and Characterization of Hypersurfaces 133
5.4.1 Reduction of Hypersurfaces to Tree-type Graphs 133
5.4.2 Topologization of Nuclear Configuration Space 137
5.4.3 Bounds for the Number of Stationary Points on Energy Hypersurfaces . . . 138
5.5 Symmetry Properties of Stationary Points 140
5.5.1 Symmetry Properties of Rigid and Non-rigid Molecules 141
5.5.2 Transition States and Symmetry 144
5.6 Summary 148
References 149

6 Isomerism and the Theory of Chemical Reactivity 157
6.1 Calculation of the Characteristics of Equilibrium and Rate Processes 158
6.1.1 Calculation of Equilibrium Thermodynamic Characteristics 159
6.1.2 Calculation of Rate Characteristics in the Framework of the Activated Complex Theory 172
6.1.3 Examples of Quantum-Chemical Calculations of Equilibrium and Rate Constants 177
6.2 Equilibrium Processes with Isomerism of the Components 181
6.2.1 A General Equilibrium Problem 184
6.2.2 Formulation at a Microscopic Level 190
6.2.3 Limiting Properties of Summary Characteristics 191
6.2.4 Examples of Quantum-Chemical Study of Chemical Equilibria with Isomerism of a Reaction Component 192
6.2.5 Inversion (Deconvolution) Equilibrium Problem 195
6.3 Rate Processes with Parallel Isomerism of the Activated Complex 196
6.4 Rate Processes with Sequential Isomerism of the Activated Complex 202
6.5 Isomerism of Multiparticle Clusters 213
6.5.1 Examples of Theoretical Study of Gas-Phase Complexes 213
6.5.2 Clusters and Real Gases 221
6.5.3 The Cluster Concept of the Liquid State 222
6.5.4 Isomerism of Adsorption Complexes 224
6.6 Summary 227
References 228

7 Résumé 243

8 Postscript 245

References 245

Subject Index 250

PREFACE

This rather limited study is an attempt to provide a survey of the theoretical approaches to the study of chemical isomerism and to summarize the results obtained using these methods. The work is based partly on the previous author's experience with an earlier pamphlet on a similar subject published by Academia in 1981 in Czech (and in a Russian translation by Mir Publishers in 1984); however, the treatment is broader by a factor of about two and completely modernized. Nonetheless, even this broader study can hardly be considered as an exhaustive, unbiased collection of the results and facts so far obtained in this field. This is partly due to the fact that the size of the text was limited to about 300 typed pages and, primarily, because the common approach of the work reflecting the author's own research interests has been adopted. Nonetheless, I hope that this study will provide a reasonably accurate picture of the state of the problem at the beginning of the 1980s. This work appears in connection with the research that the author is carrying out at the J. Heyrovský Institute of Physical Chemistry and Electrochemistry of the Czechoslovak Academy of Sciences in Prague and thus the author wishes to thank Academician A. A. Vlček and Dr. J. Vojtík for their constant interest, support and kind words of encouragement.

The final form of the work passed through several stages and the author is very thankful to all those whose constructive criticism and discussion contributed to this study. At the various stages of completion of the work these especially included the friends and/or colleagues, Profs. and/or Drs. S. Beran, E. Dalgaard, S. Fraga, D. Heidrich, B. F. Minaev, D. Papoušek, K. Rasmussen, M. M. Szczęśniak, J. Vojtík, R. G. Woolley, R. Zahradník and others, as well as many (unknown) reviewers of the author's works published in specialized journals. Special thanks are due to Profs. V. Kvasnička and L. Valko, whose numerous critical comments helped to eliminate many of the shortcomings of the manuscript as a whole.

Although the bibliography includes more than 1500 citations, this alone does not ensure its completeness, or even the inclusion of all the important works. Many colleagues assisted by sending preprints and reprints of their works or by providing unpublished results. In this connection the author would wish to thank especially Profs. R. F. W. Bader, A. T. Balaban, K. Balasubramanian, R. S. Berry, O. P. Charkin, J. R. de la Vega, P. G. Mezey, J. N. Murrell, L. Radom, M. Randić, P. v. R. Schleyer, N. Trinajstić, I. Ugi, A. van der Avoird, R. G. Woolley, and others.

Important results and works appearing or collected in the period between completion of the manuscipt (November 1982) and reading of the proofs will be condensed and summarized in the last chapter – Postscript.

The author wishes to thank Dr. M. Štulíková for the great care with which she translated the Czech text into English. Thanks are also due to the late Ing. E. Caltová for typing the Czech manuscript and to the workers of the Publishing Houses for their constant interest and care that they devoted to the production of the final form of this book.

The author also wishes to thank the following organizations for kindly permitting the reprinting of copyright material: Academia, The American Chemical Society, The American Institute of Physics, The Chemical Society, Elsevier Scientific Publishing Company, J. Wiley and Sons, Inc., McGraw-Hill, Inc., The National Academy of Sciences, Nauka, North-Holland Publishing Company, Pergamon Press Ltd., Plenum Press, Springer-Verlag, Verlag Chemie, W. A. Benjamin, Inc., The Weizmann Science Press.

Prague, November 1982 Z. S.

1 Introduction

Theoretical chemistry is a part of all fields of chemistry – independent theoretical branches of the individual basic chemical disciplines are gradually formed. Theoretical inorganic and organic chemistry, theoretical biochemistry and pharmacology have become matter-of-course concepts. Theoretical chemistry is applied in individual areas through the methods of quantum chemistry yielding numerical data for testing particular mathematical models of experimentally observed phenomena. The numerical solution of the Schrödinger equation has become the most common method in the search for relationships between a given chemical structure and its properties. The rapid development of numerical quantum chemistry is, however, based primarily on the enormous development of computer technology. The basic methodological approach of quantum chemistry has been completed for decades and one of the founders of modern theoretical chemistry, Wilson[1], recently stated that relatively few really new ideas have appeared in the last twenty years. In spite of the approximative nature of the numerical results of quantum chemistry, the use of this method has become a standard, and in some cases even a justifiable complement to experimentally obtained microscopic information concerning the problem under study. The time factor is gradually included in quantum chemical studies, permitting a transition between static and dynamic descriptions. A second factor that is being included with ever increasing frequency is temperature, connected with a transition from description in terms of the potential energy (corresponding to absolute zero temperature) to consideration of the problem in terms of the Gibbs energy. This is reflected (in contrast to the earlier distinct separation during the last two or three decades) in the contemporary reapproach or even linking of quantum chemistry and statistical mechanics. Simultaneously, quantum chemistry itself has been characterized by an attempt to free it from a number of classical concepts that had so far been considered as indispensable, intrinsic quantum chemical approximations. If this process is carried out completely and successfully, then it could lead to a rebirth of quantum chemistry as the quantum mechanics of atoms and molecules rid of all classical elements (which should be connected with a simultaneous fusion with the theory of molecular spectroscopy, which is also approaching a point necessary for the attainment of a new character free of some conventional concepts).

In addition to the numerical side of the subject, based on classical mathematics, especially analysis, contemporary theoretical chemistry has another aspect which is connected with the attempt to utilize mathematics as a theory of logical structures to obtain direct insight into the intrinsic logical structure behind chemical problems without a numerical intermediate step. For this purpose, algebraic or mathematical chemistry became a new field.

Modern algebra is no longer a set of rules for the handling of sums or products of real or complex numbers. In the 1920s, systematic use of abstract and axiomatic methods become important. At present, algebra works with very generally introduced sums and products of any kinds of elements. As in other branches of mathematics, algebra is based on set theory: modern mathematics studies sets and set mapping, especially mapping considering particular given information – structures. It has been found that individual mathematical theories considered in this way have a common characteristic. The theory of mathematical structures can be used for a more complete understanding of the relationships between the individual fields of mathematics. This new concept of mathematics is described systematically in the works of a group of French mathematicians published under the pseudonym N. Bourbaki (see e.g.[2]). This process of generalization and unification occurring in the field of mathematics simultaneously universalizes its use in other branches of science. Mathematics based on the concepts of algebraic structures, ordered sets and topological spaces is used widely in areas where classical mathematics was not applicable. At present, mathematics yields a general formalism for relationships in sets based on complete abstraction from the nature of the elements forming the sets. On the other hand, most sciences study the relationships between objects rather than the objects themsleves. This fact leads to the potential usefulness of abstract mathematical structures in the study of the characteristic properties of the real world. Thus the tools of theoretical chemistry have gradually been expanded to include non-numerical methods such as group theory formalism, graph theory, topology or information theory.

The methods of quantum and algebraic chemistry can be usefully combined in the study of concrete chemical problems. The problem of isomerism is a good example of a chemical phenomenon that was elucidated by combination of these two theoretical approaches. Although the first application[3] of mathematical means for study of isomerism is more than a hundred years old, at present this region is still full of open problems and is the subject of considerable interest.

Although the term isomerism was originally introduced for chemical species by Berzelius[4] in 1830, this is a phenomenon with far more general applications. This fact was originally anticipated in an adumbration given by ancient Greek atomists in the 5th century B. C. – in his Metaphysics, Aristotle[5] states that "Leucippus however, and his disciple Democritus hold ... that the 'differences' are the causes of everything else. These differences, they say, are three: shape, arrangement and position; because they hold that what is differs only in contour, inter-contact and inclination. (Of these, contour means shape, inter-contact means arrangement, and inclination means position.) Thus, for example, A differs from N in shape, AN from

NA in arrangement, and Z from N in position".* The phenomenon of isomerism has also been the subject of philosophical interest in more modern times. In 1879, Engels[6] interpreted it as a transition from quantity to quality. Contemporary results in physics and biology have actually indicated that chemical isomerism is part of a more general phenomenon.

It has long been known in the field of nuclear physics that two or more forms of a given atomic nucleus with the same atomic and mass numbers can exist, differing in their properties. It has been shown that this atomic nucleus isomerism (see e.g.[7–9]) is connected with metastable states – excited states with a measureable lifetime varying within a very broad range. In the theoretical description of this phenomenon the concept of isomerism of the atomic nucleus shape was introduced[7–9], connected with the existence of at least two equilibrium positions on the curve of the dependence of the potential energy on the nucleus deformation parameter. Both potential energy minima – corresponding to the nuclear form of an elongated and, in the other case, to a flattened rotational ellipsoid – are separated by a potential energy barrier. The potential energy minima are very pronounced for some isotopes. For other isotopes, two minima are found with similar energies separated by a very low barrier. The limiting case is then a nucleus vibrating with a large vibrational amplitude around the spherical form. A broad analogy between atomic nucleus isomerism and chemical isomerism is clearly apparent from these facts. The existence of atomic nucleus isomerism implies the possibility of the formal generalization of the concept of molecular isomerism, based on replacement of at least one atomic structural centre in the molecule by an isomeric nucleus. Then, for example, molecule X_2, whose nucleus X could be one of two isomers, could be formally considered as a mixture of three isomers.

The phenomenon of isomerism has been described not only for lower but also for higher forms of organization of matter than chemical compounds, this is termed biological isomerism (see e.g.[10–12]). Living systems containing identical components and differing in the relative positions of these components in space or in the relative positions and numbers of these components are called biological isomers. Components are understood to be supermolecular species fulfilling defined physiological functions and structurally connected. Enantiomorphism is a frequent manifestation of biological isomerism; this is the existence of living systems with left-handed or right-handed asymmetry. The first observation of this phenomenon was probably the description of left- and right-handedness in barley. Fig. 1-1 depicts enantiomorphism in tobacco. The discovery of enantiomorphism in microorganisms was very important. Attempts to find the reasons for biological isomerism in molecular structures or stereochemical specificity of protoplasm have not yet found direct confirmation, although genetic control of all types of bioisomerism has been demonstrated. The hypothesis of enantiomorphism being the result of the stereochemically antithetical

---

* However, cf. note 4 on p. 30 in ref.[5].

character of molecules of compounds entering the structures of living organisms is based on factual material; nonetheless, all these proofs are to date indirect.

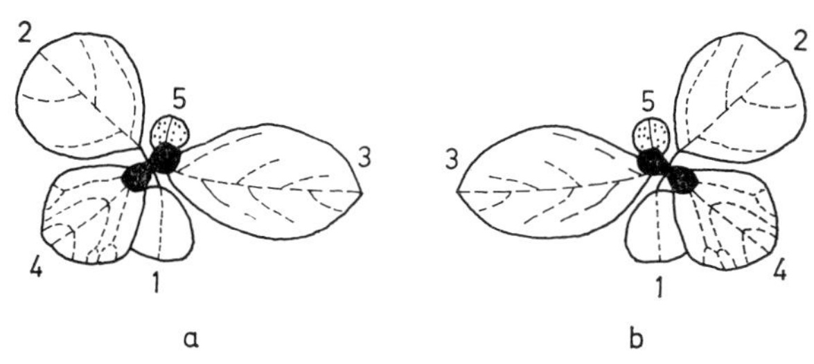

Figure 1-1. An example of biological isomerism — enantiomorphism of tobacco; the numbers indicate the order of leaf budding (from Ref.[10])

The above discussion indicates that the phenomenon of isomerism is a broad concept, with chemical isomerism bordering on atomic nucleus isomerism and bioisomerism. So far a unifying concept is lacking in this broad spectrum of phenomena, although some partial relationships are becoming clear. An example is Ulbricht's[13] discussion of possible relationships between asymmetry at the level of elementary particles and at a molecular level; more recently, Goldanskii[14] has considered the need for this unifying concept. The optical asymmetry of biological molecules[11] and its origins is one of the most important problems in the concept of chemical evolution. It is very well known that amino acids in natural proteins belong overwhelmingly to the L-series, while natural sugars are constructed almost entirely of the D optical isomers. Ulbricht[13] proposed a new mechanism for the explanation of the reasons for this asymmetry, relating the phenomenon of asymmetry in biomolecules with parity nonconservation in weak interactions at the level of elementary particles. For example, it has been postulated[13] that an explanation could be based on longitudinally polarized β-decay electrons that generate circularly polarized Bremsstrahlung on interaction with matter that could lead to photochemical asymmetric synthesis. So far, however, a convincing experimental proof of this hypothesis is lacking[15]. Recently, it has also been pointed out[16,17] that these weak interactions may contribute (in more complicated systems) to the stability of the optical isomers themselves. For the sake of completeness it should be noted that an attempt[18] was recently made to explain the problem of asymmetry of biomolecules on the basis of a completely different physical principle, viz. the effect on prochiral reactions by the gravitational or other chiral physical field[19]. This concept was, however, immediately refuted[20–22]. These broad relationships, including a wide range of objects studied by modern natural science, emphasize the need for deeper study of the phenomenon of chemical isomerism itself.

As far as the author knows, no monograph exists at present that is devoted to chemical isomerism alone. Although some specific types of chemical isomerism have been treated in monographs, these deal primarily with stereoisomerism (see, for example, [23-27,40,41]) and partially with valence isomerism[28], which became more important after the formulation of the Woodward-Hoffmann rules[29]. Selected types of processes with isomer participation – for example isomerization[30-33] or the reactions of optically active compounds[34] – have also been treated in monographs. In most of these studies theoretical considerations are treated only briefly.

The above text describes the changes occurring in contemporary theoretical chemical concepts, such as the complementary roles of quantum chemistry and quantum mechanics, classical mechanics in part, statistical thermodynamics and algebraic chemistry. In the study of a concrete chemical problem, usually one of these techniques plays a predominant role; in less frequent circumstances a combination of these methods is employed for simultaneous study of a particular problem. Nonetheless, anticipating future developments, Löwdin[35] recently described a unified combination of these procedures called trace algebra; the work of Primas[36,37] has also been an important contribution to the unification of chemical theory. In the present work, the broad concept of theoretical chemistry will also be retained.

This short study was carried out in order to consider actual theoretical problems relating to chemical isomerism and to summarize the most important results. For this purpose, the methods of quantum chemistry and quantum mechanics, statistical mechanics, and algebraic chemistry will be used. Although the final purpose of any theory is a deeper understanding and interpretation of observed phenomena, a number of concepts and quantities must often be introduced for the sake of the completeness of the theoretical framework, that (either temporarily or in principle) cannot be studied experimentally. From the point of view of the observations and measurements themselves, this conceptual superstructure may seem superfluous; thus we would like to recall Einstein's ideas as communicated by Heisenberg[38]: "It is never possible to introduce only observable quantities in a theory. It is the theory which decides what can be observed." However, for a more complete picture of Einstein's approach to the relationship between theory and experiment, this point of view as expressed in the Spencer lecture should also be included[39]: "Pure logical thinking cannot yield us any knowledge of the empirical world; all knowledge of reality starts from experience and ends in it. Propositions arrived at by purely logical means are completely empty as regards reality." A particular illustration of these statements is the evolution of the contemporary concept of isomerism as described in the next chapter.

Special attention will be paid to the following four problem regions: the relationship of the potential energy hypersurface as a central concept in the modern theory of chemical reactivity to the phenomenon of isomerism, the picture of isomerism in the rigorous quantum mechanical approach, the algebraic description of isomerism and, finally, the implications of isomerism of the components of chemical reactions for the calculation and interpretation of the characteristics of equilibrium and rate processes. The size of this work does not permit discussion of all the relevant related problems or the consideration of all the concepts discussed to the same degree.

It was also necessary to limit the number of illustrative examples, especially in the chapters on isomer enumeration and isomerism of the components of chemical processes. Naturally, it was possible to consider more standard methodical questions in relationship to quantum chemistry, vibrational analysis, statistical thermodynamics and chemical dynamics only orientatively. Consequently, references are given wherever the particular subject is reviewed in the literature.

REFERENCES

(1) Wilson, E. B.: 1976, Pure Appl. Chem. 47, pp. 41—47.
(2) Bourbaki, N.: 1966, Elements of Mathematics. General Topology, Addison—Wesley, Don Mills, Ontario.
(3) Cayley, A.: 1874, Phil. Mag. *47*, pp. 444—446.
(4) Berzelius, J. J.: 1830, Ann. Phys. Chem. *19* (der ganzen Folge 95), pp. 305—335.
(5) Aristotle: 1933, The Metaphysics (with an English translation by H. Tredennick), W. Heinemann Ltd., London, paragraph 985^b.
(6) Engels, F.: 1946, Dialectics of Nature, Lawrence and Wishart, London, p. 32.
(7) Polikanov, S. M.: 1972, Usp. Fiz. Nauk 107, pp. 685—704.
(8) Polikanov, S. M.: 1977, Isomerism of Atomic Nucleus Shape, Atomizdat, Moscow, (in Russian).
(9) Grinberg, A. P.: 1980, Usp. Fiz. Nauk *132*, pp. 663—678.
(10) Kasinov, V. B.: 1973, Biological Isomerism, Nauka, Leningrad, (in Russian).
(11) Galaktionov, S. G.: 1978, Asymmetry of Biological Molecules, Vysheishaya Shkola, Minsk, (in Russian).
(12) Kizel, V. A.: 1980, Usp. Fiz. Nauk *131*, pp. 209—238.
(13) Ulbricht, T. L. V.: 1959, Quart. Rev., Chem. Soc. *13*, pp. 48—60.
(14) Goldanskii, V. I.: 1976, Usp. Fiz. Nauk *118*, pp. 325—338.
(15) Jean, Y., and Ache, H. J.: 1977, J. Phys. Chem. *81*, pp. 1157—1162.
(16) Harris, R. A., and Stodolsky, L.: 1978, Phys. Lett. *78 B*, pp. 313—317.
(17) Harris, R. A., and Stodolsky, L.: 1981, J. Chem. Phys. *74*, pp. 2145—2155.
(18) Dougherty, R. C.: 1980, J. Am. Chem. Soc. *102*, pp. 380—381.
(19) Edwards, D., Cooper, K., and Dougherty, R. C.: 1980, J. Am. Chem. Soc. *102*, pp. 381—382.
(20) Mead, C. A., and Moscowitz, A.: 1980, J. Am. Chem. Soc. *102*, pp. 7301—7302.
(21) Peres, A.: 1980, J. Am. Chem. Soc. *102*, pp. 7389—7390.
(22) Edwards, D., Cooper, K., and Dougherty, R. C.: 1980, J. Am. Chem. Soc. *102*, p. 7618.
(23) Eliel, E. L.: 1962, Stereochemistry of Carbon Compounds, McGraw-Hill, New York.
(24) Mislow, K.: 1966, Introduction to Stereochemistry, W. A. Benjamin, New York.
(25) 1971, Conformational Analysis. Scope and Present Limitations, Ed. G. Chiurdoglu, Academic Press, New York.
(26) Sokolov, V. I.: 1979, Introduction to Theoretical Stereochemistry, Nauka, Moscow, (in Russian).
(27) Nógrádi, M.: 1981, Stereochemistry. Basic Concepts and Applications, Akadémiai Kiadó, Budapest.
(28) Maier, G.: 1972, Valenzisomerisierungen, Verlag Chemie, Weinheim.
(29) Woodward, R. B., and Hoffmann, R.: 1970, Die Erhaltung der Orbitalsymmetrie, Verlag Chemie, Weinheim.

(30) CONDON, F. E.: 1958, in Catalysis, Vol. VI, Ed. P. H. Emmett, Reinhold Publ. Comp., New York, pp. 43–189.
(31) KOPTYUG, V. A.: 1963, Isomerization of Aromatic Compounds, Ed. N. N. Vorozhtsov, Jr., Sib. Branch Acad. Sci. USSR, Novosibirsk, (in Russian); 1965, Oldbourne Press, London.
(32) KOCHLOEFL, K., and KRAUS, M.: 1967, Isomerization and Degradation, Academia, Prague, (in Czech).
(33) GAJEWSKI, J. J.: 1981, Hydrocarbon Thermal Isomerizations, Academic Press, New York.
(34) MORRISON, J. D., and MOSHER, H. S.: 1971, Asymmetric Organic Reactions, Prentice-Hall, Englewood Clifs, New Jersey.
(35) LÖWDIN, P.-O.: 1977, Int. J. Quantum Chem. *12*, Suppl. 1, pp. 197–266.
(36) PRIMAS, H.: 1980, in Quantum Dynamics of Molecules, Ed. R. G. Woolley, Plenum Press, New York, pp. 39–113.
(37) PRIMAS, H.: 1981, Chemistry, Quantum Mechanics and Reductionism, Springer-Verlag, Berlin.
(38) HEISENBERG, W.: 1974, Phys. Bull. *25*, pp. 231–235.
(39) EINSTEIN, A.: 1933, On the Method of Theoretical Physics, Clarendon Press, Oxford.
(40) DASHEVSKII, V. G.: 1974, Conformations of Organic Molecules, Khimiya, Moscow, (in Russian).
(41) DASHEVSKII, V. G.: 1982, Conformational Analysis of Organic Molecules, Khimiya, Moscow, (in Russian).

2 Basic Concepts in Chemical Isomerism

2.1 Definition and Chemical Significance of the Concept of Isomerism

In the first few decades of the nineteenth century, experimental discoveries of identical elemental compositions for substances with apparently different physical and chemical properties began to appear. These surprising findings indicated the existence of a new phenomenon to chemistry. Table 2-1 gives a survey of the most important observations made at that time[1,2]. It is apparent in the light of modern molecular formulae that identical elemental compositions in the first works stemmed from experimental imprecision. If interference from the concept of polymerism is excluded, the first isomeric compounds described were *a*- and *b*-stannic acid[3,4] and silver fulminate and cyanate[5-7]. The evident difference in the properties of silver fulminate and cyanate led to discussions[8] in which the authors publishing the composition of fulminate[5,6] (Gay-Lussac and Liebig) and cyanate[7] (Wöhler) accused each other of incorrectly carried out analyses. In a note on Wöhler's work[7], Gay-Lussac[8] suggests an explanation in terms of various combinations of elements. It is interesting that the problem of the structure of fulminic acid has not yet been satisfactorily solved[9]. In 1828 Wöhler carried out the conversion[10,11] of ammonium cyanate to urea, causing the death of the doctrine of a vital force and contributing greatly to the crystallization of the concept of substances with the same elemental compositions and different properties. This is reflected in Wöhler's note in a letter to Berzelius dated February 22, 1828: "Dies wäre also ein unbestreitbares Beispiel, das zwei ganz verschiedene Körper dieselbe Proportion von denselben Elementen enthalten können, und dass nur die ungleiche Art der Vereinigung die Verschiedenartigkeit in den Eigenschaften hervorbringt". However, as late as 1830 Berzelius used the criterion of molecular mass to separate phenomena belonging to the field of polymerism and introduced the new term* isomerism for compounds with identical chemical composition and identical molecular mass with different properties. Deeper understanding of the structure of isomeric compounds was provided (see, for example, the discussion in[14]) by Butlerov in the formulation of the basic principles of the modern structural theory

* It is interesting that, in the selection of the designation for compounds with these properties, Berzelius chose between the names isomeric and homosynthetic, and that the final name was selected[13] on the basis of its brevity and pleasant sound.

Table 2-1. The most important indications[a] preceding the description of the phenomenon of chemical isomerism by Berzelius[13] in 1830

Year(s)	Author(s)	Species
1807	Thomson	acetic acid $C_2H_4O_2$ succinic acid $C_4H_6O_4$
1811	Dalton	albumin gelatin
1811	Gay-Lussac	saccharose $C_{12}H_{22}O_{11}$ starch $(C_6H_{10}O_5)_n$ gum arabic
1816/17	Gay-Lussac[3]; Berzelius[4]	a — stannic acid b — stannic acid $SnO_2.n\,H_2O$
1823/24	Liebig[5], Liebig and Gay-Lussac[6] Wöhler[7]	silver fulminate AgCNO silver cyanate AgNCO
1825	Faraday	ethylene C_2H_4 butene C_4H_8
1828	Wöhler[10,11]	ammonium cyanate NH_4NCO urea CH_4N_2O
1830	Berzelius[13]	tartaric acid uvic acid $C_4H_6O_6$

[a] According to Refs.[1,2]

of organic chemistry and was refined[15-17] by Pasteur, Kekulé, van't Hoff* and LeBel in the development of the theory of the three-dimensional structure of molecules, and by Cayley and Körner in their pioneering studies of enumeration problems. Nonetheless, the actual definition of the phenomenon of chemical isomerism has undergone only a single change since the time of Berzelius' first version, this change was connected with the necessity of excluding relatively unstable or fictive structures (mesomers) from this concept. According to the definition proposed by Rouvray[17], isomers are now considered to be individual chemical species with an identical molecular formula which display at least some differing physico-chemical properties and which are stable for periods of time that are long in comparison with those during which measurements of their properties are made.

* Interesting information given in the works of Bláha[15] and Riddell and Robinson[16] indicates that this was not a simple process. For illustrative purposes two sentences will be quoted from the criticism of the idea of an asymmetrical carbon atom and the tetrahedral arrangement of substituents around it, incurred from one of the leaders of chemistry at the time: "Ein Dr. J. H. van't Hoff, an der Thierarzneischule zu Utrecht angestellt, findet, wie es schient, an exakter chemischer Forschung keinen Geschmack. Er hat es bequemer erachtet, den Pegasus zu beisteigen, offenbar der Thierarzneischule entlehnt und in seiner "La chimie dans l'espace" zu verkünden, wie ihm auf dem durch kühnen Flug erklommenen chemischen Parnass die Atome in Weltenraume gelagert erschienen sind".

2.2 Classification of Isomerism

The literature contains a number[18-29] of proposed classification schemes for isomerism. The difficulty of the problem depends both on the great number of individual forms of isomerism (more than thirty types have been recognized[17]) as well as on the fact that differentiation into the individual forms is frequently not exclusive – the individual concepts overlap considerably. The basic separation of isomerism into structural isomerism and stereoisomerism is generally accepted. Structural isomers (also called constitutional isomers[30]) differ in their structures, i.e. in the manner of bonding the atoms in the molecule[31,32]. Stereoisomers have identical structures but differ in configuration or conformation, i.e. in spatial atomic arrangement[31].

The field of structural isomerism is often divided into chain isomerism, place isomerism and functional group isomerism. Chain (or skeletal) isomerism is that form of structural isomerism in which two or more substances with the same composition differ in the manner of bonding of the atoms forming the main structure of the substance. This type of isomerism is considered most often for hydrocarbons with open chains. *n*-Butane and isobutane are textbook examples of chain isomerism. Place isomers have an identical skeleton with one or more different substituent positions (atoms or groups). Examples of such isomers with aliphatic skeletons are *n*-propylchloride and isopropylchloride; examples with cyclic skeletons include pyrocatechol, resorcinol and hydroquinone. Structural isomers differing in their functional groups form the class of functional group isomers; textbook examples are ethyl alcohol and methyl ether. Lunn and Senior[19] proposed alternative schemes for classsification of structural isomers; using the concepts of a skeleton and univalent substituents attached to it, they distinguished between skeletal and substituent isomers. A disadvantage of this system is the fact that it need not always lead to unambiguous assignment[31]. However, all the systems so far proposed for classification of structural isomers are characterized[31] by a certain awkwardness and by the fact that they are frequently unsuccessful in practice.

A special case of functional group isomerism is tautomerism (also called desmotropy, dynamic isomerism, kryptomerism, pseudomerism, etc.). The term tautomerism (see e.g.[33-37]) is used for isomers that differ in their functional groups and that are readily interconverted as a result of intramolecular rearrangement, leading to the coexistence in an equilibrium mixture. Table 2-2 gives a useful scheme for tautomerism classification, given by Jennen[26]. The concept of tautomerism expressing a certain structural non-uniformity, based on rapid and reversible motions of the nuclei in the molecule permitting transitions between the various arrangements is in principle different from the concept of resonance, where the individual structures with stable nucleus positions differ only in their electron distribution. If tautomerism is caused by the transfer of a hydrogen atom or proton, then it is termed prototropy; if caused by the transfer of an anion it is termed anionotropy. The terms dyadic, triadic and ring-chain tautomerism are related to the topology of the atoms among

Table 2-2. The Jennen method[26] of tautomerism classification

Tautomerism	Prototropy	Dyadic prototropy Triadic prototropy Ring-chain tautomerism
	Anionotropy	Ring-chain tautomerism Allyl-group tautomerism
	Geometrical tautomerism	

which the transfers occur. The most widespread type of tautomerism is triadic prototropy, connected with the transfer of hydrogen according to the general scheme (2-1):

$$\underset{\mathrm{H}}{\underset{|}{\mathrm{X}}}-\mathrm{Y}=\mathrm{Z} \rightleftarrows \mathrm{X}=\mathrm{Y}-\underset{\mathrm{H}}{\underset{|}{\mathrm{Z}}}, \qquad (2\text{-}1)$$

in which X, Y and Z are, in general, multivalent atoms. The most important types of triadic prototropy are given in Table 2-3. Pentadic tautomeric systems have also been described, e.g.[39]:

$$-\underset{|}{\mathrm{C}}=\underset{|}{\mathrm{C}}-\underset{|}{\mathrm{C}}=\underset{|}{\mathrm{C}}-\mathrm{OH} \rightleftarrows -\overset{|}{\underset{\mathrm{H}}{\underset{|}{\mathrm{C}}}}-\underset{|}{\mathrm{C}}=\underset{|}{\mathrm{C}}-\underset{|}{\mathrm{C}}=\mathrm{O}\,. \qquad (2\text{-}2)$$

The concept of tautomerism overlaps partially with the concept of metamerism, which is sometimes considered to be a separate type of structural isomerism. Then metamerism is a type of isomerism connected with a difference in the position of one or more atoms or groups or double bonds. Another type of isomerism that overlaps partially with tautomerism is valence isomerism, based on structures differing in pairs of bonded atoms. In addition to a different bond distribution, valence isomers must have different bond lengths and angles. Otherwise, the pairs of compounds would be only resonance hybrids. Instead of the term valence isomerism, Maier[40] suggested the term bond isomerism because of the fact that the total number of bonds is retained in this type of isomerism. A textbook example is benzene with six possible valence isomers[41]: benzene itself, Dewar benzene, benzvalene, bicyclopropenyl, benzprismane, and benzMöbiusstripane (in theory a total of 217 isomers can exist[42] with the empirical formula C_6H_6). If a tautomer-type equilibrium is formed among valence isomers, then it is called valence tautomerism[31]. Classification of valence isomers as a separate class[40] became useful when the Woodward-Hoffmann rules rationalized the study of valence isomerization.

Several types of isomerism have been described for coordination compounds[43–45] that can be formally considered as structural isomerism: ionic, hydrate, coordination, ligand, and linkage isomerism. The first three types of isomerism are connected

Table 2-3 The most important types of triad prototropic tautomeric systems $HX-Y=Z \rightleftarrows X=Y-ZH$[a]

Tautomerism type	X Y Z	Scheme of tautomeric equilibrium
Keto-enol	C C O	$-CH_2-C(=O)- \rightleftarrows -CH=C(OH)-$
Nitro-isonitro	C N O	$>CH-N(=O)(O) \rightleftarrows >C=N(OH)(O)$
Nitroso-isonitroso (oximino)	C N O	$>CH-N=O \rightleftarrows >C=N-OH$
Lactam-lactim Amide-imidol	N C O	$-NH-C(=O)- \rightleftarrows -N=C(OH)-$
Imino-enamine	C C N	$>CH-C(-)=NH \rightleftarrows >C=C(-)-NH_2$
Nitrile-enimine	C C N	$>CH-C\equiv N \rightleftarrows >C=C=NH$
Methylene-azomethine	C N C	$>CH-N=C< \rightleftarrows >C=N-CH<$
Amidine	N C N	$-N=C(-)-NH_2 \rightleftarrows -NH-C(-)=NH$
Azo-hydrazone	C N N	$-N=N-CH< \rightleftarrows -NH-N=C<$
Diazoamino	N N N	$-N=N-NH- \rightleftarrows -NH-N=N-$
Diazohydroxide-nitrosamino	O N N	$-N=N-OH \rightleftarrows -NH-N=O$
Three carbon	C C C	$-CH_2-CH=CH- \rightleftarrows -CH=CH-CH_2-$
Allene-acetylene	C C C	$-CH=C=CH- \rightleftarrows -C\equiv C-CH_2-$
P-amide-P-imidol	N P O	$>P(=O)-NH_2 \rightleftarrows >P(-OH)=NH$
P-amidine	N P N	$>P(=N-)-NH_2 \rightleftarrows >P(-NH-)=NH$
Thiol-thion	S C O	$-C(-SH)=O \rightleftarrows -C(=S)-OH$

Table 2-3. (cont.)

Tautomerism type	X Y Z	Scheme of tautomeric equilibrium
Thiol-thion (N)	S C N	—C(SH)=N— ⇄ —C(=S)—NH—
Thiol-thion (P)	S P N	>P(SH)=O ⇄ >P(=S)—OH

[a] According to Refs.[31,38,39]

with different distributions of the ligands between the cationic and anionic components of the compound. Ionization isomerism involves the exchange of acidic groups between the anion and the complex cation; a classical example is the pair of complexes[43] $[Co(NH_3)_5SO_4]Br$ and $[Co(NH_3)_5Br]SO_4$, differing in, among other things, their colours. Hydrate isomerism can be considered[45] as a special case of ionization isomerism, where various distributions of the water molecule between the two parts of the complex compound participate in the differences between the isomers; examples include the three compounds $[Cr(H_2O)_6]Cl_3$, $[Cr(H_2O)_5Cl]Cl_2.H_2O$, $[Cr(H_2O)_4Cl_2]Cl.2\,H_2O$. In coordination isomerism[43] the ligand distribution between the complex cation and complex anion changes: $[Co(NH_3)_6]\,[Cr(CN)_6]$ and $[Cr(NH_3)_6]\,[Co(CN)_6]$ or $[Pt^{II}(NH_3)_4]\,[Pt^{IV}Cl_6]$ and $[Pt^{IV}(NH_3)_4Cl_2]\,[Pt^{II}Cl_4]$. It should be noted that Wheland[31] does not consider ionization, hydrate and coordination isomerism as isomerism but classifies the various compounds as stoichiomers, simplifying the classification of isomerism. The possibility of isomerism in the ligand itself leads to ligand isomerism.[45] This type includes, for example, two different compounds with the common formula $[Co(A)_2Cl_2]Cl$, where A designates either 1,3-diaminopropane or isomeric trimethylenediamine. The ability of the ligand to coordinate in several ways forms a basis for formation of linkage isomerism[46], first observed for the pair of nitro- and nitrito-complexes: $[Co(NH_3)_5NO_2]X_2$ and $[Co(NH_3)_5ONO]X_2$, where X = Cl or $\frac{1}{2}SO_4$.

In addition to the older structural isomerism, spatial isomerism was also gradually distinguished, i.e. stereoisomerism[30,31,47–51]. This is separated classically into optical and geometrical isomerism and is treated in terms of the fundamental stereochemical triad: chirality – configuration – conformation. The conformational analysis of organic substances has yielded a number of examples of various types of stereoisomerism – see[52–57]. Optical isomerism is most frequently understood as resulting from the phenomenon of chirality[58–61] to the extent introduced by Kelvin, i.e. the structure and its planar mirror image are not superimposable. While the great majority of optically active substances contain a carbon atom, its presence is not a necessary condition, as demonstrated by the first three examples of inorganic optical activity[62]. The region of optical isomerism can be considered to be most thoroughly treated in monographs; see, for example, [30,47,49–51,63]. It is interesting in this respect that the chirality criterion permits unambiguous classification of all stereoisomers, i.e. symmetry classification[30] into enantiomers or antipodes (the pairs of the object and its nonsuperimposable mirror image) and diastereoisomers (all the other stereoisomers, i.e. those that are not in the latter relationship).

Geometrical isomerism is connected with the arrangement of substituents on sites connected by a double bond or at sites on planar rings. A textbook example is the existence of the *cis*-form (maleic acid) and *trans*-form (fumaric acid) of the dicarbonic acid, HOOC—CH=CH—COOH.

Classical classification of stereoisomerism is criticized[28] as confusing and overlapping. Examples can be given[31] of isomers that are both optical and geometrical isomers. In addition, this classification does not include the important field of isomers depending on rotation around formally single bonds. Consequently, Noyce[28] suggested an alternative classification of stereoisomers as inverse and rotational. Inverse isomers are stereoisomers that can be interconverted only through inversion of at least one carbon atom, connected with the breaking of one or more bonds and the simultaneous formation of the corresponding number of new bonds. Rotational isomers are stereoisomers that can be interconverted by simple rotation of part of the molecule using some bonds as the rotation axes. Noyce[28] further classifies rotational isomers according to the type of hindrance to internal rotation (π-bond hindrance, σ-bond hindrance, hindrance depending on non-bonding interactions).

Isomers that can be interconverted by rotation around formally single bonds are called rotamers, conformers or conformational isomers. Mislow[30] recommends the term torsional single bond isomers. Because of the low energy barrier separating isomers of this type, they can rarely be separated at room temperature; this, however does not prevent them from being detectable in mixtures. The most important isomers depending on rotation around single bonds include the conformational isomers of cyclohexane. A survey of the relationships between the individual important structures is given in terms of a one-dimensional potential curve[30,64] in Fig. 2-1. The chair form with point group of symmetry D_{3d} is energetically most favourable; the second energy minimum corresponds to the twist boat form with D_2 symmetry. The boat form with C_{2v} symmetry is a transition state for pseudo-rotation that can lead to the D_2 form enantiomer. The structure with C_s symmetry then represents a transition state for conversion between the D_{3d} and D_2 forms. It was recently demonstrated[64] that this interconversion can also occur through the transition state with C_2 symmetry.

The concept of geometrical isomerism (*cis-trans* isomerism, alloisomerism) overlaps with the concept of torsional isomerism[65] or with the term torsional double bond isomerism proposed by Mislow[30]. Because of the height of the energy barrier separating them, isomers that can be converted by rotation around a formally double

Table 2-4. Possible geometrical isomers[30] for an asymmetrical chain with *n* double bonds

n	Configuration[a]
1	c t
2	cc ct tc tt
3	ccc cct ctc ctt tcc tct ttc ttt

[a] c-*cis*; t-*trans*.

bond can mostly be separated and are stable at room temperature. The presence of a greater number of double bonds in the structure rapidly increases the number of possible geometrical isomers; a total of 2^n diastereoisomers is possible for nonsymmetrical chains with *n*-double bonds. This is illustrated in Tab. 2-4 for the first three members of this series. Empirical experience has shown that *trans* isomers are usually more stable; however, for naturally occurring non-conjugated fatty acids (oleic, linoleic, linolenic, etc.), structures with all-*cis* conformation predominate[30]. *Syn-anti* isomerism[31] is a special case of geometrical isomerism; this is analogous to *cis-trans* isomerism but rotation occurs around the C=N or N=N bond rather than around the C=C bond (with *cis-syn* and *trans-anti* correspondence).

Isomerism depending on rotation around bonds can lead to the formation of enantiomers. When a formally double bond is involved, then these are called geometrical enantiomers[30]; when rotation occurs around a formally single bond, the optical isomers formed are called atropisomers[47]. Atropisomerism is a general term designating any kind of isomers formed by hindered rotation around a single bond where the individual isomers can actually be separated; these are often enantiomers.

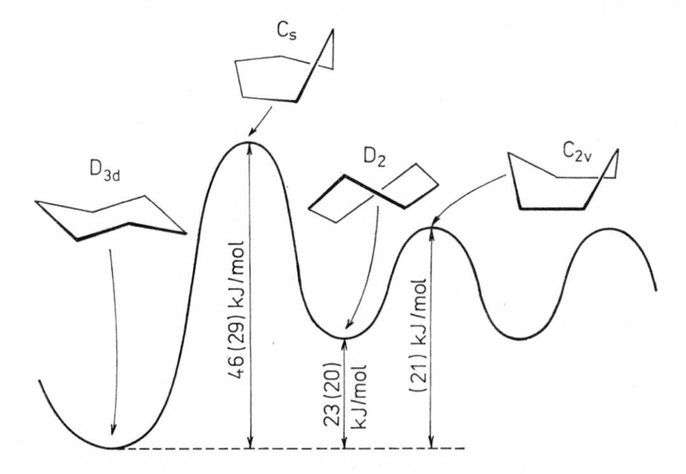

Figure 2-1. Conformational isomerism of cyclohexane (from Ref.[30]); the MINDO/2 energies[64] are given in parentheses

The best known examples of atropisomerism are[47] diphenyl isomers, ansa compounds and paracyclophane derivatives (see Fig. 2-2). Their chirality is dependent on stabilization of the nonplanar structure by bulky substituents. The possibility of separation into stable antipodes depends on the size of the substituents and the length of the methylene bridges. More recently, Mislow et al.[66,67] studied the stereochemistry of 'molecular propellers', chiral systems, good examples of which are triaryl molecules in which the aryl rings are bonded to a single atomic centre and twisted in the same sense.

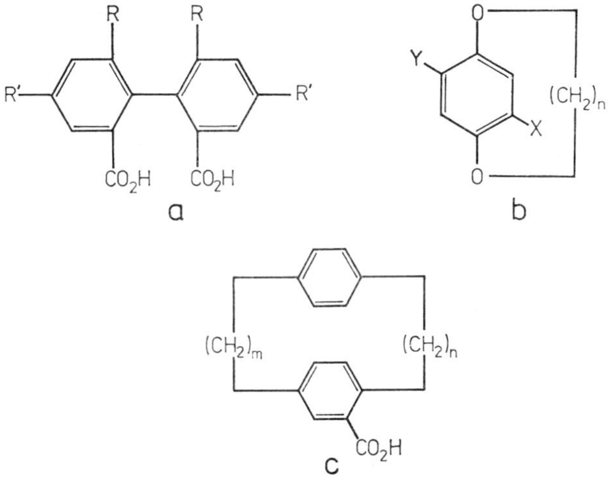

Figure 2-2. Examples of atropisomeric compounds[47]: a — derivatives of diphenic acid, b — ansa compounds, c — paracyclophane derivatives

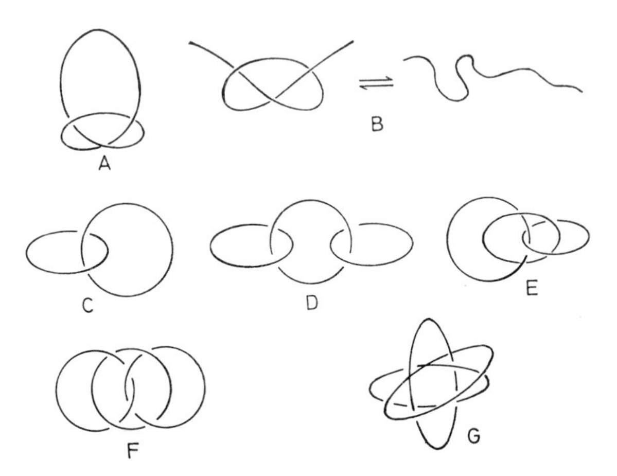

Figure 2-3. Examples of topological isomerism: A — trefoil, B — open knot, C — two-ring catenane, D—G — three-ring catenanes; only one of the enantiomers of A, B, E and F is shown (from Ref.[30])

Topological isomerism[30,68–70] can also be included in stereoisomerism. This is depicted in Fig. 2–3 by closed and open knots (A, B) and interlocking rings – catenanes (C–G). Model studies have demonstrated[30] that knotted cycloparaffin A can be formed from a chain more than fifty carbon atoms long. Knotted paraffin A is a topological isomer of unknotted cycloparaffin; they are thus stereoisomers but cannot be interconverted without breaking one C—C bond. On the other hand, open knot B is simply another conformation of acyclic paraffin. Models of catenanes[30] have demonstrated the possibility of formation of chains of twenty or thirty carbon atoms in such compounds. It should be noted that structures A, B, E and F exhibit chirality.

The process of distinguishing new types of isomerism is a continuous one[71–79]. An interesting example was described for gear-shaped molecules[74–79]. An example is[75,77,78] *bis*(4-chloro-1-triptycyl)ether (Fig. 2-4). It has been shown for *bis*(1-triptycyl)ether[75] that rotation around two C—O bonds (during which the movement of the six equivalent benzene rings is reminescent of fitted gears) is not under any apparent restriction. It can be demonstrated[77,78] that, when one of the cogs on each wheel is labelled, then if the number of cogs is given by $n = m = 3$, a pair of phase isomers is formed. These are depicted in Fig. 2-4 as the 4-chloroderivatives. While the two labelled cogs fit together once per complete rotation cycle in one isomer, in the other isomer they are out of phase and never fit together. In this new phase isomerism there is thus rapid rotation around both single bonds that is, however, internally coupled, so that separate isomers are formed (one formed by the *dl* pair and the other achiral

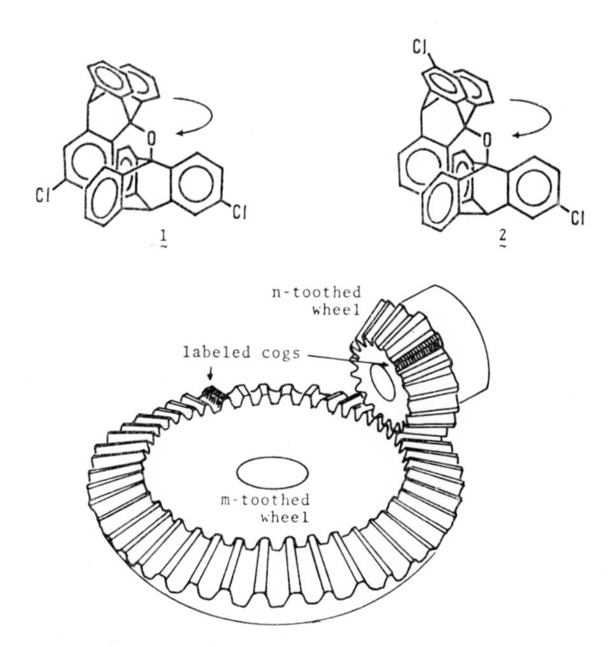

Figure 2-4. Phase isomerism in gear-shaped molecules: phase isomers meso (1) and dl (2) of *bis*-(4-chloro-1-triptycyl) ether (from Ref.[78])

(meso)). Another example of unconventional isomers is the problem of nonclassical ions[80,81].

Similarly, in inorganic chemistry the number of known types of isomers cannot be considered as complete. For example, Gažo and coworkers (see e.g.[71,147,148]) have described a new type of isomerism of coordination compounds, called distortion isomerism, based on a varying degree of distortion of the coordination polyhedron in Cu(II) compounds. The coordination sphere of Cu(II) octahedral complexes, in contrast with most other central atoms, clearly demonstrates some pronounced properties of non-rigidity (in the field also called[147] plasticity) having several stable configurations which differ in their metal-ligand distances. These plasticity properties were shown to be due to the Jahn-Teller or pseudo-Jahn-Teller effect, and hence they may occur in other metal complexes with degenerate and pseudo-degenerate electronic states.

In addition to new types of isomerism, less stable isomers have recently been found, or at least indicated, experimentally[82–88]. For example, this is the case even for the

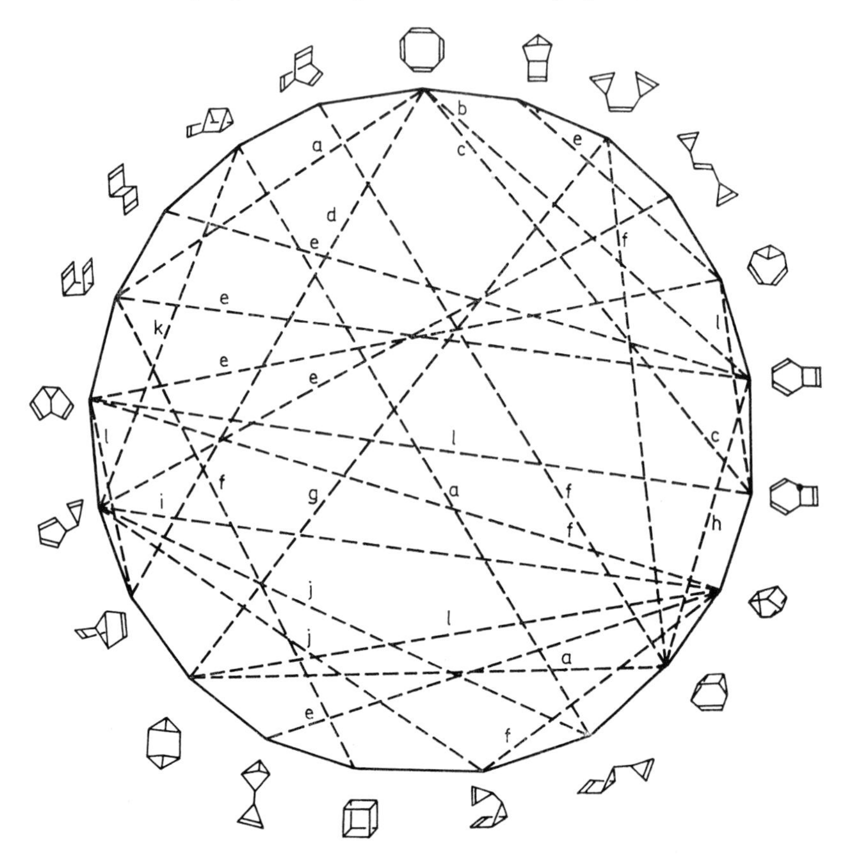

Figure 2-5. Examples of conceivable conversions within 21 selected $(CH)_8$ structures (the indicated types of conversions are explained in Ref.[91])

following very simple species: ClO_2, N_2O_4 (in addition to the symmetrical structure, also the nonsymmetrical $ONONO_2$ structure), H_2SO_4, or the smallest amino acid (glycine) where recently a conformer of lower energy than that of the earlier structure was found[88]. Discoveries of many unusual isomers are the result of the interplay between molecular spectroscopy and astronomy[89,90].

The above survey of isomerism and its usual classification exhibit two inadequacies: variety of classification criteria and the fact that these criteria are frequently not important. From a chemical point of view the convertibility of isomers is important (Fig. 2-5), which also offers the possibility of unification of classification criteria by division of isomers according to the nature and height of the energy barriers separating them[29,30]; this has already been mentioned in passing for conformational and torsional isomerism and for the conformations of cyclohexane. In ref.[29] it is proposed that isomerism be classified according to decreasing height of the energy barrier: tension barriers, flexion barriers and torsion barriers. Tension barriers imply constitutional isomerism, flexion barriers permit the existence of distinguishable stereoisomers and, the lowest, torsion barriers result in distinguishable conformers of a given stereoisomer. Data on the height of the energy barrier alone, however, are incomplete without information concerning the temperature. Properly speaking, the ratio of the barrier height and the absolute temperature should, for example, be considered; in this existing barrier classification of isomers, room temperature is

Table 2-5. The Jennen method[26,27] of isomerism classification

I. Stable isomers	1. Structural isomerism	a) Chain isomerism b) Place isomerism c) Functional group isomerism, especially tautomerism
	2. Stereoisomerism	a) Optical isomerism b) Geometrical isomerism c) Orientation isomerism
	3. Polymorphism (Physical isomerism)	
II. Labile isomers	1. Rotational isomerism 2. Inversion isomerism 3. Nuclear spin isomerism 4. Vector isomerism	
III. Metastable isomers	1. Triplet structures 2. Pseudomonotropy 3. Nuclear isomers	

implicitly assumed. It should be noted that Eliel[92] recently proposed a temperature-dependent approach to the phenomenon of isomerism, in which the notion of isomers is reserved for two substances with identical composition separated by an energy barrier greater than kT, where k is the Boltzmann constant. The idea of barrier criteria was also anticipated in the isomer classification proposed by Jennen[26,27], surveyed in Tab. 2-5. Jennen separates isomers into three groups: stable, labile and metastable, where the concept of isomerism is broader than usual. The general term orientation isomers includes the isomers of coordination compounds among stereoisomers, while rotational and inversion isomers are considered to be labile isomers. Inversion isomerism is identical with the concept of geometrical tautomerism[93] and designates isomerism based on substitution of cyclohexane in the axial or equatorial position. Jennen also includes examples of 'physical' isomerism in his classification, represented by polymorphism and pseudomonotropy. Polymorphism (see for example[44,94,95]) is related to the existence of the solid phase in a larger number of crystalline forms yielding identical solutions or vapours after conversion into the liquid and gaseous phases. In pseudomonotropy all these forms, except one, are unstable or metastable. Labile isomers also include nuclear spin isomers, i.e. isomerism dependent on the symmetry properties of the complete wave function and the wave functions of rotation and nuclear spin, isomerism based not on differences in the positions of the nuclei but on differences in their quantum states. For molecule X_2, when the nuclei have non-zero nuclear spin these symmetry requirements result[96] in two series of states, differing in the parity of their rotational quantum numbers and degeneracy factor values, where transitions between the two states are quite rare. This permits formation of two modifications of the X_2 molecule with generally different properties. A textbook example of nuclear spin isomerism[96] is the system of *ortho*- and *para*-H_2. In principle, however, this type of isomerism is possible for many polyatomic molecules – e.g. for tetrahedral molecules whose equivalent nuclei have non-zero nuclear spin, theoretical analysis[97] indicates the existence of three modifications of these molecules. The fourth type of labile isomers in Jennen's classification – vector isomerism – contains unstable electromers. This concept overlaps with that of mesomerism. The inclusion of triplet structures among isomers correctly anticipates the fact that every excited electronic state with nuclei in the equilibrium positions can be considered as a structure isomeric with the ground state.

In addition to the above-mentioned barrier classification, attempts to unify criteria for classification of the unusually complicated and wide-spread phenomenon of chemical isomerism also include more abstract approaches based on modern algebraic methods – set theory, graph theory and group theory. Ege[98] formulated relationships among isomers using set theory and pointed out the importance of equivalence relations. Ruch *et al.*[99,100] describe an algebraic analogy to configurations of permutational isomers for a common molecular skeleton on which a set of ligands is rearranged. Mislow[101] employed edge-weighted molecular graphs for classification of relationships between isomeric structures. A systematic algebraic study of the problem of chirality and chirality functions was carried out[58-61,102-113] and its possibly more general validity was pointed out[114,115].

2.3 Stereochemical Non-rigidity of Molecules

The existence of tautomerism assumes transitions between various structures and thus non-rigidity of the molecular skeleton. This is particularly important for non--rigid or fluxional molecules[116-124]. In the oldest systems of this type (ammonia with respect to inversion[124], cyclopentane with respect to pseudorotation[125], trigonal bipyramids with respect to pseudorotation[117], rearrangements of bullvalene[118] – see Fig. 2-6), transitions occurred between identical structures. A classical example is bullvalene (tricyclo[3.3.2.$0^{4.6}$]deca-2.7.9-triene) exhibiting rapid transition of the position of the cyclopropane ring and double bonds, as depicted in Fig. 2-6. Theoretical considerations have demonstrated[118,126] that there is a total of 1 209 600 ways in which this transition can occur. Averaging of the positions of the hydrogen atoms can be made possible by these transitions, so that at higher temperatures the PMR spectrum of bullvalene exhibits[118,126] only a single peak. Resonance methods, in particular, allowed the demonstration[126-129] of the fact that there are large numbers of non-rigid systems that are continuously interconverted under experimental conditions. Diffraction methods[130-132] are also very useful for the study of stereochemically non-rigid molecules. The first field in which the large-amplitude motions and internal rearrangements were extensively explored was that of inorganic and organometallic molecules[133-137]. More recently, the non-rigidity and fluxional behaviour of biomolecules have also been systematically studied[138-144].

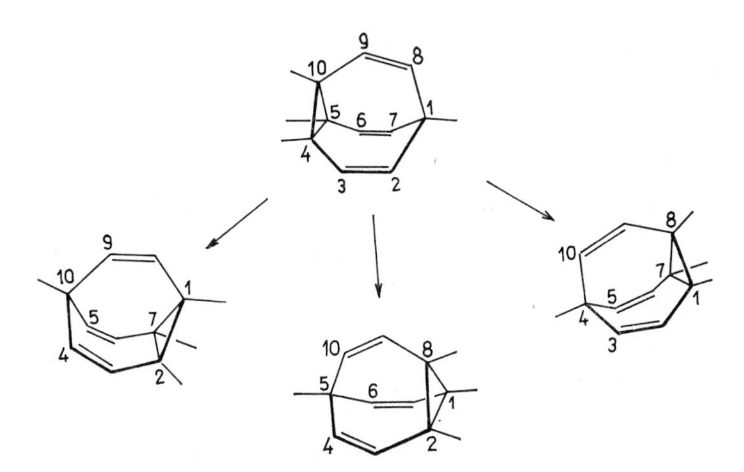

Figure 2-6. An example of a fluxional molecule — bullvalene; 4 of 1 209 600 equivalent structures are depicted together with interconversions (from Ref.[118])

The importance of the time factor for determination of molecular structure, and thus for the discovery of isomerism, was actually recognized and accepted only recently in connection with the systematic treatment of molecules as non-rigid structures. It was demonstrated[145] that not only the observed molecular configuration itself, but also the coordination number may be dependent on the type of physical technique used for their study. A survey of the time scales of the individual physical techniques used to determine molecular structures is given[119] in Tab. 2-6. If the time scale of a process involving permutating nuclei and leading to a non-degenerate stereoisomerization is longer than the time scale of the phenomenon on which the experimental technique employed is based, then the individual isomeric forms can, in principle, be distinguished. However, it is no longer possible to distinguish these forms if the time scale of the stereoisomerization process is comparable to, or shorter than, that of the particular observation[119]. The system is then termed (for both the non-degenerate and the degenerate rearrangements) a stereochemically non-rigid or fluxional molecule[118,119] or the term permutational isomerism is sometimes also employed[133]. It is apparent that the concept of stereochemical non-rigidity thus introduced is connected with a particular physical technique with a time scale relative to which the permutation process is not slow.

It is thus apparent that whether an isomeric system appears as a mixture of isomers or as a single species in an observation may depend on the experimental technique employed and on the temperature. While one technique may yield the characteristics of individual structures for a given system, another technique may yield only time--averaged values over all the configurations of the same system. Thus the concept

Table 2-6. Time scales of structural techniques as given by Muetterties[119]

Technique	Approximate time scale (s)
Electron diffraction	10^{-20}
Neutron diffraction	10^{-18}
X-Ray diffraction	10^{-18}
Ultraviolet spectroscopy	10^{-15}
Visible spectroscopy	10^{-14}
Infrared and Raman spectroscopy	10^{-13}
ESR[a]	$10^{-4}-10^{-8}$
NMR[a]	$10^{-1}-10^{-9}$
NQR[a]	$10^{-1}-10^{-8}$
Mössbauer spectroscopy (Fe)	10^{-7}
Molecular beams	10^{-6}

[a] Time scale sensitively defined by the chemical systems under investigation.

of residual isomers was introduced[92]; these are isomers that can be observed using a given observation technique under the given experimental conditions.

In the light of molecular non-rigidity, the conventional definition of the concept of chirality is also insufficient. Similar to the possibility of distinguishing isomers, manifestation of chirality also depends on the conditions of the particular physical measurement, on the time scale of the particular observation[146]. Intramolecular motions may exist in non-rigid structures, permitting removal of chirality in the sense of the Kelvin definition. A change in the temperature alone may result in a single structure appearing as chiral in one experiment and achiral in another[146]. This fact is taken into consideration in the recently proposed definition of chemical chirality[111]: a molecule is chemically chiral under the given observational conditions if there is one momentary geometry of that molecule which cannot be superimposed on its mirror image by using only rotations, translations and those intramolecular motions that can occur under the observation conditions.

It should also be noted that the time factor included in the definition of the phenomenon of chemical isomerism according to Rouvray[17] ensures only the experimental detectability of the given isomer in principle and not its separability. In order for the individual isomeric forms to be separable stronger criteria must be fulfilled. For example, Muetterties[119] states that the time scale analogous to the values in Tab. 2-6 for the experimental separability of isomers should be greater than 10^2 s. Alternatively, however, the distinguishability and separability of isomers can be discussed[30] in terms of the energy barrier separating them and of the temperature. A similar approach can be taken to the partially overlapping (but more general) concept of relative or absolute isomer stability.

2.4 Summary

Since the time of Berzelius' historic definition, intense experimental and, more recently, theoretical studies have yielded an unusually large amount of information concerning chemical isomerism; in addition, analogous phenomena have been found in atomic nuclei and biological systems. Isomerism has been demonstrated to be a general characteristic of chemical systems. Because of the remarkable variety of isomers, no unambiguous and functional classification scheme has yet been found. Concepts of the nature of isomerism have developed in close relationship and cooperation with the development of concepts of molecular structure itself. At the very beginning of organic chemistry, the concept of isomerism seemed to be very well defined. However, the recently developed concept of stereochemically non-rigid or fluxional molecular systems throws new light on isomerism. The latter is now operationally understood to be a phenomenon whose observability depends on the conditions and methods of observation. Contemporary approaches treat isomers in several ways: as clearly defined, individual species, as coexisting (but experimentally readily distinguishable) structures, and also as phenomena that can be demonstrated only under certain observational conditions and using a certain technique, with limiting transition to fluxional behaviour without the possibility of distinguishing among the individual isomers. In spite of the fact that experimental chemical techniques are increasingly more and more specific, under certain conditions not only the separability but also the distinguishability of the individual isomers is, not only temporarily but sometimes in principle, impossible. Under such conditions, the isomeric system considered must be understood as a single, indistinguishable entity.

REFERENCES

(1) Snelders, H. A. M.: 1964, Chem. Weekbl. *60*, pp. 217—221.
(2) Slanina, Z.: 1981, Chem. Listy *75*, pp. 1121—1147.
(3) Gay-Lussac, J. L.: 1816, Ann. Chim. Phys. *1*, pp. 32—45.
(4) Berzelius, J. J.: 1817, Ann. Chim. Phys. *5*, pp. 149—165.
(5) Liebig, J.: 1823, Ann. Chim. Phys. *24*, pp. 294—317.
(6) Liebig, J., and Gay-Lussac, J. L.: 1824, Ann. Chim. Phys. *25*, pp. 285—311.
(7) Wöhler, F.: 1824, Ann. Chim. Phys. *27*, pp. 196—200.
(8) Gay-Lussac, J. L.: 1824, Ann. Chim. Phys. *27*, pp. 199—200.
(9) Winnewisser, M., and Winnewisser, B. P.: 1976, Chem .Listy *70*, pp. 785—807.
(10) Wöhler, F.: 1828, Ann. Phys. Chem. *12* (der ganzen Folge *88*), pp. 253—256.
(11) Wöhler, F.: 1828, Ann. Chim. Phys. *37*, pp. 330—334.
(12) Wallach, O.: 1901, Briefwechsel zwischen J. Berzelius und F. Wöhler, Erster Band, Verlag von W. Engelmann, Leipzig, p. 207.
(13) Berzelius, J. J.: 1830, Ann. Phys. Chem. *19* (der ganzen Folge *95*), pp. 305—335.
(14) Temnikova, T. I.: 1959, Course of Theoretical Bases of Organic Chemistry, GNTICHL, Leningrad, (in Russian).
(15) Bláha, K.: 1974, Chem. Listy *68*, pp. 884—898.
(16) Riddell, F. G., and Robinson, M. J. T.: 1974, Tetrahedron *30*, pp. 2001—2007.
(17) Rouvray, D. H.: 1974, Chem. Soc. Rev. *3*, pp. 355—372.
(18) Hackh, I. W. D.: 1920, Chem. News *121*, p. 85; Chem. Abstr. *14*, 3405 (1920).
(19) Lunn, A. C., and Senior, J. K.: 1929, J. Phys. Chem. *33*, pp. 1027—1079.
(20) Tellegen, F. P. A.: 1935, Chem. Weekbl. *32*, p. 3; Chem. Abstr. *29*, 3295^7 (1935).
(21) Tellegen, F. P. A.: 1935, Chem. Weekbl. *32*, p. 40; Chem. Abstr. *29*, 3295^8 (1935).
(22) Bargalló, M.: 1950, Ciencia (Mex.) *10*, p. 257; Chem. Abstr. *45*, 7833i (1951).
(23) Bent, R. L.: 1953, J. Chem. Educ. *30*, pp. 220—228.
(24) Bent, R. L.: 1953, J. Chem. Educ. *30*, pp. 284—290.
(25) Bent, R. L.: 1953, J. Chem. Educ. *30*, pp. 328—340.
(26) Jennen, J. J.: 1954, Ind. Chim. Belg. *19*, pp. 1051—1062.
(27) Jennen, J. J.: 1955, Ind. Chim. Belg. *20*, pp. 1067—1091.
(28) Noyce, W. K.: 1961, J. Chem. Educ. *38*, pp. 23—27.
(29) Ugi, I., Gillespie, P., and Gillespie, C.: 1972, Trans. New York Acad. Sci. *34*, pp. 416 to 432.
(30) Mislow, K.: 1966, Introduction to Stereochemistry, W. A. Benjamin, New York.
(31) Wheland, G. W.: 1960, Advanced Organic Chemistry, Wiley, New York.
(32) Epiotis, N. D., Larson, J. R., Yates, R. L., Cherry, W. R., Shaik, S., and Bernardi, F.: 1977, J. Am. Chem. Soc. *99*, pp. 7460—7464.
(33) Kabachnik, M. I.: 1956, Usp. Khim. *25*, pp. 137—161.

(34) Katritzky, A. R.: 1970, Chimia *24*, pp. 134—146.
(35) Pullman, B., and Pullman, A.: 1971, Advan. Heterocycl. Chem. *13*, pp. 77—159.
(36) Kwiatkowski, J. S., and Pullman, B.: 1975, Advan. Heterocycl. Chem. *18*, pp. 199—335.
(37) Minkin, V. I., Olekhnovich, L. P., and Zhdanov, Yu. A.: 1981, Accounts Chem. Res. *14*, pp. 210—217.
(38) Wichterle, O.: 1955, Organic Chemistry, NČSAV, Prague, (in Czech).
(39) 1967, Compendious Chemical Encyclopaedia, Vol. 5, Ed. I. L. Knunyane, Soviet Encyclopaedia, (in Russian).
(40) Maier, G.: *1972*, Valenzisomeriesierungen, Verlag Chemie, Weinheim.
(41) Balaban, A. T.: 1966, Rev. Roum. Chim. *11*, pp. 1097—1116.
(42) Kent, J. E., Harman, P. J., and O'Dwyer, M. F.: 1981, J. Phys. Chem. *85*, pp. 2726—2730.
(43) Hückel, W.: 1950, Structural Chemistry of Inorganic Compounds, Vol. I, Elsevier, New York.
(44) Wells, A. F.: 1962, Structural Inorganic Chemistry, Oxford University Press, Oxford.
(45) Cotton, F. A., and Wilkinson, G.: 1966, Advanced Inorganic Chemistry, Wiley, New York.
(46) Fraser, R. T. M.: 1967, Advan. Chem. Ser. *62*, pp. 295—305.
(47) Eliel, E. L.: 1962, Stereochemistry of Carbon Compounds, McGraw-Hill, New York.
(48) Mislow, K., and Raban, M.: 1967, Top. Stereochem. *1*, pp. 1—38.
(49) 1977, Stereochemistry. Fundamentals and Methods, Vols. 1—4, Ed. H. B. Kagan, G. Thieme Publishers, Stuttgart.
(50) Sokolov, V. I.: 1979, Introduction to Theoretical Stereochemistry, Nauka, Moscow, (in Russian).
(51) Nógrádi, M.: 1981, Stereochemistry. Basic Concepts and Applications, Akadémiai Kiadó Budapest.
(52) Hanack, M.: 1965, Conformation Theory, Academic Press, New York.
(53) 1971, Conformational Analysis. Scope and Present Limitations, Ed. G. Chiurdoglu, Academic Press, New York.
(54) Bastiansen, O., Seip, H. M., and Boggs, J. E.: 1971, Persp. Struct. Chem. *4*, pp. 60—165.
(55) Wilson, E. B.: 1972, Chem. Soc. Rev. *1*, pp. 293—317.
(56) Golebiewski, A., and Parczewski, A.: 1974, Chem. Rev. *74*, pp. 519—530.
(57) Eliel, E. L.: 1975, J. Chem. Educ. *52*, pp. 762—767.
(58) Ugi, I., Marquarding, D., Klusacek, H., Gokel, G., and Gillespie, P.: 1970, Angew. Chem., Int. Ed. Engl. *9*, pp. 703—730.
(59) Brewster, J. H.: 1974, Top. Curr. Chem. *47*, pp. 29—71.
(60) Mead, C. A.: 1974, Top. Curr. Chem. *49*, pp. 1—88.
(61) Farina, M., and Morandi, C.: 1974, Tetrahedron *30*, pp. 1819—1831.
(62) Gillard, R. D., and Wimmer, F. L.: 1978, Chem. Commun., pp. 936—937.
(63) Morrison, J. D., and Mosher, H. S.: 1971, Asymmetric Organic Reactions, Prentice-Hall, Englewood Clifs, New Jersey.
(64) Komornicki, A., and McIver Jr., J. W.: 1973, J. Am. Chem. Soc. *95*, pp. 4512—4517.
(65) Epiotis, N. D., Cherry, W. R., Shaik, S., Yates, R. L., and Bernardi, F.: 1977, Fortschr. Chem. Forsch. *70*, pp. 1—242.
(66) Mislow, K., Gust, D., Finocchiaro, P., and Boettcher, R. J.: 1974, Top. Curr. Chem. *47*, pp. 1—28.
(67) Mislow, K.: 1976, Accounts Chem. Res. *9*, pp. 26—33.
(68) Schill, G.: 1971, Catenanes, Rotaxanes, and Knots, Academic Press, New York.
(69) Boeckmann, J., and Schill, G.: 1974, Tetrahedron *30*, pp. 1945—1957.
(70) Frank-Kameneckii, M. D., and Vologodskii, A. V.: 1981, Usp. Fiz. Nauk *134*, pp. 641 to 673.

(71) Gažo, J.: 1974, Pure Appl. Chem. *38*, pp. 279—301.
(72) Purcell, K. F., and Zapata, J. P.: 1978, Chem. Commun., pp. 497—499.
(73) Manohar, H., Schwarzenbach, D., Iff, W., and Schwarzenbach, G.: 1979, J. Coord. Chem. *8*, pp. 213—221.
(74) Hounshell, W. D., Johnson, C. A., Guenzi, A., Cozzi, F., and Mislow, K.: 1980, Proc. Natl. Acad. Sci. U.S.A. *77*, pp. 6961—6964.
(75) Kawada, Y., and Iwamura, H.: 1980, J. Org. Chem. *45*, pp. 2547—2548.
(76) Cozzi, F., Guenzi, A., Johnson, C. A., Mislow, K., Hounshell, W. D., and Blount, J. F.: 1981, J. Am. Chem. Soc. *103*, pp. 957—958.
(77) Kawada, Y., and Iwamura, H.: 1981, J. Am. Chem. Soc. *103*, pp. 958—960.
(78) Kawada, Y., and Iwamura, H.: 1981, Tetrahedron Lett. *16*, pp. 1533—1536.
(79) Johnson, C. A., Guenzi, A., and Mislow, K.: 1981, J. Am. Chem. Soc. *103*, pp. 6240 to 6242.
(80) 1975, Contemporary Problems of Chemistry of Carbonium Ions, Ed. V. A. Koptyug, Nauka, Novosibirsk, (in Russian).
(81) Brown, H. C. (with comments by P.v.R. Schleyer): 1977, The Nonclassical Ion Problem, Plenum Press, New York.
(82) Benson, S. W., and Buss, J. H.: 1957, J. Chem. Phys. *27*, pp. 1382—1384.
(83) Fateley, W. G., Bent, H. A., and Crawford Jr., B. C.: 1959, J. Chem. Phys. *31*, pp. 204—217.
(84) Hisatsune, I. C., Devlin, J. P., and Wada, Y.: 1960, J. Chem. Phys. *33*, pp. 714—719.
(85) Bolduan, F., and Jodl, H. J.: 1982, Chem. Phys. Lett. *85*, pp. 283—286.
(86) Kuczkowski, R. L., Suenram, R. D., and Lovas, F. J.: 1981, J. Am. Chem. Soc. *103*, pp. 2561—2566.
(87) Schäfer, L., Sellers, H. L., Lovas, F. J., and Suenram, R. D.: 1980, J. Am. Chem. Soc. *102*, pp. 6566—6568.
(88) Suenram, R. D., and Lovas, F. J.: 1980, J. Am. Chem. Soc. *102*, pp. 7180—7184.
(89) Green, S., and Herbst, E.: 1979, Astrophys. J. *229*, pp. 121—131.
(90) Herzberg, G.: 1980, Highlights of Astronomy *5*, pp. 3—26.
(91) Smith, L. R.: 1978, J. Chem. Educ. *55*, pp. 569—570.
(92) Eliel, E. L.: 1976/77, Isr. J. Chem. *15*, pp. 7—11.
(93) Pitzer, K. S., and Beckett, C. W.: 1947, J. Am. Chem. Soc. *69*, pp. 977—978.
(94) Furmanova, N. G.: 1981, Usp. Khim. *50*, pp. 1491—1516.
(95) Bar, I., and Bernstein, J.: 1982, J. Phys. Chem. *86*, pp. 3223—3231.
(96) Woolley, H. W., Scott, R. B., and Brickwedde, F. G.: 1948, J. Res. Nat. Bur. Stand. *41*, pp. 379—473.
(97) Herzberg, G.: 1960, Molecular Spectra and Molecular Structure, Vol. II, Van Nostrand Company, Princeton.
(98) Ege, G.: 1971, Naturwissenschaften *58*, pp. 247—257.
(99) Ruch, E., Hässelbarth, W., and Richter, B.: 1970, Theor. Chim. Acta *19*, pp. 288—300.
(100) Hässelbarth, W., and Ruch, E.: 1973, Theor. Chim. Acta *29*, pp. 259—268.
(101) Mislow, K.: 1977, Bull. Soc. Chim. Belg. *86*, pp. 595—601.
(102) Ruch, E., Schönhofer, A., and Ugi, I.: 1967, Theor. Chim. Acta *7*, pp. 420—432.
(103) Ruch, E., and Schönhofer, A.: 1968, Theor. Chim. Acta *10*, pp. 91—110.
(104) Ruch, E.: 1968, Theor. Chim. Acta *11*, pp. 183—192.
(105) Ruch, E., and Schönhofer, A.: 1970, Theor. Chim. Acta *19*, pp. 225—287.
(106) Ruch, E.: 1982, Accounts Chem. Res. *5*, pp. 49—56.
(107) Hässelbarth, W., and Ruch, E.: 1976/77, Isr. J. Chem. *15*, pp. 112—115.
(108) Keller, H., Langer, E., Lehner, H., and Derflinger, G.: 1978, Theor. Chim. Acta *49*, pp. 93—100.

(109) DERFLINGER, G., and KELLER, H.: 1978, Theor. Chim. Acta *49*, pp. 101—105.
(110) RUCH, E.: 1978, Theor. Chim. Acta *49*, pp. 107—112.
(111) DUGUNDJI, J., KOPP, R., MARQUARDING, D., and UGI, I.: 1978, Top. Curr. Chem. *75*, pp. 165—180.
(112) DRESS, A. V. M.: 1979, in The Permutation Group in Physics and Chemistry, Ed. J. Hinze, Springer-Verlag, Berlin, pp. 215—230.
(113) MEAD, C. A.: 1980, Theor. Chim. Acta *54*, pp. 165—168.
(114) ATKINS, P. W., and GOMES, J. A. N. F.: 1976, Chem. Phys. Lett. *39*, pp. 519—520.
(115) BARRON, L. D.: 1981, Chem. Phys. Lett. *79*, pp. 392—394.
(116) BENFEY, O. T.: 1957, J. Chem. Educ. *34*, pp. 286—288.
(117) BERRY, R. S.: 1960, J. Chem. Phys. *32*, pp. 933—938.
(118) DOERING, W.v.E., and ROTH, W. R.: 1963, Angew. Chem. *75*, pp. 27—35.
(119) MUETTERTIES, E. L.: 1965, Inorg. Chem. *4*, pp. 769—771.
(120) COTTON, F. A.: 1968, Accounts Chem. Res. *1*, pp. 257—265.
(121) MUETTERTIES, E. L.: 1970, Accounts Chem. Res. *3*, pp. 266—273.
(122) BERRY, R. S.: 1979, in The Permutation Group in Physics and Chemistry, Ed. J. Hinze, Springer-Verlag, Berlin, pp. 92—120.
(123) BERRY, R. S.: 1980, in Quantum Dynamics of Molecules, Ed. R. G. Woolley, Plenum Press, New York, pp. 143—195.
(124) DENNISON, D. M., and UHLENBECK, G. E.: 1932, Phys. Rev. *41*, pp. 313—321.
(125) KILPATRICK, J. E., PITZER, K. S., and SPITZER, R.: 1947, J. Am. Chem. Soc. *69*, pp. 2483 to 2488.
(126) SERGEEV, N. M.: 1973, Usp. Khim. *42*, pp. 769—798.
(127) STEIGL, A.: 1978, in Dynamic NMR Spectroscopy, Ed. P. Diehl, E. Fluck, R. Kosfeld, Springer-Verlag, Berlin, pp. 1—53.
(128) JARDETZKY, O.: 1980, Biochim. Biophys. Acta *621*, pp. 227—232.
(129) COTTON, F. A., and HANSON, B. E.: 1980, in Rearrangements in Ground and Excited States, Vol. 2, Ed. P. de Mayo, Academic Press, New York, pp. 379—421.
(130) SPIRIDONOV, V. P., ISCHENKO, A. A., and ZASORIN, E. Z.: 1978, Usp. Khim. *47*, pp. 101—126.
(131) BASTIANSEN, O., KVESETH, K., and MØLLENDAL, H.: 1979, Top. Curr. Chem. *81*, pp. 99—172.
(132) BIANCHI, R., PILATI, T., and SIMONETTA, M.: 1981, J. Am. Chem. Soc. *103*, pp. 6426—6431.
(133) GILLESPIE, P., HOFFMAN, P., KLUSACEK, H., MARQUARDING, D., PFOHL, S., RAMIREZ, F., TSOLIS, E. A., and UGI, I.: 1971, Angew. Chem., Int. Ed. Engl. *10*, pp. 687—715.
(134) UGI, I., MARQUARDING, D., KLUSACEK, H., GILLESPIE, P., and RAMIREZ, F.: 1971, Accounts Chem. Res. *4*, pp. 288—296.
(135) HOLMES, R. R.: 1972, Accounts Chem. Res. *5*, pp. 296—303.
(136) MUETTERTIES, E. L.: 1974, Tetrahedron *30*, pp. 1595—1604.
(137) ALBRIGHT, T. A.: 1982, Accounts Chem. Res. *15*, pp. 149—155.
(138) OVCHINNIKOV, YU. A., and IVANOV, V. T.: 1974, Tetrahedron *30*, pp. 1871—1890.
(139) SUNDARALINGAM, M., and WESTHOF, E.: 1979, Int. J. Quantum Chem., Quantum Biol. Symp. *6*, pp. 115—130.
(140) BOLTON, P. H., and JAMES, T. L.: 1980, J. Am. Chem. Soc. *102*, pp. 25—31.
(141) VENKATACHALA M, C. M., KHALED, M. A., SUGANO, H., and URRY, D. W.: 1981, J. Am. Chem. Soc. *103* pp. 2372—2379.
(142) GUPTA, V. D.: 1981, Int. J. Quantum Chem. *20*, pp. 9—21.
(143) 1981, Biomolecular Stereodynamics, Vols. I and II, Ed. R. H. Sarma, Adenine Press, New York.
(144) OLSON, W. K., and SUSSMAN, J. L.: 1982, J. Am. Chem. Soc. *104*, pp. 270—278.
(145) HARGITTAI, M., and HARGITTAI, I.: 1977, The Molecular Geometries of Coordination Compounds in the Vapour Phase, Akadémiai Kiadó, Budapest.

(146) Mislow, K., and Bickart, P.: 1976/77, Isr. J. Chem. *15*, pp. 1–6.
(147) Gažo, J., Bersuker, I. B., Garaj, J., Kabešová, M., Kohout, J., Langfelderová, H., Melník, M., Serátor, M., and Valach, F.: 1976, Coord. Chem. Rev. *19*, pp. 253–297.
(148) Melník, M.: 1982, Coord. Chem. Rev. *47*, pp. 239–261.

3 Quantum Chemical Description of Isomerism

This chapter is concerned with a concept lying at the very basis of the contemporary theory of chemical reactivity – the concept of a potential energy hypersurface. This concept is an immediate consequence of the separation of nuclear and electronic motions proposed in 1927 by Born and Oppenheimer[1]. In terms of the properties of the potential energy hypersurface, classical key characteristics of molecular systems are introduced, e.g.: molecular structure, energy barriers, force constants, and the frequencies of vibrational and rotational transitions. The belief that all experimental observations for molecular systems can be explained in terms of the molecular structure and related notions is so widespread that it has been called[2] the central dogma of molecular science. The concept of a potential energy hypersurface also permits rationalization of the concept of chemical isomerism.

3.1 The Born-Oppenheimer Approximation — The Concept of a Potential Energy Hypersurface

In order to obtain information on a system (in the absence of external fields) consisting of N atomic nuclei and n electrons, the time-independent Schrödinger equation (3-1) must be solved:

$$\hat{H}\, \Psi(\boldsymbol{r}, \boldsymbol{R}) = E\, \Psi(\boldsymbol{r}, \boldsymbol{R}) \tag{3-1}$$

where $\boldsymbol{r} = \{\boldsymbol{r}_i\}$ is the set if positional vectors $\boldsymbol{r}_i$ of the individual electrons and $\boldsymbol{R} = \{\boldsymbol{R}_\alpha\}$ is the set of positional vectors $\boldsymbol{R}_\alpha$ of the individual atomic nuclei. The non-relativistic, spinless Hamiltonian $\hat{H}$ for this multi-particle system involving only Coulomb interactions has the form:

$$\hat{H} = \hat{T}_e + \hat{T}_n + \hat{V}(\boldsymbol{r}, \boldsymbol{R})\,, \tag{3-2}$$

where the operators $\hat{T}_e$ and $\hat{T}_n$ of the kinetic energy of the electrons with mass m and of the nuclei with masses M_α are given by the relationships:

$$\hat{T}_e = -\frac{\hbar^2}{2m} \sum_{i=1}^{n} \nabla_i^2\,, \tag{3-3}$$

$$\hat{T}_n = -\frac{\hbar^2}{2}\sum_{\alpha=1}^{N}\frac{\nabla_\alpha^2}{M_\alpha}, \tag{3-4}$$

and $\hat{V}(\boldsymbol{r}, \boldsymbol{R})$ is the potential energy operator.

Equation (3-1) can be solved using the following approach. Firstly, the characteristic problem (3-1) is solved for a hypothetical state of the nuclei fixed in positions described by selected values of coordinates $\boldsymbol{R}$, stepwise for all choices of $\boldsymbol{R}$ values; i.e. the characteristic equation (3-5) is used to find all the eigenvalues $U_k(\boldsymbol{R})$ and the corresponding eigenfunctions $\Phi_k(\boldsymbol{r}, \boldsymbol{R})$ of operator $\hat{H}_e$, introduced by relationship (3-6):

$$\hat{H}_e \Phi_k(\boldsymbol{r}, \boldsymbol{R}) = U_k(\boldsymbol{R})\, \Phi_k(\boldsymbol{r}, \boldsymbol{R}), \tag{3-5}$$

$$\hat{H}_e = \hat{T}_e + \hat{V}(\boldsymbol{r}, \boldsymbol{R}). \tag{3-6}$$

Assume that the set of functions $\Phi_k(\boldsymbol{r}, \boldsymbol{R})$ forms a complete orthonormal set; then it is possible (see for example[3-5]) to express the solution of the Schrödinger equation (3-1) as expansion (3-7) (the justifiability of this premise is considered below):

$$\Psi(\boldsymbol{r}, \boldsymbol{R}) = \sum_k \Theta_k(\boldsymbol{R})\, \Phi_k(\boldsymbol{r}, \boldsymbol{R}), \tag{3-7}$$

yielding Eq. (3-1) in form (3-8):

$$\sum_k [\hat{T}_n + U_k(\boldsymbol{R})]\, \Theta_k(\boldsymbol{R})\, \Phi_k(\boldsymbol{r}, \boldsymbol{R}) = E \sum_k \Theta_k(\boldsymbol{R})\, \Phi_k(\boldsymbol{r}, \boldsymbol{R}). \tag{3-8}$$

In order to convert Eq. (3-8) into a more manageable form, it will be multiplied from the left by function $\Phi_l^*(\boldsymbol{r}, \boldsymbol{R})$ and integrated with respect to the electron coordinates; the orthonormality of function $\Phi_k(\boldsymbol{r}, \boldsymbol{R})$ and action of operators ∇_α^2 will be employed; thus we obtain

$$[\hat{T}_n + U_l(\boldsymbol{R}) + \hat{C}_{ll}(\boldsymbol{R}) - E]\, \Theta_l(\boldsymbol{R}) = -\sum_{k \neq l} \hat{C}_{lk}(\boldsymbol{R})\, \Theta_k(\boldsymbol{R}), \tag{3-9}$$

where it holds for operator $\hat{C}_{lk}(\boldsymbol{R})$ that

$$\hat{C}_{lk}(\boldsymbol{R}) = -\frac{\hbar^2}{2}\sum_{\alpha=1}^{N}\frac{1}{M_\alpha}\left[2\int \Phi_l^*(\boldsymbol{r}, \boldsymbol{R})\, \nabla_\alpha \Phi_k(\boldsymbol{r}, \boldsymbol{R})\, \mathrm{d}\boldsymbol{r}\, \nabla_\alpha + \int \Phi_l^*(\boldsymbol{r}, \boldsymbol{R})\, \nabla_\alpha^2\, \Phi_k(\boldsymbol{r}, \boldsymbol{R})\, \mathrm{d}\boldsymbol{r}\right]. \tag{3-10}$$

When $l = k$, the first integral in Eq. (3-10) equals zero, so that the diagonal element $\hat{C}_{ll}(\boldsymbol{R})$ becomes simply the multiplicative operator $C_{ll}(\boldsymbol{R})$. In this way the solution of the Schrödinger equation (3-1) was converted (through solution of the electronic problem for fixed nuclei, (3-5), as an intermediate stage) into an equivalent set of an infinite number of coupled equations (3-9) for functions $\Theta_k(\boldsymbol{R})$. In the decoupling of Eqs. (3-9) it is necessary to neglect the non-diagonal operators $(l \neq k)\ \hat{C}_{lk}$, in other words to limit the infinite expansion of (3-7) to a single term. This approximation proposed by Born[6] and Born and Huang[7] is called the adiabatic approximation[8] (solution of the complete set of equations (3-9) is termed non-adiabatic). Introduction of the adiabatic approximation not only removed coupling in set of equations (3-9) but also converts it into Eq. (3-11) of the Schrödinger type:

$$[\hat{T}_n + U_l(\boldsymbol{R}) + C_{ll}(\boldsymbol{R})]\, \Theta_l(\boldsymbol{R}) = E\, \Theta_l(\boldsymbol{R})\,. \tag{3-11}$$

Eq. (3-11) has an apparent physical significance – it describes the behaviour of the nuclei in the potential field given by the function $U_l(\boldsymbol{R}) + C_{ll}(\boldsymbol{R})$. After separation of the translational motion by transfer to coordinates related to the centre of mass of the system, Eq. (3-11) describes the rotations and vibrations of the system[8-11]. It should be noted here that the standard introduction of the adiabatic approximation[6,7] is formally not entirely correct, as separation of translation is not carried out (the translational continuum). However, the derivation for prior separation of translational motion occurs in a very similar manner; in addition, mass polarization terms may appear, depending on the selection of internal coordinates[8]. Nonetheless, the resultant potential for nuclear motion is the same in both cases. The standard derivation method was retained here, in consideration of the nature of this chapter.

In addition to the adiabatic approximation, an additional, cruder approach, known as the Born-Oppenheimer approximation (also the clamped nuclei approximation[8]) was introduced[1] in the solution of Eq. (3-1) or (3-9). This approximation introduces the omission of the contribution of $C_{ll}(\boldsymbol{R})$ to the potential term of Eq. (3-11) so that, in this framework, the nuclear motion is determined by a potential given simply as the eigenvalue of the electronic Hamiltonian with stepwise fixing of the nuclei (3-6). It is thus apparent that the energy hypersurface, generated by solution of the electronic problem (3-5) for all spatial arrangements of the atomic nuclei, controls the nuclear motion within the nuclear characteristic problem (3-12) in the framework of the Born-Oppenheimer approximation:

$$[\hat{T}_n + U_l(\boldsymbol{R})]\, \Theta_l(\boldsymbol{R}) = E\, \Theta_l(\boldsymbol{R})\,. \tag{3-12}$$

In contrast with the adiabatic and Born-Oppenheimer approximations, the concept of the potential energy hypersurface no longer has any sense in the non-adiabatic approximation, as assignment of the potential terms in Eq. (3-9) that does not have the form of the Schrödinger equation cannot be carried out as it could for Eqs. (3-11) and (3-12).

Separation of the electronic and nuclear motions not only simplifies the equations that must be solved for a description of the system, but also represents an intermediate step between the strictly quantum and classical approaches. This feature appears most strongly during discussion of the concept of molecular structure. It is also apparent, for example in the calculation of the derivatives with respect to the nuclear coordinates for the potential hypersurface $U_k(\boldsymbol{R})$ given by solution of the electronic problem (3-5). The potential energy can be readily converted into an alternative form:

$$U_k(\boldsymbol{R}) = \int \Phi_k^*(\boldsymbol{r}, \boldsymbol{R})\, \hat{H}_e \Phi_k(\boldsymbol{r}, \boldsymbol{R})\, \mathrm{d}\boldsymbol{r}\,. \tag{3-13}$$

Calculation of the derivative with respect to coordinate R_α should lead to the following three terms:

$$\frac{\partial U_k(\boldsymbol{R})}{\partial R_\alpha} = \int \frac{\partial \Phi_k^*(\boldsymbol{r}, \boldsymbol{R})}{\partial R_\alpha} \hat{H}_e \Phi_k(\boldsymbol{r}, \boldsymbol{R}) \, d\boldsymbol{r} + \int \Phi_k^*(\boldsymbol{r}, \boldsymbol{R}) \frac{\partial \hat{H}_e}{\partial R_\alpha} \Phi_k(\boldsymbol{r}, \boldsymbol{R}) \, d\boldsymbol{r} + $$
$$+ \int \Phi_k^*(\boldsymbol{r}, \boldsymbol{R}) \hat{H}_e \frac{\partial \Phi_k(\boldsymbol{r}, \boldsymbol{R})}{\partial R_\alpha} \, d\boldsymbol{r} \, . \tag{3-14}$$

If the fact that $\hat{H}_e$ is the Hermitian operator is employed, then incorporation of Eq. (3-5) permits the third term in Eq. (3-14) to be rewritten as

$$\int \Phi_k^*(\boldsymbol{r}, \boldsymbol{R}) \hat{H}_e \frac{\partial \Phi_k(\boldsymbol{r}, \boldsymbol{R})}{\partial R_\alpha} \, d\boldsymbol{r} = \int \frac{\partial \Phi_k(\boldsymbol{r}, \boldsymbol{R})}{\partial R_\alpha} U_k(\boldsymbol{R}) \, \Phi_k^*(\boldsymbol{r}, \boldsymbol{R}) \, d\boldsymbol{r} \, , \tag{3-15}$$

enabling combination of the first and third terms in Eq. (3-14):

$$\int \frac{\partial \Phi_k^*(\boldsymbol{r}, \boldsymbol{R})}{\partial R_\alpha} \hat{H}_e \Phi_k(\boldsymbol{r}, \boldsymbol{R}) \, d\boldsymbol{r} + \int \Phi_k^*(\boldsymbol{r}, \boldsymbol{R}) \, \hat{H}_e \frac{\partial \Phi_k(\boldsymbol{r}, \boldsymbol{R})}{\partial R_\alpha} \, d\boldsymbol{r} =$$
$$= U_k(\boldsymbol{R}) \int \left[\frac{\partial \Phi_k^*(\boldsymbol{r}, \boldsymbol{R})}{\partial R_\alpha} \Phi_k(\boldsymbol{r}, \boldsymbol{R}) + \Phi_k^*(\boldsymbol{r}, \boldsymbol{R}) \frac{\partial \Phi_k(\boldsymbol{r}, \boldsymbol{R})}{\partial R_\alpha} \right] d\boldsymbol{r} =$$
$$= U_k(\boldsymbol{R}) \frac{\partial}{\partial R_\alpha} \int \Phi_k^*(\boldsymbol{r}, \boldsymbol{R}) \, \Phi_k(\boldsymbol{r}, \boldsymbol{R}) \, d\boldsymbol{r} = 0 \, . \tag{3-16}$$

Thus Eq. (3-14) is finally simplified to yield

$$\frac{\partial U_k(\boldsymbol{R})}{\partial R_\alpha} = \int \Phi_k^*(\boldsymbol{r}, \boldsymbol{R}) \frac{\partial \hat{V}(\boldsymbol{r}, \boldsymbol{R})}{\partial R_\alpha} \Phi_k(\boldsymbol{r}, \boldsymbol{R}) \, d\boldsymbol{r} \, , \tag{3-17}$$

considering that only the second term in Hamiltonian (3-6) depends on the nuclear coordinates. Relationship (3-17) represents the well-known Hellmann[12]-Feynman[13] theorem, i.e. if wave function $\Phi_k(\boldsymbol{r}, \boldsymbol{R})$ is available (or some approximation to it[14]) the calculation of the forces acting on the nuclei is essentially reduced to a procedure analogous to the classical electrostatic interaction formula. For this purpose, according to Eq. (3-17), the classical term for the force component acting on the nucleus, $-\partial V(\boldsymbol{r}, \boldsymbol{R})/\partial R_\alpha$, weighted over the electronic wave function, is sufficient. As will be discussed later, for wave functions satisfying the Hellmann-Feynman theorem, result (3-17) considerably simplifies calculation of the first and second derivatives of the potential energy $U_k(\boldsymbol{R})$.

The Born-Oppenheimer approximation introducing separation of the electronic and nuclear motions demarcates two only slightly overlapping fields in contemporary theoretical chemistry: solution of Eq. (3-5) practically limits the region in which quantum chemistry is effective, while molecular spectroscopy is mainly concerned with solution of Eq. (3-12). Works beyond the limitations of the Born-Oppenheimer approximation are rare[8,15] and always limited to the simplest systems because of the numerical magnitude of the problem.

The work of Born and Oppenheimer[1] was initiated[16] by the need to theoretically explain the experimentally observed separation of the energy levels of the molecules into rotational, vibrational and electronic levels. Using perturbation expansion of the wave functions and operators using powers of the (small) parameter $\varkappa$, given by

$$\varkappa^4 = m/M\,, \tag{3-18}$$

where M is the average nuclear mass, they demonstrated that the second-order terms in the perturbation expansion correspond to the vibrational energy, and the fourth-order terms to the rotational energy and corrections to the vibrational energy. Later, Longuet-Higgins[17] came to the same conclusions on the basis of the variational approach. Recently, Essén[16] produced a new conception of the Born-Oppenheimer approximation demonstrating that the smallness of coefficient $\varkappa$ is not responsible for the separability of the motions in systems with Coulomb interaction, but rather the form of the interaction between the particles. This has been shown for the example of a hypothetical system of particles differing from the usual system of electrons and atomic nuclei in that all the interactions are attractive when the possibility of separation is no longer actual. It should be noted, however, that the separation of molecular energy has been demonstrated to be derivable also in the framework of the non-adiabatic approach, using the generator coordinate method[18,19].

While the Born-Oppenheimer approximation is used in most works, very little is known about the relationships between the values yielded by this approximation and the values from the adiabatic or non-adiabatic approximations. On one hand, the works of Kołos and Wolniewicz[8,20-24] have demonstrated the great usefulness of the Born-Oppenheimer as well as the adiabatic approximation in the calculation of the molecular parameters of the hydrogen molecule. Table 3-1 compares[23] the nonrelativistic dissociation energies of the ground states of the H_2, HD and D_2 molecules

Table 3-1. Comparison of nonrelativistic dissociation energies D_0 of H_2, HD, and D_2 molecules in the ground states calculated[a] in the Born-Oppenheimer, adiabatic, and non-adiabatic approximation

Approximation	$10^{-2}\,.\,D_0(\mathrm{m}^{-1})$		
	H_2	HD	D_2
Born-Oppenheimer	36112.2	36401.5	36745.6
Adiabatic	36118.0	36405.7	36748.3
Nonadiabatic	36114.7	36402.9	36746.2
Experiment	36113.6 ± 0.3	36400.5 ± 1.0	36744.2 ± 0.5

[a] From Ref.[23]

calculated by the Born-Oppenheimer, adiabatic and non-adiabatic approximations, together with experimental data. On the other hand, Czub and Wolniewicz [25] have demonstrated for the electronic ground state of the H_2 molecule in vibrationally excited states that the non-adiabatic pseudopotential and adiabatic potential behave quite differently in the vicinity of each zero of the adiabatic vibrational wave function. The conditions under which the Born-Oppenheimer approximation fails and calculation of the corresponding correction terms are a subject of constant interest (see for example[26-29]). Important application fields lie in the study of vibronic coupling for spectroscopic interpretations[15,30-35] and in the theory of elementary rate processes[36-38].

Separation of the nuclear and electronic motions can typically not be used for degenerate or near-degenerate electronic states when the coupling terms (3-10) neglected in the adiabatic approximation become important[39-41]. It is thus necessary[15,17,40-42] in the study of the Renner effect[43] and of the Jahn-Teller effect[44] to consider the marked dependence of the electronic wave function on the nuclear motion and to abandon the adiabatic approximation. In addition to cases involving intersection or very close approach of the energy hypersurfaces corresponding to different electronic states, the Born-Oppenheimer approximation also fails in regions with very rapid changes of the potential function upon a change in the nuclear coordinates. The recent results of a study[45-48] of the general properties of potential energy hypersurfaces have shown that, in general, it is always necessary to assume the presence of regions on the given hypersurface where the adiabatic hypothesis fails completely.

Longuet-Higgins[45] derived an interesting theorem in this connection. Consider electronic wave function $\Phi(\boldsymbol{r}, \boldsymbol{R})$ that is real everywhere and a configuration of atomic nuclei $\boldsymbol{R}_0$ such that the electronic state is non-degenerate at this given point in the nuclear configuration space. Consider closed loops in this configuration space passing through point $\boldsymbol{R}_0$. Let us begin at $\boldsymbol{R}_0$ and return to it along one of the curves. Such motions permit separation of all the closed curves into two disjunctive subsets (as both functions $\Phi(\boldsymbol{r}, \boldsymbol{R})$ and $-\Phi(\boldsymbol{r}, \boldsymbol{R})$ are solutions of the electronic characteristic problem): those that do not undergo a change in the sign of the function and those along which $\Phi(\boldsymbol{r}, \boldsymbol{R})$ returns to point $\boldsymbol{R}_0$ with the opposite sign. Assume that S is a simply connected surface in the nuclear configuration space, bounded by closed loop L. Then, according to the Longuet-Higgins theorem[45], if function $\Phi(\boldsymbol{r}, \boldsymbol{R})$ changes sign when transported adiabatically around L, then there must be at least one point on S at which $\Phi(\boldsymbol{r}, \boldsymbol{R})$ is discontinuous, i.e. a point at which an intersection with the potential hypersurface of another electronic state appears. This theorem then permits detection of the presence of an intersection between potential energy hypersurfaces on the basis of the behaviour of function $\Phi(\boldsymbol{r}, \boldsymbol{R})$ moving around a closed loop remote from this intersection. Later, Stone[46] gave a generalization for a wave function that is not essentially real (i.e. cannot be rendered real simply by multiplying by a complex number independent of $\boldsymbol{r}$).

The failure of the Born-Oppenheimer approximation in the vicinity of the nuclear configuration leading to degenerate electronic states has two aspects, static and dynamic. The static aspect is related to structural distortion resulting from the Jahn-Teller theorem; the dynamic aspect is related to the non-negligible interaction itself between the motions of the nuclei and electrons in this region[49,50]. Recently, an interesting analogy to the conventional Jahn-Teller effect was suggested[51], i.e. the vibrational Jahn-Teller effect. In van der Waals molecules there is typically a 1-3 order difference in the vibrational frequencies of inter- and intramolecular motions. Analogously to the usual Born-Oppenheimer assumption of the separability of nuclear and electronic motions, the separability of the two types of nuclear motion can also be considered. The analogy does not end here as, if a degenerate molecular state is formed in a symmetric cluster with fixed intermolecular coordinates, then distortion of these coordinates must occur, leading to a decrease in the symmetry of the cluster. Once again, the static (mentioned immediately above) and dynamic aspects can be distinguished, where the latter is connected with interactions between inter- and intramolecular vibrational motions. The dynamic vibrational Jahn-Teller effect permits[51] deduction of the properties of the intermolecular interaction hypersurfaces from the spectra of the molecular complexes.

It should finally be noted that the question of the applicability of the Born-Oppenheimer approximation should be considered always in connection with the type of phenomenon to be studied. The fact that the adiabatic hypothesis permits correct description of some properties of a system does not imply that it is also useful for the, in some sense, more subtle properties of the same system. Simultaneously, however, it should be borne in mind that the precision obtainable from the Born-Oppenheimer approximation is so far quite sufficient for most chemical problems.

3.1.1 The Concept of Molecular Structure

The adiabatic and Born-Oppenheimer approximations permit assignment of a potential energy hypersurface $U(q_1, q_2, \ldots, q_{3N-6})$ to each electronic state of any N-atomic molecule, giving the energy of the system in a hypothetic state of fixed nuclei as a function of $3N$-6 ($3N$-5 for a linear structure) independent coordinates of the nuclear positions. Consider the ith local minimum of this function of $3N$-6 variables, characterized by coordinates $(q^0_{1,i}, q^0_{2,i}, \ldots, q^0_{3N-6,i})$ and the classical energy $E^{(cl)}_{0,i}$:

$$E^{(cl)}_{0,i} = U(q^0_{1,i}, q^0_{2,i}, \ldots, q^0_{3N-6,i}) \,. \tag{3-19}$$

According to the Lagrangian equations from classical mechanics, the time dependence of the classical trajectory of lowest energy passing through the considered local minimum is described by the equations[2]:

$$q_{k,i}(t) = q^0_{k,i} \,, \quad (k = 1, 2, \ldots, 3N\text{-}6) \,, \tag{3-20}$$

and is thus time-independent; the particles remain in positions corresponding to the considered local minimum with constant total energy equal to $E_{0,i}^{(cl)}$. This fact is manifested in the classical theory of molecular vibrations[52] in the existence of a trivial solution to the secular equations. It is apparent that this classical trajectory with the lowest energy represents the classical ground state of the system for a certain region around the ith minimum. At the level of classical mechanics it is natural to designate the arrangement of nuclei corresponding to the local minimum on the hypersurface as the equilibrium configuration. A convenient geometric representation of this arrangement then determines the corresponding molecular structure, a concept on which the great majority of our understanding of the physico-chemical properties of molecules is based (implicitly at least).

Ideally, complete description of the molecular structure would require (as pointed out by Liehr[53–58]) the complete potential energy hypersurface(s) or at least the information triad proposed by Bauer[59]: (i) the precise location of the minima on the potential energy hypersurface for each set of electronic quantum numbers (especially for the lowest energy set); (ii) knowledge of the curvatures of the hypersurface at these minima; and (iii) knowledge of the heights of the barriers that separate adjacent minima. The Bauer concept anticipates three interdependent characteristics of the classical concept of molecular structure: multiplicity of occurrence of local minima within each energy hypersurface, the possibility of transitions between them (even between two different hypersurfaces) and stereochemical non-rigidity of the molecules[60,61] (in this connection the terms[62–67] for fluxional molecules, degenerate intramolecular rearrangement or autoisomerization are also used). The relationships between the properties of the stationary points of the energy hypersurface and the phenomenon of stereochemical non-rigidity are discussed in a review[67]. While the molecular structure of a rigid molecule is represented by a single deep minimum on the hypersurface, non-rigid systems correspond to several equivalent or different, but energetically similar, local minima separated by relatively low potential barriers that can be readily overcome under the given conditions.

Thus, in terms of classical molecular structure, isomerism is simply and graphically interpreted as the occurrence of more than one non-equivalent local minimum on a single hypersurface or on the potential energy hypersurfaces corresponding to the different electronic states. The feasibility of transition between these minima depends on the heights of the barriers separating neighbouring minima and on the temperature (including the possibility of quantum mechanical tunnelling at lower temperatures), and possibly also on the transition probabilities between the individual energy hypersurfaces. The possibility, in principle, of separating two configurations can be expressed either by the above-mentioned Muetterties criterion[60] for their lifetimes or in terms of the characteristics of the hypersurface by the inequality suggested by Bersuker[68]

$$\Delta E \geqq 2h\nu\,, \tag{3-21}$$

where ΔE is the depth of the minimum relative to the lowest barrier and ν is the vibrational frequency associated with this direction. The factor of two results from the inclusion of tunnelling. For very low barriers separating flat minima, the static concept of equilibrium structures is no longer useful and a dynamic model of quasi-free particles must be used[69-73].

The concept of non-rigidity of the molecular structure is mostly introduced and discussed intuitively; nonetheless, it is apparent that it would be useful to introduce some quantitative measure of this property, e.g. in the form of a one-dimensional parameter. A special case of such a structural parametrization is the introduction of the parameter suggested by Yamada and Winnewisser[74], which is a measure of the deviation from linearity in polyatomic chain molecules. This parameter, which qualifies the quasilinearity, was based on the correlation of the levels that become bending vibrations in the limiting case of an ideal linear molecule with levels that become rotations in the limiting case of a bent rigid molecule. In this sense, Berry *et al.*[75-78] recently suggested generalization of the non-rigidity parameter based on a correlation diagram connecting the energy levels of the structure conceived as either rigid or non-rigid.

The selection of the description of the two limiting cases was based on the fact that the model was to be used primarily for the study of non-rigid clusters of atoms. The non-rigid limit was selected in agreement with the procedure used in a similar situation in the study of the conditions in the atomic nucleus, i.e. a set of N identical particles interacting with pairwise harmonic forces, with zero equilibrium distances.

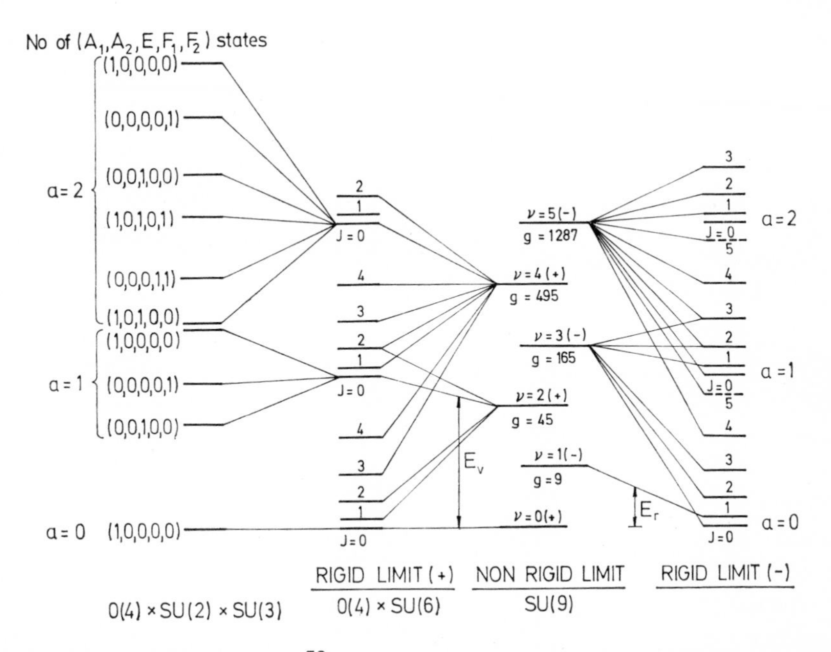

Figure 3-1. The correlation diagram[78] for the 4-body problem with a tetrahedron as the rigid limit. For clarity, the even and odd sets of levels are placed on opposite sides of the non-rigid case

This model has two important features. The energy spectrum of the system can be constructed on the basis of the knowledge of only a few parameters. The symmetry of the model is so high that it includes all the types of symmetry that can be encountered for an arbitrary degree of rigidity (the symmetry of the model is described[79] by the unitary group $U(3N\text{-}3)$). The rigid limit is then often modelled by the usual rigid rotator with vibrations of small amplitude, fulfilling the assumption of the separability of rotations from vibrations. Under certain conditions, however, it is useful to allow greater non-rigidity. For example, for the $(H_2)_2$ system the rigid limit was modelled by two rigid diatomics that were so loosely bounded that free internal rotation was possible. Most often, however, a model based on a spherical top is used, where all of its $3N$-6 vibrational modes are harmonic and degenerate. This artificial model again has the important property of sufficiently high symmetry, as already exhibited by the model for the non-rigid limit. The artificially high degeneracy of the vibrational modes can be removed by transfei to an alternative mcdel of a regular polyhedron.

The correlation diagram is constructed[75-78] in the conventional manner. The individual energy states for each of the two limits are ordered sequentially on a vertical scale with the ground state at the bottom. Because of the harmonic ocsillator approximation, the vibrational levels are naturally equidistant; the spacing of the rotational levels increases linearly with increasing rotational quantum number. Then, assignment of the levels of both limits corresponding to the same representation is carried out. Unambiguousness is attained by using the noncrossing rule for levels of the same symmetry (see Fig. 3-1).

Consider the rigid limiting case for excited states with quantum numbers $\alpha = 0$, $J = 1$ (i.e. the first rotationally excited, nonvibrational state) and $\alpha = 1$, $J = 0$ (i.e. the first non-rotating, vibrationally excited state) and designate their energies as E_r^0 and E_v^0, respectively. The rigid limit states considered are correlated with quantum states with $v = 1$ and $v = 2$ for the non-rigid limit case, where it holds for the energy within this passage that $E_r^0 \to h\omega$, $E_v^0 \to 2h\omega$ (where ω designates the frequency of harmonic vibrations in the non-rigid limit). In these terms, the non-rigidity parameter was introduced as the ratio[75-78]:

$$\gamma = \frac{2E_r^0}{E_v^0} \,. \tag{3-22}$$

It is apparent that $\gamma = 1$ for the non-rigid limit and that $\gamma \ll 1$ for the rigid limit, while for extreme rigidity (represented by infinitely large vibrational frequencies), $\gamma = 0$. It thus holds for positions on the correlation diagram between the two limits that

$$0 < \gamma < 1 \,. \tag{3-23}$$

It is, of course, apparent that the value of γ depends on the model employed; it is not dependent on the model for the non-rigid limit but rather on the representation selected for the rigid limit.

The real molecule lies somewhere between the two rigidity limits. Evaluation of its γ value requires the determination of all its states correlating with states $\alpha = 0$, $J = 1$ and $\alpha = 1$, $J = 0$ and calculation of the average energies for both these state sets (as there are lower degeneracies in the real molecule, in contrast to the idealized limits). These statistically weighted averages correspond to values E_r^0 and E_v^0 when the symmetry of the molecule is as great as that assumed in the correlation diagram. Table 3-2 gives values of γ for a number of three- and four-atom systems[75-78]. It might at first seem surprising that all the values listed are (mostly much) closer to zero than to unity. This is because of the nature of the non-rigid limit, reflecting more the conditions in atomic nuclei (if parameter γ were applied to a nucleus, typical values would lie[78] between 0.1 and 1). The high value of γ for H_2O is connected with the presence of two light hydrogen atoms (the mass (m) dependence of γ can be approximated[75] as proportional to $(1/m)^{1/2}$). The example of NH_3 is interesting, as the γ value obtained as the average over all the vibrational motions is much lower than when only the inversion motion is considered.

Parameter γ represents the first one-dimensional (but physically dimensionless) quantification of non-rigidity. Its most important characteristic is that it is not connected with the existence of the potential energy hypersurface; consequently, the

Table 3-2. Examples of the non-rigidity parameter γ taken from Berry *et al.*[75-78] for various triatomic and tetratomic molecules

System	γ
$Ar—N_2$	0.093
H_2O	0.023
Ar_3[a]	0.013
NO_2	0.0082
O_3	0.0061
SO_2	0.0035
ClO_2	0.0033
$(H_2)_2$[a]	0.40
NH_3	0.034[b]
	0.013
CH_4	0.013
Ar_4[a]	0.0077
P_4	0.00096

[a] Based on estimated parameters (otherwise on observed spectra).
[b] Inversion fundamental only.

possibility of tunnelling was not considered during its introduction. It does have a drawback in that it yields no information on the temperature dependence of the non-rigidity. It has been found that, in terms of γ, the great majority of molecular systems (in contrast to atomic nuclei) approach the rigid limit, so that collective rotational motion is retained.

Recent quantum chemical studies (see e.g.[69-73,80-92]) have yielded a number of instructive examples of non-rigid or even fluxional behaviour for relatively simple systems. For example, Liskow *et al.*[80] carried out *ab initio* calculations in large basis sets of potential curves for the bending motion of the C_3 linear structure, in connection with the existing contradictions between the theoretical and experimental data for the entropy of this molecule. Fig. 3-2 depicts the shape of this curve for the basis set choice considered to be the most reliable. The highly anharmonic nature of this potential is apparent; if it were not for the dimple between 160 and 180° the potential would be readily describable as a square well. Non-rigidity in the C_3 structure is apparent from the following data: on a change in the CCC angle from 180 to 120°, the potential energy increases[80] by only about 1.4 kJ mol^{-1}. This could also explain the anomalously low values of the lowest vibrational levels for bending in C_3 (47, 116, 155, ... cm^{-1}), clearly manifested in the entropy values. C_3 nonrigidity has been interpreted[80] in terms of the Walsh diagrams[93] for the orbital energies.

Non-rigidity to fluxional behaviour can be expected in systems of classical/non--classical carbonium ions[94]. For example, the *ab initio* SCF CI study[82] has demonstrated that, for the vinyl cation, the classical (linear) and non-classical (bridged) structures are very close in energy, while the latter is probably energetically slightly more stable. It has been demonstrated that the barrier for the mutual rearrangement of the two structures is very low, less than 4–13 kJ mol^{-1}.

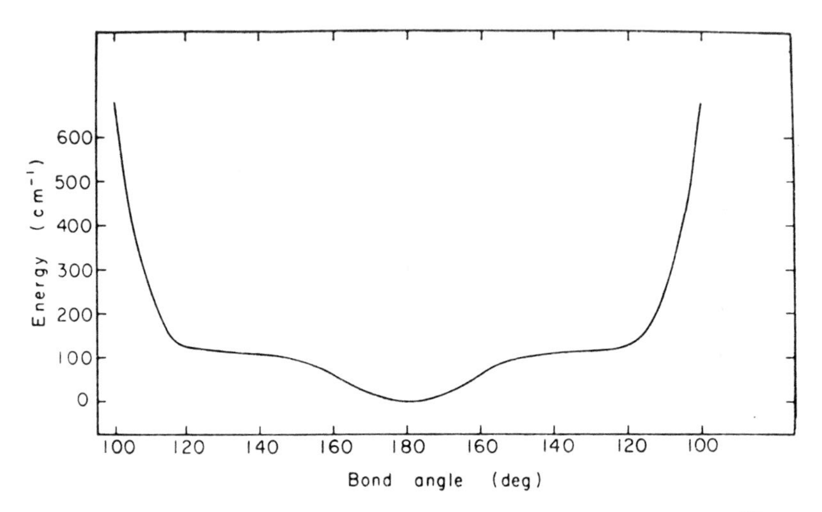

Figure 3-2. The bending potential of C_3 based on the (4s3p1d) SCF calculations[80]

Another species represented by a very shallow potential minimum has been found to be vinylidene, $H_2C{=}C$, for which a potential energy barrier for isomerization to acetylene of 17 kJ mol^{-1} was found by *ab initio* SCF CI calculations[87], while use of the perturbation approach[88] for calculation of the electron correlation led to an estimate of 9 kJ mol^{-1}. Alternatively, it has even been proposed[88] that vinylidene be considered as an effective transition structure for the degenerate rearrangement of acetylene, connected with exchange of hydrogen atoms. The ability of singlet vinylidene to be rapidly converted to acetylene leads to a paradoxical situation, i.e. that triplet vinylidene should have[89] a much longer lifetime.

Provided that certain simple thermodynamic and kinetic criteria are fulfilled, then this type of structural behaviour, i.e. non-rigidity with possible structural fluctuations, must be expected for a given tautomeric system. This should be the general approach; however, so far quantum chemical calculations have been carried out only for simple, model systems[84,91].

A great deal of data on structural non-rigidity has been collected in the quantum--chemical study of inorganic systems[69–73,85]. It has also been found here that classical concepts of a static molecular structure, of atomic valence or of spatial orientation of bonds cannot be used. A good example is the Li_2BeH_4 system. *Ab initio* SCF calculations[73] (describing only the isolated molecule) have shown that a planar cyclic structure $[LiHLi]^+ [BeH_3]^-$ lies even lower than the classical structure $Li^+[BeH_4]^{2-}Li^+$, by about 10 kJ mol^{-1}. No appreciable energy barrier was found between the two structures, so that it must be assumed that continuous conversion from one form to the other occurs in the Li_2BeH_4 system, which is connected with reorganization and destruction of its anion component. The simplest example of these polytopic molecules is the ABC ↔ BCA system, among which recent SCF calculations[95] included the LiOB, OBLi system. Simple van der Waals systems currently being studied (see e.g.[96–103]) can be considered as an extreme case of non-rigid systems.

Recently, a number of theoretical studies of non-rigidity and flexibility in biomolecules have been carried out[104–113], whose behaviour results from the multiplicity and variety of the interactions and motions involved. This flexibility and the associated structural fluctuations have often been found to be a necessary factor in explanation of the function of biomolecules, e.g. globular proteins[107]. For example, the recent discovery[104] of the extreme conformational flexibility of the furanose ring can have far reaching consequences for the structure of polynucleotides and the rigidity of DNA. Calculations by the consistent force field method have demonstrated that there are two minimum energy conformations of the furanose ring separated by a barrier with a height of only about 2.5 kJ mol^{-1}.

All these recent quantum-chemical discoveries of molecular non-rigidity and fluctuations document that, even though the classical concept of molecular structure is completely justifiable and readily useable in certain situations, there are (even in the framework of the Born-Oppenheimer approximation) cases where this is not true.

This holds especially for chemical isomers for which a distinct, clearly defined character is not a universally useful characteristic. On the contrary, under certain conditions these results emphasize the possibility not only of mutual inseparability but even indistinguishability of isomers.

Now reconsider, within the adiabatic or Born-Oppenheimer approximations, the forms of the characterization of the classical concept of molecular structure. In terms of the characteristics of the energy hypersurface, the equilibrium configuration is clearly defined and is described by the equilibrium structural parameters r_e of the corresponding minimum on the potential energy hypersurface, with the significance of the distance between the (idealized) pairs of motionless nuclei in the free molecule. While quantum chemical calculations generally lead to value of r_e, experimental results are much more complicated. So far, structural parameters r_e could be found experimentally[114-117] only for some of the simplest molecules. At every non-zero temperature many experimentally observed values are averaged over all the rotational-vibrational states of the species or even of all the isomers that may be present. Averaging occurs through the corresponding factors following from the Boltzmann distribution. The dependence of the measured values on the experimental temperature is thus apparent. Even if the experiment were to be carried out at absolute zero temperature, the values obtained would be averaged over the vibrational motions of the ground vibrational state of the most stable isomer. Thus, the experimental data employed does not generally yield the r_e values, but various modified analogues belonging to certain average structures, e.g. the values[114,115,118-123] of r^0, r_v, r_α, r_g, r_a, r_s, r_m, etc. Parameter r^0 (also called r_z) carries the significance[114,115] of the distance between average nuclear positions in the ground vibrational state; r_v designates the same quantity for the vth excited vibrational state. Parameters r_α and r_g are the values defined[122] under conditions of thermal equilibrium: the values of r_α

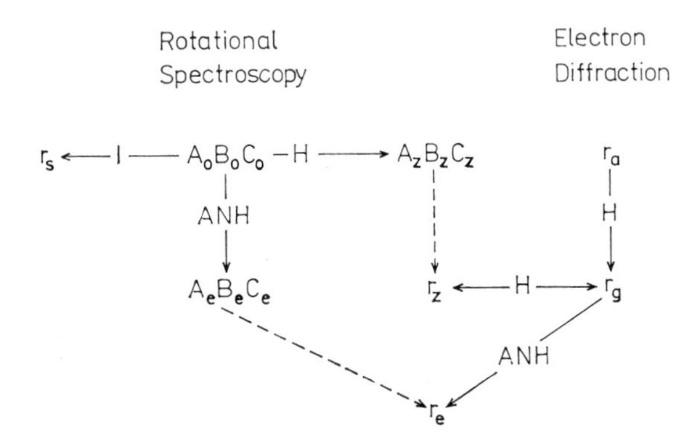

Figure 3-3. Diagram[121] showing the relationship among the rotational constants (A, B, C) and structural parameters determined by spectroscopy and gas electron diffraction. Symbols H and ANH refer to the harmonic and anharmonic corrections for the vibrational effects, respectively, and I stands for isotopic substitution

correspond to the distances between thermal average nuclear positions, and value r_g designates the thermal average value of the internuclear distance. In addition to r_g, parameter r_a is also used[118,119]; this is related to the former through the mean square of the vibrational amplitude. The substitution or r_s structure results from[118,119] the evaluation of the effect of isotopic substitution on the principal moments of inertia. Using the perturbation theory, Watson[124] obtained the mass--dependent structure, r_m, which generally represents a better approximation to the r_e values than the r_s terms. A number of expressions have been proposed for the relationships between these structural parameters[114,115,118–124]. Qualitatively, these relationships between the structural parameters obtained from rotational spectroscopy and electron diffraction are depicted in Fig. 3-3.

The existence of a number of possible representations of molecular geometry and the resultant inconsistency of the experimental information available in the literature has complicated the comparison of the quantum chemical r_e values with the experimental data. For simple and relatively rigid systems, however, the differences resulting from the use of different experimental techniques will probably be smaller than the error involved in the quantum chemical calculation. For example, the differences between the related r_g and r_s distances, taken from recent[121] precise experiments, varied mostly between 0.014 and -0.002×10^{-10} m; however, even differences as large as 0.025×10^{-10} m (the C—Cl bond in *t*-butylchloride) or -0.010×10^{-10} m (the C=C bond in acetone) cannot be considered as exceptional. On the other hand, it should be considered that, for very small molecules, the differences in the values of the structural parameters for the various possible representations are much larger than the actual experimental error in contemporary precise measurements (see Tab. 3-3). It is useful to note, for example, that, for CO_2, the observed r_g or r_a values lead[119] to disagreement with the $D_{\infty h}$ equilibrium symmetry. When experimental structural information is used for testing the quality of quantum-chemical calculations of parameters r_e, it must be borne in mind, especially for non-rigid molecules, that the experimental data are not measured directly but are derived on the basis

Table 3-3. Comparison of structural parameters r_e, r_g, and r_a for CO_2 and SO_2[a]

System	Parameter	r_e	r_g	r_a
CO_2	C—O (10^{-10} m)	1.1600	1.1652	1.1646
	O—O (10^{-10} m)	2.3200	2.3251	2.3244
	∢ OCO (°)	180.00	172.27	172.64
SO_2	S—O (10^{-10} m)	1.4308	1.4360	1.4351
	O—O (10^{-10} m)	2.4696	2.4771	2.4758
	∢ OSO (°)	119.31	119.20	119.22

[a] From Ref.[119]

of a mathematical treatment of the observed quantities using a model and theoretical formalism. The quality of the theoretical interpretation model thus affects the resulting numerical values. A good example is the study[125,126] of the pronounced asymmetry of the methyl group in compounds of the CH_3YH_2X type ($Y = C, Si; X = F, Cl, Br$) exhibited in the r_s structures found from microwave spectra. Calculations based on the experimentally determined force fields[125] and on the *ab initio* SCF approach[126] have demonstrated that this substantial asymmetry is an artefact of the substitution structure resulting from molecular vibrational effects.

The existence of vibrational or vibrational-rotational motions is also manifested in the temperature dependence of the observed structural parameters. A systematic study of this effect for simple molecules can be found in the works of Hedberg *et al.*[127-129] and Cyvin[130]. The study of the effect of pressure on structural parameters[131-135] is formally analogous: for example, the effect of pressure-produced changes in the (effective) angle of internal rotation and the corresponding changes in the equilibrium ratios between the conformational isomers. In contrast to the temperature dependence of the structural parameters, the detailed theory has not yet been developed on a molecular level for these situations.

All the above considerations concerning the relationship between the experimental and quantum-chemical structural parameters are also valid for experimental studies of mixtures of several isomers which could not be separated in the temperature interval suitable for the measurements. Utilization of diffraction methods for this purpose assumes a high structural similarity of the isomers. This permits a decrease in the number of parameters deduced from the mathematical treatment of the intensity curves. This reduction in the number of parameters is important, as for mixtures it is also necessary to determine the component ratio. Consequently, diffraction methods are used for finding the structural parameters of mixtures of rotamers[136-139]. Knowledge of the relative contents of the individual isomers is not necessary for microwave spectroscopy of mixtures of isomers[140-142]; it is, however, necessary to assign the individual transitions to the particular isomers. An example of the determination of the structural characteristics of a mixture of three isomers by microwave spectroscopy is the study of epifluorhydrine[140]. The NMR technique is an important tool in the study of molecular conformation[143,144] in solution. However, as Jardetzky[144] recently emphasized, if the individual contributing conformations are not known, the average structure deduced from NMR data is completely without physical significance. It is apparent that combination with quantum-chemical information would be useful here[145], especially when the information can be readily transferred from the isolated molecule to conditions in the condensed phase.

3.2 Molecular Energy

Construction of the potential energy hypersurface, i.e. solution of the characteristic problem (3-5) representing the centre of interest of quantum-chemical methodology, can be carried out at various approximation levels, either in the choice of the Hamiltonian $\hat{H}_e$ or (primarily) in the obtaining of the corresponding electronic wave function Φ (for recent reviews, see for example[146-155]). The most frequently used approach at present is represented by the Hartree-Fock treatment, where the many--electron wave function is approximated by a single determinant built up from the molecular orbitals. The molecular orbitals themselves are mostly represented by a linear combination of a finite set of atomic (basis) functions (mostly of the Slater or Gaussian type). The expansion coefficients required for carrying out this approximative step are obtained from the usual SCF scheme. If the applicable set of atomic functions is sufficiently large and flexible, then the dependence on the basis set disappears and the Hartree-Fock limit is attained. This is, however, not true for most molecules, where the selection[156] of a basis set compatible with the reliability requirements and the feasibility of the calculation become important. In the framework of the *ab initio* method, all the integrals involved over the basis set functions are properly calculated; if this is not so, then the method is said to be semiempirical. In the *ab initio* method, the magnitude of the minimal basis set of Slater orbitals can be considered to be a meaningful lower limit. An extension of this rudimentary basis set leads to double zero basis sets, where every atomic orbital of the minimal basis set is replaced by two orbitals with the same symmetry, but with different exponents. A further expansion of the basis set includes the polarization functions (for example the d-orbitals on carbon), permitting the description of the charge distribution in the system. Table 3-4 gives one of the possible simple classification schemes for basis sets.

Table 3-4. Illustrative classification of basis sets used in ab initio calculations

Type of basis	Examples
Perfect orbitals	Very large bases
Split shell + polarization	6-31G*, 6-31G**
Split shell	4-31G, 6-31G
Minimal	STO, STO-3G

Gradually, the approaches replacing the single determinant approximation to the wave function by a many determinant approximation are becoming more common. This procedure permits description of the correlation of motion of electrons with different spins. This is mostly carried out (see for example[157-161]) through the configuration interaction technique (either as multiconfiguration interaction (MC) carrying out complete variational optimization of all the expansion coefficients or as simple configuration interaction (CI) fixing the expansion coefficients from the SCF solution and optimizing only the coefficients for the individual configurations), electron pair approximations or using the perturbation theory. Construction of the correlated wave functions is typically realized in the framework of the *ab initio* approach to atomic integrals.

Semiempirical methods are constantly being developed for applications in the field of large systems; these methods are characterized by use of the SCF scheme, minimal Slater orbital basis set and reduction of the number of evaluated atomic integrals. This is balanced by the use of some experimental information (e.g. the atomic ionization potentials) in the framework of approximative expressions for some atomic integrals. In addition, some of the adjustable parameters can be varied in order to obtain better agreement between some of the calculated and observed quantities (heats of formation) for a reference set of substances. Typical examples are the CNDO, INDO, NDDO methods[162], followed by the MINDO and MNDO methods[163]. In all these methods, overlap between orbitals is neglected to a certain degree, with retention of rotational invariance. The SINDO method[164], working with symmetrically orthogonalized orbitals, is based on similar principles. A typical region of application of semiempirical methods involves relatively large molecules containing heavy atoms. For this purpose, parametrization including elements from H to Br was recently proposed[165,166] in the framework of the CNDO/INDO formalism.

For even larger systems, the most frequently used methods include phenomenological potentials, especially in the framework of the consistent force field approach[167-170]; the approach taken in its parametrization is analogous to that in semiempirical quantum-chemical methods.

In the subsequent text, the particular means used for the construction of the given molecular energy will mostly be disregarded. We will be primarily concerned with the general features of the molecular energy as a function of the positions of the atomic nuclei or characterization of the corresponding molecular system deductible from the potential energy hypersurface.

3.3 Localization and Identification of Stationary Points on the Energy Hypersurfaces

The present state of numerical quantum chemistry, i.e. (to a considerable extent) of computer technology, does not in general permit calculations of the potential energy hypersurfaces for systems that are sufficiently large to be chemically interesting. Consequently, the contemporary theory of chemical reactivity is concerned with representation of energy hypersurfaces in terms of selected points, namely stationary (or critical) points[169-176] introduced by relationship (3-24):

$$\left(\frac{\partial U(\boldsymbol{R})}{\partial q_i}\right)_0 = 0\,; \quad i = 1, 2, \ldots\,; \quad \boldsymbol{R} = \{q_i\}\,, \tag{3-24}$$

and concepts of chemical reactivity are formulated in terms of these points[177-179], especially in organic chemistry[171]. Further differentiation among the individual stationary points is made possible by the force constant matrix **F** (or the eigenvalues of this matrix), whose elements are introduced by the relationship

$$F_{ij} = \left(\frac{\partial^2 U(\boldsymbol{R})}{\partial q_i \, \partial q_j}\right)_0 \tag{3-25}$$

(index 0 indicates that the elements of the matrix are evaluated at the stationary points of the hypersurface). It should be noted for the sake of completeness that in the literature matrix **F** is frequently termed the Hessian matrix (see e.g.[174,175]). The eigenvalues of matrix **F** corresponding to local minima on the energy hypersurface are all positive[171], while critical points with just one negative (and otherwise positive) eigenvalue(s) of matrix **F** have the significance[171] of transition states or activated complexes. If all the eigenvalues of matrix **F** are negative, then a local maximum is involved[171]. If at least one eigenvalue is equal to zero, then a degenerate critical point is present[175]. Under certain conditions[175], a degenerate critical point may also be a minimum or a shoulder. Non-degenerate stationary points that have at least one positive and one negative eigenvalue of matrix **F** are termed saddle points[175]. It should also be noted for the sake of completeness that, while the terms transition state and activated complex are used synonymously in quantum-chemical studies of potential energy hypersurfaces (see for example[171]), the term transition state can have much broader significance in chemical kinetics.

If the stationary points on the energy hypersurface of a given system were located by the trial and error or simplex method[180-182], there would be no great difference in the computational requirements compared to the construction of the potential energy hypersurface in the analytical form. A principal change was brought in the localization of the stationary points by the gradient methods introduced into the field of quantum-chemical studies of molecular structure by the pioneering works of Pulay[183-185] and McIver and Kormornicki[186]. The application of optimization techniques in theoretical chemistry is much older[187,188]; even structural problems solved at the level of the consistent force field method[170] were solved with the same scheme [167] as that used later for quantum-chemical methods. In general, this scheme can be separated into two steps: (i) construction of the values of the components of the potential energy gradient from the analytical formulae in a chosen point of the energy hypersurface, and (ii) recurrent generation of the subsequent approach to the coordinates of the stationary point sought by an iterative optimization technique selected according to the character of the problem.

Expressions are given for the first derivatives of the total energy in the framework of the single determinant LCAO approximation with respect to a non-linear parameter (here this is the coordinates of the atomic nucleus) in the works of Moccia[189] and Gerratt and Mills[190]; however, this result was essentially already contained in the works of Hurley[191] and Bratož[192]. Even the simplest versions of the SCF method lead to the energy in terms of a function of at least two types of variables – the nuclear coordinates and the LCAO expansion coefficients; however, the latter are naturally dependent on the nuclear coordinates. Nevertheless, as the wave function is found by a variational technique (and thus the energy is stationary with respect to these linear coefficients), it is no longer necessary to differentiate the LCAO expansion coefficients with respect to the nuclear coordinates, which would be a serious problem. This very important property of the variationally determined functions that permits quite straightforward calculation of their gradients in the analytical form was originally recognized by Hurley[191]. However, the dependence of the wave function on the nonlinear parameters[193], particularly on the atomic orbital exponents, can be involved in the *ab initio* calculations. Moreover, the problem may include the Lagrangian multipliers in connection with the constraints placed on the wave function, namely the orthonormality conditions. Nonetheless, as finally pointed out by Epstein[194], Hurley's simplifying condition is valid even in the presence of Lagrangian multipliers. It thus holds that, provided that the energy is stationary with respect to both the expansion coefficients and the Lagrangian multipliers, only the explicit dependence of the energy on the atomic nuclei coordinates need be taken into consideration in construction of the first derivatives. The detailed form of these formulae for the *ab initio* SCF method for systems with closed and open shells is given, for example, in references[193,195–198]. It should, however, be noted for the sake of completeness that gradient calculation does not lead to such a simple scheme if the wave function is not variationally optimized. An example is the CI wave functions, as only the CI expansion coefficients are determined variationally. If, however, MC SCF wave functions are employed, where all the expansion coefficients are determined variationally, the simple scheme is again valid for the gradient calculation.

A critical factor in the analytical energy gradient calculation in the SCF methods is thus the determination of the derivatives of the atomic integrals over expressions built up from the basic functions with respect to the atomic nuclei coordinates. This can be reduced[186,193] to the sums of the integrals containing the derivatives of the basis functions with respect to the nuclear coordinates. The derivatives of the Gaussian functions can be readily drawn up, as described by Boys[199]. The corresponding analytical formulae for the Slater type orbitals are given in the work of Gerratt and Mills[200] for a specially selected coordinate system (this special case leads to derivatives only for the s, p basis set); this problem is solved quite generally in ref.[201]. It has been found[201] that the derivatives of an arbitrary Slater orbital with respect to the Cartesian nuclear coordinates can be expressed as a linear combination of, at most, eight Slater orbitals of various types. Thus, the results listed in the works[183,189–201] permit the derivation of the analytical formulae for the derivatives of the total energy with respect to the nuclear coordinates in an arbitrary quantum-chemical method, working in the framework of the single-determinant LCAO approximation in the Slater or Gaussian basis set.

A simplification was pointed out in Chap. 3.1, introduced by the Hellmann-Feynman theorem into the calculation of the energy gradient (see Eq. (3-17)). The theorem is primarily satisfied for exact wave functions that are, however, encountered only rarely. Nonetheless, there are a number of types of approximate wave functions that satisfy this theorem. Introduction of these function classes is useful not only for calculation of the first but also of the second derivatives of the energy (where, however, knowledge of the derivatives of the expansion LCAO coefficients is generally necessary for variationally obtained wave functions). First of all, it can be demonstrated that the Hellmann-Feynman theorem is fulfilled at the level of the Hartree-Fock limit[191,202,206], although this fact can hardly be used in practice. This theorem is not generally satisfied for the approximative SCF wave functions: if the formal classification given by Pulay[193] is maintained, the Hellmann-Feynman force is, in addition, accompanied by the density force and integral force terms. If the basis functions are, however, optimal with respect to all the non-linear parameters (i.e. floating functions[191]) then the wave functions formed from them (called stable functions[202]) fulfill the Hellmann-Feynman theorem. Stable functions are not very suitable from the computational point of view as, even though they simplify the construction of the gradient, optimization of the nonlinear parameters makes the overall calculation too complicated[193].

Recently, Nakatsuji *et al.*[207–209] suggested a simple theorem valid for both the SCF and MC SCF approaches as well as for other types of theories, to test whether the Hellmann-Feynman theorem is satisfied. A sufficient condition for the wave function constructed within basis functions χ_r to satisfy this theorem is that the basis set includes the derivative χ_r' for any basis element χ_r. Such a set is $\{\chi_r, \chi_r', \chi_r'', \ldots\}$, that can become finite for a recurrent basis, i.e. when $\chi_r = \chi_r^{(n)}$. In addition, it has been shown[207] that, when sequence $\{\chi_r, \chi_r', \chi_r'', \ldots\}$ is artificially terminated sooner (e.g. when only the first derivatives of the basis set are included), the degree to which the Hellmann-Feynman theorem is fulfilled can be very satisfactory. Similarly, the Hellmann-Feynman force can be used[210] as an approximation of the exact energy gradient in some cases, as a time-saving tool for the first few iterations in the geometry optimization. It should be noted that, even though this theorem is not satisfied for the CI wave functions, nonetheless it can be useful[211] for construction of the exact energy gradient in this approximation even for a finite-dimensional CI matrix.

If the analytical formulae permitting calculation of the components of the total energy gradient at an arbitrary point on the energy hypersurface are available, then solution of step (ii) can be begun. Quantum-chemical optimization of the molecular geometry is most often carried out using methods based on the general class of variable metric minimization methods (for reviews, see for example[170,172,187,188,212–214]), first proposed by Davidon[215] and later extended by Fletcher and Powell[216]. The Davidon-Fletcher-Powell method is considered[170,198] to be the most effective general minimization procedure requiring knowledge of the first derivatives; for quantum-chemical purposes it is extensively used in the Murtagh and Sargent version[217]. In this optimization procedure, the $(n + 1)$th approximation to the coordinates of stationary point $\boldsymbol{R}_{n+1}$ is generated from the values of the nth approximation $\boldsymbol{R}_n$ using recursion formula (3-26):

$$\boldsymbol{R}_{n+1} = \boldsymbol{R}_n - \alpha_n \mathbf{A}_n \,\mathrm{grad}_n\, U\,, \tag{3-26}$$

where $\mathrm{grad}_n\, U$ designates the total energy gradient at point $\boldsymbol{R}_n$, $\mathbf{A}_n$ is a certain square matrix and α_n is a constant, selected according to the following proposed rules: α_n is chosen as unity provided that step (3-26) leads to a decrease in the potential energy, otherwise the value of α_n is halved and the process is repeated until a decrease does occur. In the first calculation step matrix $\mathbf{A}_n$ is chosen as an identity matrix (i.e. optimization is carried out in the same way as in the steepest descent method[170]). In the subsequent steps, matrix $\mathbf{A}_n$ is constructed according to the recursion formula:

$$\mathbf{A}_n = \mathbf{A}_{n-1} + z_n z_n^+ / c_n\,, \tag{3-27}$$

where superscript '+' designates transposition and vector z_n is constructed according to the relationship

$$z_n = -\mathbf{A}_{n-1}[\mathrm{grad}_n\, U - (1 - \alpha_{n-1})\, \mathrm{grad}_{n-1}\, U]\,, \tag{3-28}$$

while the scalar term c_n is found from the relationship

$$c_n = (\mathrm{grad}_n\, U - \mathrm{grad}_{n-1}\, U)^+ \, z_n\,. \tag{3-29}$$

It can be demonstrated[217] for matrix $\mathbf{A}_n$ that it converges to the inverse of the force constant matrix at the stationary point in question.

The described optimization method is symmetry conservative[171,186,218-220], i.e. when the numerical calculations are sufficiently precise the point group of symmetry of the optimized structure is retained in each step. This fact is a result[186] of the circumstance that the gradient in Cartesian coordinates transforms according to the totally symmetric representation of the corresponding point group of symmetry. This fact can simplify the search for all the minima on the given hypersurface, or at least all the minima with the given symmetry, and also affects the applicability[219,220] of the variable metric method for localization of saddle points on the studied hypersurface. Similar to most other optimization procedures, the variable metric method cannot prefer the global minimum over localization of the other local minima. Nonetheless, it should be noted for the sake of completeness that an iterative technique is available in the literature for global optimization[221].

While methods of the variable metric type employ only first derivative values, the literature also contains procedures employing explicit formulae for the second derivatives. However, at present these techniques are used rather more in connection with the methods of molecular mechanics[170,222] than for quantum chemical structure optimization. An example of a method working with both the first and second derivatives of the minimized function is the Newton-Raphson method, e.g. in a modification[223]. On the other hand, however, even the simple method of the steepest descent can be useful[170]; in this method matrix **A** in relationship (3-26) is replaced by the identity matrix. It has been found that, for the various situations that can be important in structural optimization, it is preferable to select the optimization procedure according to the momentary nature of the problem rather than to employ a single technique. For example, for conformation studies it was recommended[170] for situations where the starting structure is a very bad estimate of the minimum that the calculation be begun with the method of steepest descent and then that the final optimization be carried out by the Newton-Raphson method. Between these two extremes lies a region that is typically convenient for the application of the Davidon-Fletcher--Powell method. In conclusion, it should be noted for the sake of completeness that it has recently been demonstrated[224] that there are certain differences in the symmetry properties of the Newton-Raphson method compared to the remaining two procedures discussed here, which contributes to the differences in the convergence rates.

Techniques based on energy minimization are intended primarily for localization of stationary points of the minimum type. Saddle points, and especially transition states can, under certain conditions (when certain symmetry requirements are fulfilled or when the starting structure accidentally almost overlaps with this saddle point) also be localized by energy minimization; however, this procedure is not universally useful (cf.[171,225,226]). McIver and Komornicki[171] suggest determination of the positions of the transition states in terms of the minimization of function $\sigma(\boldsymbol{R})$:

$$\sigma(\boldsymbol{R}) = \operatorname{grad} U(\boldsymbol{R}) \cdot \operatorname{grad} U(\boldsymbol{R}), \tag{3-30}$$

i.e. the square of the Euclidian norm of the energy gradient. Function (3-30) was minimized using a modification of the least squares method, as described by Powell[227], that does not require knowledge of the derivatives of the gradient components. Minimization of function (3-30) can also be used for localization of minima on the potential energy hypersurfaces; it has been found[171] that this procedure is as efficient as the above use of the variable metric method. The search for the transition states can nonetheless be treated by the energy minimization method, provided that a single suitable coordinate is selected and the optimization is carried out in a subspace with one less dimension. Recently, Rothman and Lohr[228] found rules for the suitable selection of this coordinate and conditions so that the search led to localization of the transition state. The Cerjan and Miller[229] algorithm for localization of the transition state employs the matrix of the second derivatives of the energy explicitly in each step. On the other hand, however, techniques not requiring construction of the energy gradient[230-234] have also been proposed for this purpose. Such an alternative to methods based on gradient minimization is represented, for example, by the synchronous transit technique[231], the 'X-method'[232] or the promising procedure proposed by Müller and Brown[233] (see below). The problem of localization of transition states and methods useful for this purpose overlap considerably with the construction of (in a certain sense) optimal curves on the energy hypersurfaces (reaction paths), and thus this problem will be reconsidered in the subsequent discussion. This is also true of the problem of the symmetry properties of stationary points, especially of transition states.

After localization of the stationary points on the hypersurface, numerical construction of the force constant matrix is usually carried out[171] (3-25); after diagonalization, this matrix can be used in a reliable test of the type of each stationary point of the studied potential energy hypersurface. Recently, Scharfenberg[235] pointed out that the eigenvalues of the **F** matrix yield more information than is absolutely necessary for classification purposes (i.e. knowledge of these values is sufficient but not necessary). Instead of the solution of the characteristic problem of matrix **F**, he proposed its factorization in terms of two triangular matrices and a diagonal matrix within a very simple algorithm. Information on the types of stationary points can be very readily extracted from this decomposition.

It should be noted for the sake of completeness that, in addition to optimization techniques for the localization of stationary points (especially minima) employing the analytically constructed values of the components of the potential energy gradient, procedures have also been developed[236-241] that employ the gradient values constructed by the numerical finite difference method. This approach is particularly useful in connection with a carefully chosen iterative optimization technique (cf.[236-238]).

The effectiveness of all the optimization procedures is, *inter alia*, dependent on the selection of the starting structure; consequently, it has been proposed[198] that optimization at the *ab initio* level be commenced using structures that have been previously optimized at a semiempirical level. An improvement in the convergence was also attained[242] by using information from spectroscopic experiments or empirical potential fields in the construction of matrix **A** in relationship (3-26). For estimation of a suitable starting structure for the localization of transition states (especially for isomerizations) it has been proposed[243] that the linear internal coordinate path method be employed. A set of internal coordinates shared between the reactant and the product is selected and they are all varied linearly and simultaneously from their reactant to their product values. This results in a continuous pathway connecting the reactant and product. It has been demonstrated[243] for this pathway that a secondary minimum of the square of the gradient norm may exist (this value is essentially zero for the reactant and product geometries); the structure of this secondary minimum is then used as an initial estimate for optimization of the transition state.

However, none of the numerical optimization methods yields a criterium that would determine the number of stationary points on the given hypersurface with any certainty. The only way to ensure the localization of all the stationary points with relative certainty is thus systematic variation of the starting structures. The optimization of the conformation of N-acetyl-N′-methyl amides of aminoacids is an example of the extent of such studies; here up to 20 000 starting structures had to be optimized[244] for the localization of the minima in some cases. Similarly, in the study of cyclohexaglycyl, a total of 2160 starting conformations was used for the localization of nine minima with C_2 symmetry[245]. For some applications it may be sufficient[246] or desirable to localize only the global (absolute) minimum on the hypersurface, disregarding all other stationary points. Under these circumstances, the Bremermann technique[221] mentioned above is useful. This may be the case of the empirical potentials that can exhibit both realistic and artificial minima (cf.[247,248]).

Since the pioneering works of Pulay[183-185] and McIver and Komornicki[186], a number of papers have been published[170,198,222,242,249-280] dealing with the automatic optimization of the molecular structure for various types of energy function using analytical formulae for the first derivatives (for reviews, see[193,281,282]). At present, efficient computer programs are available for all the important types of theoretical energy functions, beginning with empirical potentials[170,222,249-251] and the EHT method[252-254], through semiempirical[186,255-266] and *ab initio* SCF[183-185,198,242,267-271] methods, to various types of *ab initio* correlated wave functions (CI[211,272-277], MC[278,279] and those based on perturbation theory[280]). Illustrative examples are given in Table 3-5.

The existence of efficient computer programs for automatic optimization of the molecular geometry naturally considerably simplifies the systematic study of the quality of the structural parameters yielded by the individual types of energy functions (cf.[193,281-284]). Without going into detail, some contemporary problems will be mentioned here. The prediction ability of quantum chemical methods for the study of small organic molecules is well documented[281-284]. This is illustrated by the

Table 3-5. Selected illustrative examples of automatic geometry optimization techniques with analytic evaluation of the energy gradient

Methodical level of $U(\boldsymbol{R})$	References
Empirical (incl. EHT)	Lifson and Warshel (1968)[167], Niketić and Rasmussen (1977)[170], Ferguson and Beckel (1973)[252], Slanina *et al.* (1978)[253], Beran *et al.* (1981)[254]
Semiempirical	McIver and Komornicki (1971)[186], Zeiss and Whitehead (1971)[255], Rinaldi and Rivail (1972)[256], Pulay and Török (1973)[257], Pancíř (1973)[258], Grimmer and Heidrich (1973)[260], Rinaldi (1976)[261], Dewar and Komornicki (1977)[262], Khalil and Shanshal (1977)[263], Scharfenberg (1978)[264], Dewar and Yamaguchi (1978)[265], Leška *et al.* (1979)[266]
Ab initio SCF	Pulay (1969)[183], Schlegel *et al.* (1975)[267], Huber *et al.* (1976)[268], Komornicki *et al.* (1977)[198], Dupuis and King (1978)[269], Pople *et al.* (1978)[270], Pulay (1979)[271]
Ab initio CI SCF (incl. perturbation treatment)	Tachibana *et al.* (1978)[211], Pople *et al.* (1979)[280], Yamashita *et al.* (1980)[272], Brooks *et al.* (1980)[273], Krishnan *et al.* (1980)[274], Osamura *et al.* (1981)[276]
Ab initio MC SCF	Kato and Morokuma (1979)[278], Goddard *et al.* (1979)[279]

Table 3-6. The mean absolute differences[a,b] between theoretical (INDO, *ab initio* SCF in the STO-3G and 4-31G basis sets) and experimental values of selected structural parameters in simple organic molecules

Structural type	INDO	STO-3G	4-31G
A-H Bonds	0.046	0.021	0.010
A-B Bonds	0.071	0.032	0.011
Bond angles	3.1	3.3	4.1

[a] According to Ref.[283]
[b] Bond lengths in 10^{-10} m, bond angles in °.

data[283] in Tab. 3-6 that also demonstrate that, under some circumstances, the results of the semiempirical methods may be of better quality than those obtained by the *ab initio* SCF approach. It is, however, necessary to critically consider the selection of the basis set in *ab initio* methods, as demonstrated by more recent comparative studies[285-288]. A number of examples of the necessity of including correlation effects for correct description of the structural characteristics of some systems have been described[289-292]. In some situations, geometric optimization is carried out only partially with fixation of some degrees of freedom (e.g. intramolecular internal coordinates in molecular complexes). It has, however, been demonstrated[293], that the transition from a partially rigid to a completely relaxed structure may lead to non--negligible effects on the values of all the structural parameters.

3.4 Curves on the Potential Energy Hypersurfaces

In addition to localization of stationary points on the given potential energy hypersurface, some of the curves lying on it can be constructed; the meaning of the reaction pathways or reaction coordinates is then attributed to such a set of points (these two terms are not necessarily identical – see for example[294]). The physical significance of these curves is often problematical – they can depend on the choice of coordinate system and need not have any apparent relationship with classical trajectories[36-38,295] that (in the framework of the classical description of the problem) represent the physically correct solution. On the other hand, the construction of these curves (in contrast with the classical trajectory problem) does not necessarily require *a priori* knowledge of the whole potential energy hypersurface. Thus the concept of reaction pathways can be considered to be a compromise between representation of the hypersurface in terms of stationary points alone and exhaustive (mostly hypothetical) numerical point-by-point definition of the hypersurface. The concept of reaction pathways naturally yields more information on the shape and behaviour of the hypersurface than follows from the characteristics of the stationary points alone. In addition, this concept is useful in visualizing the relationships between some stationary points.

The concept of reaction coordinates is frequently encountered, converting the problem to the finding of the steepest descent path, familiar from textbooks on differential geometry. Fukui[296] suggested that the reaction coordinate be conceived as a curve that connects the reactants and products, passes through a transition state, and is orthogonal to the equipotential contours; the more sophisticated intrinsic reaction coordinate is also based on this concept[296-301]. It is an easy matter to express the curves defined above in terms of the components of the energy gradient. Let us describe the positions of the nuclei of the reactive system in terms of $n = 3N\text{-}6$ arbitrary, independent coordinates $\xi_1, \xi_2, \ldots, \xi_n$, that are sufficient for the description of the system and determination of its potential energy $U = U(\xi_1, \xi_2, \ldots, \xi_n)$. A set of equipotential points with energy c is described by the equation (corresponding to a plane curve in $(n + 1)$-dimensional space):

$$c = U(\xi_1, \xi_2, \ldots, \xi_n) . \tag{3-31}$$

It is well known from differential geometry that the normal vector to hypersurface (3-31) at point $(\xi_1, \xi_2, \ldots, \xi_n)$ is described by the equation

$$\frac{\Xi_1 - \xi_1}{\dfrac{\partial U}{\partial \xi_1}} = \frac{\Xi_2 - \xi_2}{\dfrac{\partial U}{\partial \xi_2}} = \ldots = \frac{\Xi_n - \xi_n}{\dfrac{\partial U}{\partial \xi_n}} , \tag{3-32}$$

where Ξ_i designates the coordinate of a general point on the normal. If the reaction path is expıessed parametrically in terms of parameter t:

$$\xi_i = \xi_i(t) , \quad (i = 1, 2, \ldots, n) , \tag{3-33}$$

then the equation of the tangent to curve (3-33) at point $(\xi_1, \xi_2, \ldots, \xi_n)$ is given by the equation familiar from differential geometry:

$$\frac{\Xi_1 - \xi_1}{\dfrac{d\xi_1}{dt}} = \frac{\Xi_2 - \xi_2}{\dfrac{d\xi_2}{dt}} = \ldots = \frac{\Xi_n - \xi_n}{\dfrac{d\xi_n}{dt}} . \tag{3-34}$$

The requirement of orthogonality of the reaction coordinate to the equipotential contours leads directly to a set of differential equations

$$\frac{\dfrac{d\xi_i}{dt}}{\dfrac{\partial U}{\partial \xi_i}} = a , \quad (i = 1, 2, \ldots, n) , \tag{3-35}$$

where a is an arbitrary constant. If finite differences are employed, then the magnitude of the shifts $\Delta\xi_i^0$ leading from arbitrary point $(\xi_1^0, \xi_2^0, \ldots, \xi_n^0)$ on the reaction coordinate to a new point is given by the relationship

$$\Delta\xi_i^0 = a\left(\frac{\partial U}{\partial \xi_i}\right)_0, \quad (i = 1, 2, \ldots, n). \tag{3-36}$$

Relationships (3-36), however, do not permit the determination of the direction of movement from a stationary point, as relationships (3-36) predict a zero shift in this point (see stationary point determination (3-24)). If, however, Taylor expansion of the first derivative is carried out at this point, then neglecting of higher terms leads to Eq. (3-36) in the form

$$\Delta\xi_i^0 = a\sum_{j=1}^{n}\left(\frac{\partial^2 U}{\partial \xi_i\, \partial \xi_j}\right)_0 \Delta\xi_j^0, \quad (i = 1, 2, \ldots, n). \tag{3-37}$$

The condition of solvability of the set of homogeneous equations (3-37) leads to the secular problem for the force constant matrix at the stationary point in question. It is apparent that the direction of motion in the stationary point is then given by an eigenvector of the force constant matrix.

The reaction pathway introduced in this manner would, however, in general be dependent on the choice of coordinates (see, for example, the discussion in ref.[219]). A solution of this problem was described in the classical work of Murrell and Laidler[302] and was extended by Fukui *et al.*[297], Schaeffer[303] and Stanton and McIver[219]. Analysis of the physical significance of the reaction pathway introduced by equations (3-36) and (3-37) must begin with the physically clear concept of the classical reaction trajectories, following from the solution of the corresponding set of Lagrangian equations:

$$\frac{\mathrm{d}}{\mathrm{d}t}\frac{\partial L}{\partial \dot{\xi}_i} - \frac{\partial L}{\partial \xi_i} = 0, \quad (i = 1, 2, \ldots, n), \tag{3-38}$$

where, for the conditions given here, the Lagrangian L is given by

$$L = T - U = \tfrac{1}{2}\sum_{i=1}^{n}\sum_{j=1}^{n} a_{ij}\dot{\xi}_i\dot{\xi}_j - U. \tag{3-39}$$

Specification of the initial conditions permits selection of the particular reaction trajectory from the set of all possible solutions of Eq. (3-38). The discussion will now be limited to the trajectory corresponding to zero initial velocity; in addition, the requirement of infinitely slow nuclear motion at each point on this trajectory will be included (assuming continuous dissipation of the nuclear kinetic energy). This then yields the hypothetical trajectory of infinitely slow nuclear motion; Fukui *et al.*[297] demonstrated that this trajectory permits the rewriting of general equation (3-38) in the form (3-40) determining the shift $\Delta\xi_j^0$ from one trajectory point to another trajectory point:

$$\frac{\sum_{j=1}^{n} a_{ij}\,\Delta\xi_j^0}{\left(\frac{\partial U}{\partial \xi_i}\right)_0} = a\,, \quad (i = 1, 2, \dots, n)\,, \tag{3-40}$$

where a is again an arbitrary constant. Comparison of Eqs. (3-36) and (3-40) permits the determination of a sufficient condition under which this trajectory becomes equivalent to the path of steepest descent on the potential energy hypersurface, U. As demonstrated e.g. in work[298], the sufficient condition has form (3-41) requiring the

$$T = \tfrac{1}{2}\sum_i a_{ii}\dot{\xi}_i^2\,, \quad (a_{11} = a_{22} = \dots = a_{nn}) \tag{3-41}$$

selection of a coordinate system that diagonalizes the expression for the kinetic energy of the system and the corresponding expansion coefficients are the same for all the coordinates (provided these coefficients are equal to units, then the condensed expression $T = \frac{1}{2}\sum_i P_i^2$ can be used, given in terms of the components of the linear momentum P_i). The shifts at the stationary points are then determined by an eigenvector of the corresponding vibrational problem, i.e. by one of the normal coordinates. While the early Fukui definition[296] ensured only uniqueness of the reaction path within a given coordinate system, nonetheless it was dependent on the choice of this system of coordinates (cf.[219,304]), then passage to the trajectory of infinitely slow motion[297,298,303] permits elimination of this dependence on the coordinates. It is thus justifiable to term this hypothetical but clearly physically defined reaction path as the intrinsic reaction coordinate[297]. The fact that the intrinsic reaction coordinate becomes equivalent to the path of steepest descent for the special choices of coordinates mentioned represents a connection with the less rigorous concepts of the reaction path. Now the coordinate dependence of the steepest reaction path itself can be clarified[219]. Assume that the transition from coordinate system q to system q' in matrix form is described in terms of matrix $\mathbf{A}$ by Eq. (3-42) and assume also that the inverse transformation is described by Eq. (3-43)

$$q' = \mathbf{A}q\,, \tag{3-42}$$

$$q = \mathbf{A}^{-1}q'\,. \tag{3-43}$$

While transformation of force constant matrix $\mathbf{F}$ is given by relationship (3-44), similarity transformation (3-45) holds for transformation of the $\mathbf{GF}$ vibrational problem (that represents a solution of the Lagrangian equations (3-38)). As Eq. (3-44) is generally not a similarity transformation, matrices $\mathbf{F}$ and $\mathbf{F}'$ have different characteristic problems, while the vibrational problem itself is not affected by the transformation of coordinates from q to q'.

$$\mathbf{F}' = (\mathbf{A}^{-1})^{+}\,\mathbf{F}\mathbf{A}^{-1} \tag{3-44}$$

$$(\mathbf{GF})' = \mathbf{A}(\mathbf{GF})\,\mathbf{A}^{-1} \tag{3-45}$$

Requirement (3-41) is fulfilled, for example, by the mass-weighted Cartesian coordinates η_i introduced by Eq. (3-46) in terms of the Cartesian coordinates of atoms q_i (x, y or z) and their masses m_i:

$$\eta_i = (m_i)^{1/2}\,q_i\,. \tag{3-46}$$

(Condition (3-41) is also fulfilled by normal coordinates.) The number of degrees of freedom here is $n = 3N$; otherwise coordinates η are a special case of coordinates ξ in relationship (3-31). It thus holds in relationship (3-41) that $a_{11} = a_{22} = \ldots = = a_{nn} = 1$. While Cartesian mass-weighted coordinates are a type of coordinates ensuring equivalence between the intrinsic reaction coordinate and the path of steepest descent, in applications it may be a drawback that the translational and rotational motions of the system as a whole are not separated. It may then be more useful to abandon this equivalence and to pass to a system of internal coordinates, connected with a reduction of the dimensions of the problem to $3N$-6. Transformation problems connected with this step are discussed in ref.[301]. In any case, the dependence on the atomic masses is retained, a phenomenon that was first clearly pointed out by Murrell and Laidler[302].

A numerical procedure for calculating the intrinsic reaction path is described in ref.[298], assuming that the coordinate set used fulfills conditions (3-41), i.e. for equi-

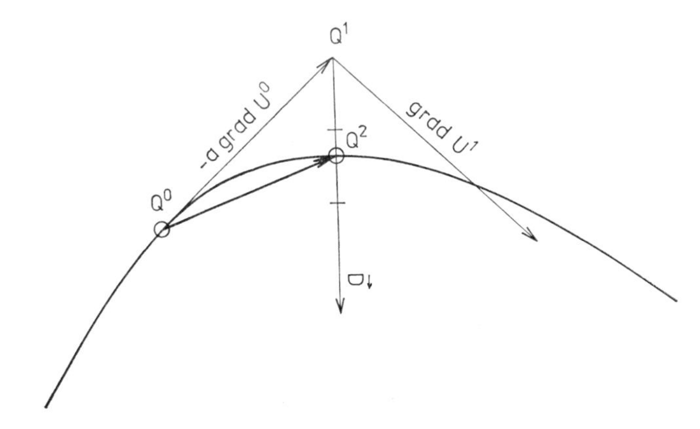

Figure 3-4. The unit step for intrinsic reaction coordinate determination (a hypothetical classical reaction trajectory of infinitely slow nuclear motion) yielding an auxiliary point Q^1 from the energy gradient at Q^0, and then giving the new point Q^2 of the coordinate from the energy gradients at Q^0 and Q^1 and a linear search in the $\boldsymbol{D}$ direction (from Ref.[298])

valence of the trajectory of infinitely slow nuclear motion with the path of steepest descent. Assume that point $Q^0 = (\xi_1^0, \xi_2^0, \ldots, \xi_n^0)$ is attained on the constructed intrinsic reaction coordinate (Fig. 3-4). Provided that this point is not a stationary point on the energy hypersurface, the corresponding shift is determined by relationship (3-36). For an arbitrary choice of parameter a (in ref.[298] this choice is carried out in connection with the curvature of the hypersurface in the given region), the generated point Q^1 is not necessarily a point on the constructed curve; correction is carried out using the energy gradient, grad U^1, at point Q^1. Vector $\boldsymbol{D}$ is constructed:

$$\boldsymbol{D} = -\frac{\text{grad } U^0}{|\text{grad } U^0|} + \frac{\text{grad } U^1}{|\text{grad } U^1|}, \qquad (3\text{-}47)$$

bisecting the angle formed by vectors grad U^0 and grad U^1 (Fig. 3-4), and point Q^2 with minimal energy is sought along direction $\boldsymbol{D}$ (from a parabolic fit using three points). The thus-constructed point is now considered as a new starting point for construction of the hypothetical reaction trajectory of infinitely slow nuclear motion. Provided that a saddle point is reached that is a transition state, then the path direction is determined by the normal vibrational coordinate with imaginary vibrational frequency. It is then necessary to carry out numerical construction of the force constant matrix and subsequent vibrational analysis in this particular point. The unit step (i.e. procedure from Q^0 to Q^2) in the construction of the trajectory can be realized using only two gradient calculations and two (or more, if a linear search using three points does not lead to the discovery of a minimum) additional energy calculations. The selection of parameter a in Eq. (3-36), determining the magnitude of the step from Q^0 to Q^1, is essential for the optimal construction of the intrinsic reaction coordinate. Ishida *et al.*[298] estimated that 20 to 25 points should be sufficient for construction of the reaction coordinates for the systems that they studied, however, only for the qualitative determination of the reaction coordinate. The procedure for plotting the intrinsic reaction coordinate has been termed[305] the reaction ergodography. It is apparent from a general point of view that the set of equations for the intrinsic reaction coordinate has an infinite number of solutions, as the starting point can be selected as an arbitrary general point on the hypersurface. All these solutions are termed[300] meta-intrinsic reaction coordinates. The intrinsic reaction coordinate can then be selected from this set as the particular solution that is connected with passage through the transition state.

The concept of the intrinsic reaction coordinate has also been employed in the study of some chemical systems[297,298,305–309]. These calculations result not only in the energetics and geometry of the reacting system, but also in more detailed information on the shape of the potential energy barrier on the hypersurface cross section given by the intrinsic reaction coordinate. An interesting example is, e.g., the study[297,305] of isotopically differentiated reactions (see Eqs. (3-48)–(3-51)). For such reactions the

$$CH_4 + T \rightarrow CH_3T + H \qquad (3\text{-}48)$$

$$CD_4 + T \rightarrow CD_3T + D \tag{3-49}$$

$$CH_4 + T \rightarrow CH_3 + HT \tag{3-50}$$

$$CD_4 + T \rightarrow CD_3 + DT \tag{3-51}$$

intrinsic reaction coordinates are in general different (even though their potential hypersurface is essentially common as a result of the Born-Oppenheimer approximation). Stationary points (that are common for the otherwise different intrinsic reaction coordinates) are an exception to this rule. Consequently, the potential barriers resulting from these reaction paths for isotopically related reactions have different widths. While, for example, for substitution processes (3-48) and (3-49) barrier broadening was remarkable[305] primarily in the region following passage through the transition state, for abstraction processes (3-50) and (3-51) this was primarily a characteristic of the region before reaching the transition state. It can thus be concluded[305] that the mass effect is mainly related to the C—H stretching mode.

It should also be noted, for the sake of completeness, that the loss of a clear physical significance for the steepest descent path in passage to coordinates that are not related to the mass weighted Cartesian coordinates by an orthonormal transformation would be removed if the requirement to follow grad U were replaced by a condition to follow vector $\mathbf{g}$. grad U, where $\mathbf{g}$ is a certain suitably chosen metric tensor[233,299,300]. A special case has also been found[173] where the gradient-following reaction path remains invariant under a non-orthonormal transformation of the mass-weighted Cartesian coordinates to an internal coordinate system. The concept of the intrinsic reaction coordinate (or, as more recently suggested, the intrinsic reaction path[310]) is the subject of constant refinement and formalization, especially using differential geometry and classical mechanics[300,311–315].

The concept of the intrinsic reaction coordinate can be considered as a compromise between the requirements of the classical theory of reaction trajectories and the possibilities of description of the potential energy hypersurfaces of real systems. Instead of knowledge of the whole hypersurface, construction of the intrinsic reaction coordinate requires only the ability to generate the energy gradient at a general point in configurational space. The curve can be considered[301] as a 'centre line' in the reaction path region, because of its vibrationless and rotationless character.

In addition to the concept of the intrinsic reaction path, a number of other approaches to the reaction path have been proposed. The construction of the steepest descent reaction paths in a general coordinate system (i.e. not fulfilling condition (3-41)) is of especial importance. Alternatively, the term 'minimum energy reaction path' is used; however, this term can sometimes have a more general meaning. A useful technique for calculation of the steepest descent path was recently proposed by Miller and Brown[23[illegible]]; most importantly, it does not require calculations of the gradient or matrix of the second derivatives of the energy, but only calculation of the energy itself. This technique is realized in the framework of a constrained simplex optimization. Assuming that two points are known on the sought gradient path, P_1 and P_2, then a further point Q can be obtained by minimizing the energy on the hypersphere centred around the energetically higher of the two points, P_1 and P_2, where the radius of this hypersphere is selected as less than the distance between points P_1 and P_2. Energy minimization begins at the intersection point of the line connecting points P_1 and P_2 with the hypersphere considered, and continues with the constraint that all the subsequent points remain on the hypersphere.

Another concept, not identical with the steepest descent path, can be formulated on the basis of the requirement that[316], along the hypersurface valley, the energy gradient remains parallel to one of the eigenvectors of the force constant matrix. Such a pathway has interesting geometric properties[317], including the possibility of climbing the valley walls[233]. This concept represents a further illustration of the diversity of ways of introducing various reaction paths and is also useful as a possible alternative means for localizing transition states[317].

A further type of reaction path is represented by the frequently employed minimum energy reaction paths[234,318–322], obtained by the optimization of all the geometric parameters in the studied system with respect to the potential energy minimum for gradual fixing of the values of a single geometric parameter (called the reaction coordinate[318] or mapping parameter[172]). A reaction path constructed in this manner with a given choice of the fixed parameter can lead to different curves, depending on whether the calculation is begun in the region of the reactants or of the products[171,243,320] (chemical hysteresis[320]). In addition, various selections of the fixed parameter may lead[231] to completely different reaction paths. It has also been demonstrated that this type of reaction path may be discontinuous[171,234,243,320,323,324]. Moreover, the reaction path constructed in this way need not pass through a transition state at all[171,321] and need not even lead to stable products[320]. It should be noted that other ways of selecting reaction paths have been proposed. These include, for example, the concept of linear and quadratic synchronous transit paths[230,231] or a procedure based on the Rice and Teller principle of least motion[327,328].

A list of more general analyses of the properties of reaction paths should also include the Mezey concept[310] of the reaction path stability based on the partitioning of the configurational space into domains and classification of their importance from the point of view of the theory of chemical reactions. In this connection, the possibility of an intuitively straightforward definition of the chemical notion of structure was pointed out, based on passage from a rigid nuclear arrangement corresponding to a given critical point to an open set of points given by a suitable geometric neighbourhood. This versatile approach reflects better the features of the common concept of structure – for example all the vibrational states of a given species are considered to also be this species in spite of the fact that their time-averaged geometries vary and differ from the equilibrium structure.

The above discussion emphasizes the wide variety possible in the choice of reaction paths as soon as the rigorous (within classical mechanics) level of classical reaction trajectories is abandoned. The considerable literature in this field includes a large number[296–338] of interesting properties of curves on potential energy hypersurfaces that are often more interesting from a geometrical than from a physical or even chemical point of view.

The concept and construction of reaction coordinates permits the visualization of the fate of the reactants in the region up to or after the attainment of the transition state, i.e. yields a specific kind of information that cannot be obtained from the theory of absolute reaction rates alone. This is sometimes considered to be justification for the concept of a selected reaction path. It should, however, be noted here that the theory of absolute reaction rates does not require such a specification (because of its statistical-mechanical basis), and on the contrary admits the existence of any path leading from the reactants to the transition state. To select a single reaction path from the infinite number of possible paths would imply negation of an elementary assumption lying at the basis of the theory of absolute reaction rates, that of equilibrium distribution of states. The concept of selected reaction paths is, however, relevant for the theory of absolute reaction rates as a tool for safe assignment[171] of reactants and products to the transition state (as the procedure using the normal coordinate corresponding to the imaginary vibrational frequency of the activated complex need not be sufficient). Although no direct relationship was found between one of the proposed realizations of the reaction path and the dynamics of the chemical reaction, it is felt[301,323] that a selected reaction path could lie close to the average or most probable classical trajectory (given by the most probable values of the initial conditions). This assumption permits the analytical fitting of the potential energy hypersurface for the study of classical trajectories to be limited to a certain region around the selected reaction path[339] alone.

3.5 Calculation of the Potential Energy Hypersurfaces

It is generally known that, except for the simplest models, the solution of the Schrödinger equation in quantum mechanics cannot be obtained in a closed, analytical form. This is also completely true for calculations of the potential energy hypersurfaces by solution of the characteristic problem (3-5) of the electronic Hamiltonian (3-6). It is thus necessary to carry out stepwise numerical solution of the Schrödinger equation (3-5) for electrons in a field of fixed nuclei for the individual discrete nuclear configurations. The region in which the nuclear coordinates are systematically varied in selected steps depends on the purpose that the generated potential is to serve. If a universally useful potential is to be obtained, however, it is necessary to ensure a reasonable precision in the whole coordinate space of the studied system. In the study of spectroscopic problems knowledge of the behaviour in a certain region around the relevant potential minimum should be sufficient, while its use for kinetic purposes requires correct asymptotic behaviour of the potential with respect to each of the dissociation limits. The precision of the potential would increase, in general, with the refinement of the division of the intervals for values of the individual coordinates; this refinement is, of course, limited by the number of points for which the numerical solution of Eq. (3-5) can be carried out with reasonable consumption of computer time. Moreover, sufficient fineness of the division into intervals is also necessary to prevent exchange of the hypersurfaces corresponding to the individual electronic states in connection with possible hypersurface crossing. This is especially important in the absence of distinct symmetry labels.

In problems in which the potential energy hypersurfaces are employed, it is useful to express the hypersurfaces not only in tabular form but also (and even preferably) in the form of an analytical function obtained by fitting free parameters to the energy for selected discrete nuclear configurations. The functions selected must be sufficiently flexible in order to permit reliable reproduction of the tabulated energy data. Simultaneously, its form for various functional expressions appearing in the solution of a concrete physical problem should permit analytical treatment, e.g., of the definite integrals involved. The actual quantum chemical solution of Eq. (3-5) and the form of the Hamiltonian (3-6) is always approximative[9,146]; the Hamiltonian mostly has classical (nonrelativistic) form, some types of interactions are neglected and possibly only the valence electrons are included. Simultaneously, the solution of the characteristic problem of this incomplete Hamiltonian (mostly carried out in the framework of the LCAO approximation) is only approximate – questions of the finiteness of the expansion basis set in the LCAO approximation, of the consequences of possibly neglecting some types of integrals between the AO's as a result of the introduction, for example, of the ZDO approximation, of single-determinant wave functions and of the problem of the correlation energy, etc., must be considered. Nonetheless, an accuracy of greater than 4 kJ mol^{-1} (termed the chemical accuracy) at least in the critical regions (e.g. around the tops of the potential barriers) is usually necessary for the energy function from this approximative approach, if it is to be reasonably useful for the solution of a given physical problem. In addition, when it is necessary to abandom the Born-Oppenheimer hypersurfaces in favour of adiabatic hypersurfaces, it is necessary in the calculation of elements $C_{ll}(\boldsymbol{R})$ from Eq. (3-11) to generate the corresponding electronic wave function $\Phi_l(\boldsymbol{r}, \boldsymbol{R})$ explicitly and with sufficient precision. It is understandable under these conditions that purely quantum-chemical calculations of the potential energy hypersurfaces are relatively rare and are limited to quite simple systems.

The most rigorous applications of potential energy hypersurfaces include studies[178] of the dynamics of chemical reactions, either at the level of the classical Hamiltonian equations of motion (the classical trajectory methods[36–38,295]) or (more recently) at the level of the quantum-mechanical scattering theory[3,36,37]. In the application of these theories of elementary chemical processes, the empirical potential energy hypersurfaces are often employed[295]. The LEPS method[295] (London, Eyring, Polanyi,

Sato) method will be mentioned here; it is based on the historical LEP method[340] (London, Eyring, Polanyi) for triatomic systems. Another important method of this kind is the BEBO method[341,342] (Bond Energy – Bond Order). The number of systems for which quantum-chemical (at least) *ab initio* SCF potential energy hypersurfaces with sufficient precision are available is gradually increasing[37,295]. A critical review is given in the survey article of Bader and Gangi[343]. Calculation of the hypersurfaces by the DIM method[344] (Diatomics in Molecules) is also increasingly popular in this field.

Potential energy hypersurfaces can also be generated[345] from vibrational-rotational data; they are then expressed in the form of conventional expansion (3-52) in terms of displacements $R_i = r_i - r_{i,e}$ of coordinates r_i from the equilibrium configration $r_{i,e}$:

$$U = \frac{1}{2}\sum_{i,j} f_{ij}R_iR_j + \frac{1}{3!}\sum_{i,j,k} f_{i,j,k}R_iR_jR_k + \dots, \tag{3-52}$$

where $f_{ij}, f_{ijk}, \dots$ are the quadratic, cubic, ... force constants. A deficiency of potential functions of type (3-52) lies in the fact that they satisfactorily describe only the region around the equilibrium point, while they fail drastically in the determination of the dissociation energy values[346], even when the fourth term of expansion (3-52) is included. Simons[346] demonstrated that expansion of (3-52) expressed in terms of (dimensionless) coordinates ϱ_i does not have this drawback:

$$\varrho_i = \frac{r_i - r_{i,e}}{r_i}. \tag{3-53}$$

In addition, it has been found[347] that the use of coordinates (3-53) in expansion (3-52) leads to a marked improvement in the quality of the reproduction in fitting the energy--geometry data and to the stability of the expansion coefficients. Consequently, expansion in terms of Eq. (3-53) has been recommended[347] for the analytical representation of quantum-chemical hypersurfaces. Another approach to the generation of potential energy hypersurfaces on the basis of spectroscopic data can be found in the works of Murrell *et al.*[348–353]. The type of potential employed in this approach combines a polynomial including quartic (biquadratic) terms with a factor of the $(1 - \tanh(\gamma_i(r_i - r_{i,e})/2))$ type, where γ_i's are parameters selected to attain agreement with the quantum chemical and kinetic data and the symbol tanh designates the hyperbolic tangent. This type of potential has been found to be reliable in the whole coordinate space of the systems studied[348–353]. In connection with isomerism it should be noted that this method has been applied[349,350–352] to the potential hypersurfaces of systems exhibiting more than a single energy minimum (O_3, HO_2, SO_2, ClO_2). Study of this type of potential hypersurface enabled[350] clarification of the existence of the SOO isomer (in addition to the more stable OSO isomer) and a relationship has been found[350] between isomeric structures OClO and ClOO.

In contrast to diatomic molecules, no generally applicable direct procedure is available for obtaining the potential functions from the observed vibrational-rotational frequencies of polyatomic molecules. For diatomics, this problem is solved in the framework of the Rydberg-Klein-Rees (RKR) method[355]. Recently, an approximative RKR scheme has been proposed[354] that would also be valid for polyatomics. In addition, an approximative inversion method has been obtained, based on a transformation expressible in closed form that, in principle, permits direct construction of the hypersurface from spectral data.

Vibrational and rotational spectra are also a source of information for construction (parametrization) of potential energy hypersurfaces within phenomenological potentials, especially using the consistent force field method[167–170,172,356–359]. Further types of information that can be used for this purpose include, for example, thermodynamic data for the estimation of rotational barriers and non-bonding interactions, structural data and crystal properties (heats of sublimation, lattice vibrations, and lattice constants). All this data and possibly some additional information (or, on the contrary, some part thereof) can be used in an iterative process in an attempt to obtain the best agreement between the observed and calculated quantities. This results in a set of parameters appearing in the selected analytical form for the force field for the given class of compounds.

Another field in which spectral data has recently been intensively used for the construction of hypersurfaces is the formation of van der Waals molecules (for a review see[96]) or, in more general terms, of molecular complexes or weak intermolecular interactions. A large number of other sources of information can also be used for these purposes[360]. These primarily include the thermodynamic properties of moderately dilute gases, mostly condensed in the form of the second or third virial coefficient, with the familiar relationship to the interaction energy hypersurface. This can simultaneously represent an effective test of the quality of the hypersurface obtained, viz. realization of the fitting procedure using only the second virial coefficient and subsequent calculation of the third virial coefficient on the basis of the interaction hypersurface thus generated. For the purposes of comparison it would, however, be preferable if the third virial coefficient were available with sufficient precision, which is at present exceptional in the evaluation of experimental information on state behaviour[361]. In addition, the transport properties of the dilute gas can be utilized; these are most often condensed in the values of the diffusion coefficients, viscosity coefficients or coefficients of thermal conductivity. A newer technique employs molecular beams and the interaction potential is obtained from fitting of the scattering intensity. It is also possible to utilize the thermodynamic properties of the condensed phases, e.g. the heat capacity, lattice energy, or lattice spacing. A great drawback of most of the potential functions obtained in this way is their effective character in the sense that they are based on the assumption (valid only for atoms) of complete spherical symmetry of the interacting species, so that a single geometrical parameter appears in the formulae – the absolute value of the radius vector. Potential functions employing a larger number of geometrical parameters (i.e. two to six) are at present exceptional. It is also apparent that various types of measurements emphasize different regions of molecular separation, so that an adequate, generally useful interaction hypersurface should result if various types of observation techniques are combined. For example, Hutson and Howard[362] employed such an approach in obtaining the anisotropic potential surfaces of the Ar—HCl system by combining experimental data from molecular beams, rotational spectral lines and the second virial coefficient. It is important here that the linear structure Ar.HCl was found as a global minimum on the hypersurfaces thus obtained and that, simultaneously, the existence of a secondary minimum, Ar.ClH, was indicated (cf.[363]).

However, only quantum-chemical *ab initio* hypersurface calculations are completely independent of the experimental information. The most extensive quantum-chemical calculation of the potential energy hypersurface so far carried out is the study of Clementi *et al.* of water clusters[364-369] and of solvation of biomolecules[369-376]. The molecular complexes of water were studied using pair interaction potentials of the type (3-54), based on a point charge interaction model (Fig. 3-5) similar to that already employed by Bernal and Fowler[377]:

$$\begin{aligned} U = {} & q^2(1/r_{13} + 1/r_{14} + 1/r_{23} + 1/r_{24}) + 4q^2/r_{78} - \\ & - 2q^2(1/r_{18} + 1/r_{28} + 1/r_{37} + 1/r_{47}) + a_1 \exp(-b_1 r_{56}) + \\ & + a_2(\exp(-b_2 r_{13}) + \exp(-b_2 r_{14}) + \exp(-b_2 r_{23}) + \\ & + \exp(-b_2 r_{24})) + a_3(\exp(-b_3 r_{16}) + \exp(-b_3 r_{26}) + \\ & \qquad + \exp(-b_3 r_{35}) + \exp(-b_3 r_{45})), \end{aligned} \tag{3-54}$$

where q, a_1, a_2, a_3, b_1, b_2, and b_3 are fitted constants and the r_{ij}'s are the distances between the centres in Fig. 3-5. It has been demonstrated that this analytical formula represents the best combination of numerical precision with mathematical simplicity.

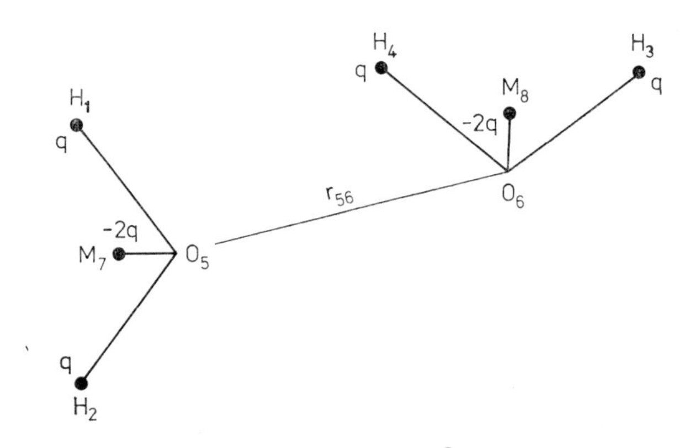

Figure 3-5. Point charge model used in derivation of the analytical hypersurface of the interaction energy for the water dimer in the studies of Clementi *et al.*[364-369]

The potential parameters in Eq. (3-54) were found[364] on the basis of 216 (or[366] 230) quantum-chemical calculations at the Hartree-Fock level for various nuclear configurations for a rigid structure of the monomer units and also[367] 66 calculations using the configuration interaction method (here potential (3-54) was expanded by four exponential terms representing the contribution of the correlation effect). This type of potential was also used[366] in a study of water oligomers $(H_2O)_n$ ($n = 3, 4, \ldots, 8$) and for a Monte Carlo simulation[364,365,369] of the liquid state of water. These applications assumed additivity of the pair potential, which cannot be used as a general assumption, as demonstrated by a study of nonadditivity effects on hypersurfaces of Ne_3 (ref.[378]) and Be_4 (ref.[379]). Pair interaction potential (3-54) was also employed[380,381] for calculation of the second virial coefficient of water vapour and for the solution of the problem of isomerism of the water dimer[382–384]. Hydration was studied[369,370,375] using the interaction potential $U(\mathrm{M}, \mathrm{W})$ between molecule M and the water molecule W:

$$U(M, \mathrm{W}) = \sum_{i} \sum_{j \neq i} (-A_{ij}^{\mathrm{ab}}/r_{ij}^{6} + B_{ij}^{\mathrm{ab}}/r_{ij}^{12} + C_{ij}^{\mathrm{ab}} q_i q_j / r_{ij}) , \tag{3-55}$$

where A_{ij}^{ab}, B_{ij}^{ab} and C_{ij}^{ab} are fitted constants, the r_{ij}'s are the distances between the ith atom in M and the jth atom in W, q_i and q_j are the charges at the centres and indices a and b permit distinguishing of types with the same atomic number with respect to their surroundings in the molecule. The constants of the interaction potential (3-55) were obtained by fitting[370] to 1690 energy values for the interaction of the water molecule in various positions around some of the studied 21 naturally occurring aminoacids, or by fitting[371] to 368 energy values corresponding to various configurations of the water molecule and one of four DNA bases. Tests[372] on phenylalanine (this aminoacid was not included in the fitting set[370]) indicated that the analytical potential can[370] readily be transferred to molecules related to those studied in ref.[370].

Expansion in terms of radial functions $U_n(R)$ and Legendre polynomials $P_n(\cos \Theta)$ has also been used for analytical expression of the interaction potential hypersurfaces for weak intermolecular interactions, e.g. for the two-dimensional problem[96,362,385] with variables R and Θ:

$$U(R, \Theta) = \sum_{n=0}^{N} U_n(R)\, P_n(\cos \Theta) , \tag{3-56}$$

based on multipole expansion[212] in terms of Legendre functions $P_n(\cos \Theta)$. Expansions of type (3-56) in Legendre polynomials are a useful means for the study of anisotropy of the interaction potential. Even though it is necessary to include many terms in expansion (3-56) for small values of R, at values of this distance typically corresponding to bonding states of van der Waals molecules this expansion becomes a rapidly converging series. This has been demonstrated, for example, by the analysis of infrared spectra of H_2 – inert gas complexes[386]. In general, the interaction potential between two arbitrary systems can be expressed in terms of the spherical expansion[387].

A new approach to the problem can be found in the technique of interpolation of hypersurfaces by polynomial roots[388,389]. This is based on the fact that the energy function most frequently results from the variational principle and is thus a root of the characteristic polynomial of a certain matrix. This indicates the possibility of using the roots of a polynomial in λ, $F(\lambda, \boldsymbol{R})$ (whose coefficients will be simple functions in terms of nuclear coordinates $\boldsymbol{R}$), for fitting such a hypersurface. The polynomial

$$F(\lambda, \boldsymbol{R}) = \sum_{i=0}^{m} P_i(\boldsymbol{R})\, \lambda^{m-} \tag{3-57}$$

was selected[388], where $P_i(\boldsymbol{R})$ designates the complete ith degree multinomial ($P_0(\boldsymbol{R}) = 1$). The coefficients in $P_i(\boldsymbol{R})$ are determined from the requirement that it hold, for all the particular points $\boldsymbol{R}_g$ used for the fitting, that

$$F(U(\boldsymbol{R}_g), \boldsymbol{R}_g) = 0\,. \tag{3-58}$$

The number of input points required for an interpolation of degree m with dimension

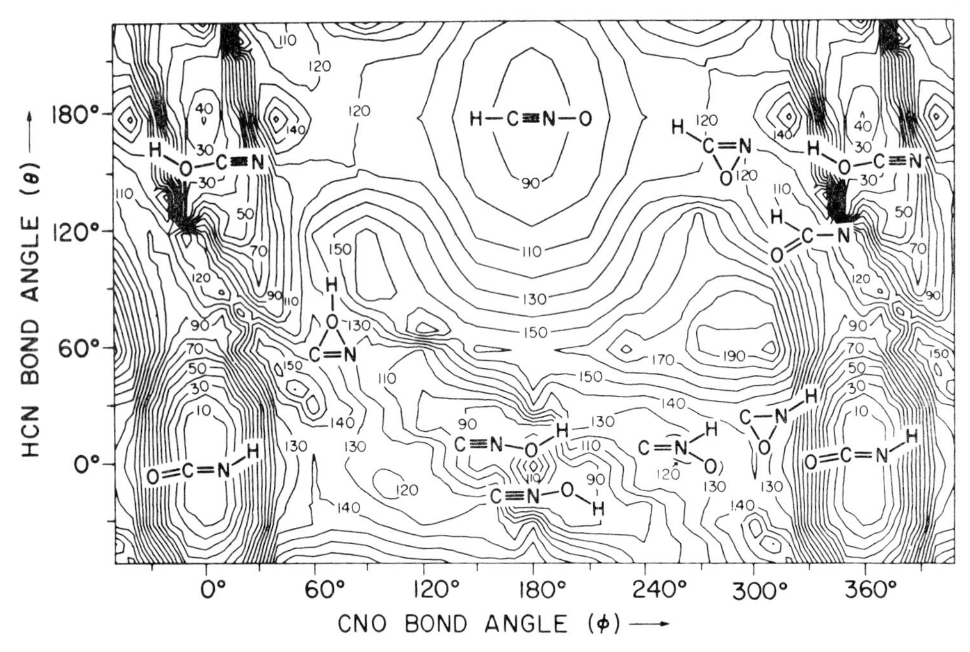

Figure 3-6. Visualization of the six-dimensional 4-31G potential energy hypersurface of CHNO by two-variable mapping in terms of bond angles $\theta = \sphericalangle$ HCN and $\phi = \sphericalangle$ CNO. Contour labels are in kcal mol^{-1} (1 kcal mol^{-1} = 4.184 kJ mol^{-1}) with respect to the lowest calculated point on the hypersurface ($\theta = \phi = 0$). (from Poppinger *et al.*[392])

of the interpolated function d is given by the expression

$$M = \binom{m + d + 1}{m} - 1 , \qquad (3\text{-}59)$$

which ensures reasonably slow growth with dimension d. Testing fits carried out using this new method exhibited very good quality (the errors varied between 10 and 1000 J mol^{-1}).

Calculations of potential energy hypersurfaces of electronically excited states are important for photochemistry (and potentially also for transition to the non-adiabatic approximation). A survey of quantum-chemical calculations of potential energy hypersurfaces of triplet states is given in the review article of Devaquet[391].

Mapping in terms of two variables is a very useful form of visualization of the conditions on a multidimensional hypersurface (especially of the relationships between its stationary points). Fig. 3-6 depicts, for illustration, a 4-31 G contour map of the CHNO system[392]. The figure is described in terms of two bond angles, where the energy was minimized with respect to the remaining four structural parameters for each point used in construction of this map.

The potential energy hypersurface can generally be considered as a particular case of property hypersurfaces[393]. Similarly to the potential energy, further properties of the molecular system (e.g. the dipole moment) can also be treated as functions $P(\boldsymbol{R})$ of the nuclear coordinates and expanded in Taylor series in these variables around the equilibrium geometry*. In this series the expansion coefficients are the first, second, ... partial derivatives $(\partial P/\partial R_i)_0$, $(\partial^2 P/\partial R_i\, \partial R_j)_0$, ... It is apparent that it is useful to study the properties of these coefficients, especially the number of independent coefficients. First, the relationships between the non-zero derivatives were studied for the potential energy[394,395] and later also for the dipole moment hypersurfaces[396]. Recently, Fowler[393] used group theory to find the number of independent coefficients describing a general property hypersurface (for a non-degenerate electronic state).

* For the sake of completeness it should be noted that, at one time, it was suggested that a global minimum (or maximum) of function $P(\boldsymbol{R})$ for general property P be sought in analogy to the potential energy (cf. also[493]). This was to serve for selection of the species with the smallest (or largest) value of the chosen property from a number of possible species, thus determining the structure (and even its geometry) suitable for synthesis (*a priori* design of a substance with the required properties). It is thus apparent that this was a trivial misunderstanding. For this purpose, only the P_e value for the equilibrium geometry of the considered species can be employed, i.e. in energy minima $U(\boldsymbol{R})$. Additional localization of stationary points on the $P \neq U$ hypersurface for this purpose would lack meaning.

3.6 Vibrational Motions of the Atomic Nuclei

Description of nuclear motion in terms of the Born-Oppenheimer approximation assumes solution of the nuclear characteristic problem (3-11) or (3-12), which is at present conventionally conceived as a problem lying beyond quantum chemistry, i.e. belonging rather to molecular spectroscopy. However, there are instructive examples indicating that combination of these two approaches can lead to valuable new results. In connection with the expected gradual abandonment of the Born-Oppenheimer approximation, the process of fusion of these, at present, distinct regions of theoretical chemistry will probably become clearer. This should be of particular importance for the theory of chemical reactivity. However, even in the framework of the Born-Oppenheimer approximation there are a number of good reasons for considerable overlapping. With the possible exception of photochemical processes, every chemical conversion is connected with motion of the atomic nuclei and, on the other hand, quantum-chemical energy hypersurfaces represent key information for molecular spectroscopic techniques (that is, however, used at present quite exceptionally). The remarkably developed theory of the vibrational-rotational Hamiltonian (recently, semi-rigid or non-rigid molecular systems have received considerable attention – see for example[123,397–401]) is used at present primarily and overwhelmingly in the framework of the inversion vibrational problem. It is, however, true that only utilization of the most accurate quantum-chemical hypersurfaces, i.e. of the *ab initio* CI or MC SCF types (cf.[402–405]) would be adequate because of the high degree of development of the theory of molecular spectroscopy. There are, however, also interesting, although rather exceptional, examples of the utilization of essentially qualitative quantum-chemical information for quantitative spectroscopic predictions[406,407].

The harmonic oscillator approximation is typical for the contemporary quantum-chemical approach to the vibrational problem. The vibrational frequencies of $3N$-6 independent quantum harmonic oscillators used in this approximation for description of the vibration of a non-linear N-atomic molecule are obtained by solution of the classical equations of motion. Solution of the Lagrangian equations for the motion of N points in a harmonic potential field is a standard problem in classical mechanics textbooks (see for example[408]). This problem has been solved by Wilson[409,410] and Elyashevich[411] for molecular vibrations. Solution of the problem in matrix notation[412], where the frequencies of the normal vibrational modes result from diagonalization of the matrix product of two symmetrical matrices, viz. the (Wilson) kinematic matrix $\mathbf{G}$ (the elements of the inverse matrix are the coefficients in the expression of the kinetic energy in the coordinates used), and the force constant matrix $\mathbf{F}$, is especially simple and elegant. The vibrational problem can be solved in an arbitrary coordinate system; mostly Cartesian or mass-weighted Cartesian, internal, and symmetry coordinates are used. The $\mathbf{G}$ matrix has an especially simple form in Cartesian coordinates; in internal coordinates it is constructed by the general (vector) method, commonly described in books on the theory of vibrational spectra[123,130,412], and for symmetry coordinates it is derived from the expression for internal coordinates. When working with Cartesian coordinates, translational and rotational motion of the system as a whole are not separated, increasing the dimension of the problem by six degrees of freedom (or by five for a linear molecule); the subsequent two coordinate systems do not involve this complication. Symmetry coordinates (completely respecting the assignment of the vibrational modes to individual representations of the point group of symmetry of the system) ensure minimal dimensions of the problem. In some choices of internal coordinates, the problem of redundant coordinates may appear[382,383,413–415]; this problem has not alway been correctly treated in the literature. Redundancy is connected with the use of a greater number of coordinates than is kinematically necessary (i.e. $3N$-6) in order, for example, to completely consider the symmetry, to obtain transferable force constants with clear significance, or for algorithmic simplicity. The presence of redundancy requires a very circumspect approach to the problem as, for example, the condition for an equilibrium nuclear configuration does not generally have a simple form (3-24). The selection of coordinates for solution of the problem leading to its smallest dimensions is especially important for solution of the inversion vibrational problem[416] (determination of the force constants on the basis of the known **G**-matrix and a set of experimental information – the vibrational frequencies, frequency shifts as an effect of isotopic substitution, Coriolis constants, centrifugal distortion constants, mean vibrational amplitudes, etc.). In conclusion, however, it should be noted that the recently introduced[494,495] Green's function method may bring a substantial change to the solution of the inversion vibrational problem.

While solution of the inversion vibrational problem represents a traditional subject of study in the classical theory of vibrational spectra, solution of the actual vibrational problem has become widely used only for the application of quantum-chemical methods to generate matrix **F** (and to replace the previously used empirical potentials, analogies or estimates). In the theoretical construction of the force constant matrix, three qualitatively different levels can be distinguished: (i) double numerical differentiation[417–422], (ii) double analytical differentiation[190,192,280,423–431] and (iii) numerical differentiation following analytical differentiation[171,183–185,257,262,272,432–435]. The double numerical differentiation technique is the simplest method of constructing matrix **F** on the basis of quantum-chemical calculations. Systematic variation of the nuclear configurations is carried out in the chosen coordinate set in the immediate vicinity of the studied energy minimum, followed by calculation of the potential energy of the configurations generated in this way. The portion of the energy hypersurface mapped in this manner, important for vibrational motion, is then used for construction of the second derivatives of the potential energy with respect to the nuclear coordinates in the studied stationary point by numerical differentiation to the second order. In case (ii), analytical formulae are obtained for the force constants by means of the differentiation of the expression for the total energy in the given quantum-chemical scheme. These formulae can be derived using[190,193,436–438] the Hellmann-Feynman theorem[12,13], provided a wave function is available that is an exact eigenfunction of the corresponding Hamiltonian. Schutte[14] and, more recently, Nakatsuji *et al.*[209] discussed the possibility of using the Hellmann-Feynman theorem for approximative wave functions. This theorem is concerned primarily with the first derivatives of the energy (see Eq. (3-17) and the discussion in Chap. 3.3), but if it is satisfied at this level, it can considerably simplify terms appearing in the calculation of the second derivatives and decrease their number. In addition, an intuitive physical significance can be attributed to the second derivatives[209] similar to that already mentioned for the first derivative in connection with its interpretation in terms of the Hellmann-Feynman force. Method (ii) of double analytical differentiation was long limited to the simplest systems, even thought the results of Bloemer and Bruner[427] indicated, in principle, the possibility of general applications. However, the recent work of Pople *et al.*[280,431] brought about a basic change in this area; these authors utilized analytical formulae for the second derivatives of the energy in the framework of the *ab initio* SCF approach for the study of small organic molecules. Technique (iii) can be considered as the most practical and, at present, most frequently used approach to the quantum-chemical construction of matrix **F**. This approach was introduced and widely used by Pulay *et al.*[183–185,193,257,439–444]. In approach (iii) the force constants are constructed by numerical differentiation of the analytical formulae for the first derivatives of the potential energy. In order to suppress the effect of potential profile asymmetry, two distortions with the same absolute value and opposite sign are considered for each coordinate. Thus calculation of the whole force constant matrix requires a number of energy gradient calculations equal to twice the dimension of this matrix. The precision of the calculation is, of course, dependent on the selection of distortion steps; in a work employing internal coordinates, Pulay recommends[193] a distortion of $\pm 0.025 \times 10^{-10}$ m for bond-length distortion and of $\pm 2^\circ$ for bond angle distortion. An increment value of $\pm 0.012 \times 10^{-10}$ m has been recommended for work in Cartesian coordinates, on the basis of an extensive analysis of smaller hydrocarbons[171]. A great deal of information on the quality of the individual types of elements of matrix **F** obtained using various quantum-chemical methods has been collected and condensed into simple rules in connection with the present extensive development of calculations of force constants (at present carried out primarily using technique (iii)). For example, the following rules[193] were derived from *ab initio* SCF calculations near the Hartree-Fock limit (or from extrapolation of results using smaller basis sets). Coupling force constants are reproduced with remarkable precision, with an error usually not greater than ± 5 Nm^{-1} or 10% (whichever is larger). Experimental data are then often less precise. Diagonal deformation force constants are consistently overestimated by about 10% with the exception of out-of-plane deformation in planar π-systems, where the deviation is about 25%. It should be noted that this type of information is usually obtained from the experiment with high precision. Stretching force constants are usually overestimated by several percent compared with the experimental values obtained from observed data relative (N. B.) to equilibrium r_e geometry. This type of data is, however, only rarely available (except for diatomic molecules), which considerably affects the quality of the comparison. Stretching force constants are very sensitive to the choice of reference geometry, as is documented by the following example. The possible error in[445] the determination of the equilibrium C—H bond length in CH_4 is only 0.005×10^{-10} m; however, this may lead to an error in the corresponding stretching force constant of about 16 Nm^{-1}, i.e. 2.8%. Knowledge of these and similar rules (i.e. the systematic nature of the error in some situations and the high precision of the theoretical values under other conditions) permits the setting up of useful correction procedures and thus to ensure the reasonable usefulness of the resultant force constant matrix for prediction purposes.

Provided that the problem studied is of such a nature that information on the type of stationary points is sufficient, it would be satisfactory to simply evaluate matrix **F** alone. As was apparent in Chap. 3.4, however, transition to, for example, the intrinsic reaction coordinate requires inclusion of the masses of the atomic nuclei in a manner that is essentially equivalent to that of vibrational analysis. In most applications, it is actually typically more useful to carry out complete vibrational analysis (considering connection with existing or potentially obtainable experimental information). It is not surprising that there has recently been a systematic increase in the number of studies based on the **F**-matrix obtained by quantum-chemical calculations and dealing with the vibrational analysis of a number of substances in order to determine the relationships between the experimental vibrational frequencies and those obtained by the studied quantum-chemical method. A number of works[442–444] have demonstrated that use of the CNDO/2 method ensures good reproducibility of a number of types of experimental frequencies and that use of a scaling factor for some types of force constants leads to complete agreement with the experimental results. In particular, the force constants describing stretching of the bond lengths are systematically overestimated by a factor of greater than two[257,432,442–444,446]. The extensive study of Dewar and Ford[447], based on the vibrational analysis of 34 molecules, yielded similar results for the MINDO/2 method. It has also been demonstrated at the *ab initio* SCF level[448–454] that reliable reproduction of the experimental vibrational data requires the use of scaling factors that have, however, been found to be transferable in a given series of substances. On the other hand, the first *ab initio* SCF CI or MC results confirmed[227] the assumption that the need for scaling factors should disappear at this level of the quantum-chemical approximation (with increasing size of the basis set) (cf. Tab. 3-7). The error in the best approximation in Tab. 3-7 is less than 4%. For the sake of completeness, a recent interesting generalization of vibrational analysis will be mentioned. Tachibana[455] gave a rigorous concept of vibrational analysis along an intrinsic reaction coordinate, i.e. at points that are not, in general, stationary.

Testing of quantum-chemical methods for the calculation of the vibrational

Table 3-7. Theoretical predictions of harmonic vibrational frequencies (cm^{-1}) of H_2CO at different *ab initio* levels with numerical differentiation of the analytic energy gradient[a]

Mode	Symmetry	DZ[b] SCF	DZ[b] CI	DZ + P[b] SCF	DZ + P[b] CI	Observed
ω_1	a_1	3223	3028	3149	3074	2944
ω_2	a_1	1878	1703	2006	1869	1764
ω_3	a_1	1651	1544	1656	1596	1563
ω_4	b_1	1324	1194	1335	1243	1191
ω_5	b_2	3315	3112	3226	3155	3009
ω_6	b_2	1349	1263	1367	1306	1287

[a] From Ref.[277]
[b] DZ — double zeta basis sets; DZ + P — DZ augmented by polarization functions.

characteristics at the level of the (harmonic) vibrational frequencies (and not at the force constant level) is of a basic nature because of the well-known general ambiguity of the inversion **GF** problem. It is interesting in this connection that information from quantum-chemical calculations can be used to eliminate sets without physical significance among the individual sets of experimental force constants compatible with the measured characteristics of the given molecule, and thus to eliminate ambiguity. Examples can be found in works[456,457] on N—F compounds and in a study of formic acid[458].

For more detailed confrontation of the theoretical and experimental results or for more minute prediction of vibrational spectra, it is necessary to calculate the intensities of the infrared and Raman spectra; thus quantum-chemical calculations of this type also tend to be more frequent[459-475]. Theoretical calculation of infrared or Raman intensities assumes knowledge of the derivatives of the components of the dipole moment μ or of components of the polarizability tensor α. The conventional approach is reminiscent of the generation of the potential energy hypersurface point by point — the μ and α values are calculated for a relatively large number of distorted geometries and the appropriate derivatives are extracted numerically from this data set. An alternative approach that elegantly employs the contemporary capabilities of analytical calculations of the energy gradient was described by Komornicki and McIver[470]. Here a molecule is considered in the presence of an external electric field (one of its components is designated as F). The first and second derivatives of component $\text{grad}_i\, U(\boldsymbol{R})$ of the energy gradient with respect to F can be obtained by numerical differentiation; six quantum-chemical calculations are sufficient for this purpose for the first derivative (two variations for each of the three Cartesian directions). Provided it is assumed for the sake of simplicity that set $\boldsymbol{R}$ is formed of the Cartesian coordinates X_i of atomic nuclei, then it holds (in abbreviated notation) that

$$\frac{\partial \mu}{\partial X_i} = - \frac{\partial}{\partial X_i} \frac{\partial U}{\partial F} = - \frac{\partial}{\partial F} \frac{\partial U}{\partial X_i} = - \frac{\partial \, \mathrm{grad}_i \, U}{\partial F}, \tag{3-60}$$

$$\frac{\partial \alpha}{\partial X_i} = \frac{\partial}{\partial X_i} \frac{\partial \mu}{\partial F} = \frac{\partial}{\partial F} \frac{\partial \mu}{\partial X_i} = - \frac{\partial^2 \, \mathrm{grad}_i \, U}{\partial F^2}. \tag{3-61}$$

The calculation is further simplified by the fact[470] that the integrals over a given basis set do not depend on the value of the applied field, so that a single integral calculation is sufficient for all these calculations. The mean square amplitudes represent another type of characteristic that is closely connected with the vibrational problem, derivable from the results of vibrational analysis or the available experimental measurements[130]. However, this characteristic has not yet been systematically evaluated on the basis of quantum-chemical methods as generators of input data.

In connection with the principal mastering solution of the problem of calculation of the second derivatives of the energy, greater interest is now being paid to the question of quantum-chemical generation of higher energy derivatives in apparent connection with the problem of the anharmonicity of vibrational motions. When analytical formulae are available for the second derivatives, the third derivatives can be readily obtained by numerical differentiation; otherwise, the problem leads to more awkward point mapping of the hypersurface, at least in a certain region around the stationary point(s). Pulay pointed out an interesting situation in the SCFapproach[193]. Although this type of quantum-chemical methods leads to erroneous behaviour of the energy function in the dissociation limit, the values of the higher energy derivatives need not be affected much, as the effect of the correlation energy on the force constant has been found to be relatively weak (or constant) around the equilibrium point. A number of quantum-chemical studies have been carried out on the anharmonic effects on the vibrational properties of simple systems[101,476–491]. At present, anharmonicity is often included for only some selected degrees of freedom (see e.g.[463,480,485]); however, both cubic and quartic force constants are sometimes included (e.g.[485] for HNO).

An interesting approach (at present again limited to simple systems) is taken in works carrying out solution of the vibrational problem for an anharmonic potential hypersurface in terms of expansion of the wave function in harmonic oscillator eigenfunctions (the 'nuclear CI method')[477–479,481,482,492]. This, however, implies a transition from the concept of a stationary point with its surroundings characterized by harmonic and higher force constants to utilization of the whole quantum-chemical hypersurface for solution of the vibrational problem (cf.[402–405]).

3.7 Summary

The potential energy hypersurface at present represents a predominant, practical instrument for the theoretical study (and a basis for interpretation of experimental results – interpretation more or less implicitly includes a certain theoretical model and the concepts following from it) of the behaviour and properties of chemical substances (at present primarily only in the gaseous phase), in particular their structure and reactivity. It thus plays a key role especially in relation to chemical isomerism – most contemporary theoretical studies in isomer chemistry leading to quantitative numerical results are realized within the framework of a potential hypersurface under full acceptance of the Born-Oppenheimer approximation. Because of the close relationship between the concepts of molecular structure and chemical isomerism (the latter is the apparent particular manifestation of the former), the close parallelism or identity in the theoretical description and discussion of these two phenomena is not surprising. The actual means of theoretical characterization of the potential energy hypersurfaces are essentially universal (in spite of the wide variety of ways of construction and quality of the individual energy functions available at present). Organically, three clearly different levels of description can be distinguished: representation in terms of isolated stationary points alone, mapping of relationships between these points in terms of various types of reaction paths and, finally, still rather rarely, exhaustive point-by-point generation of hypersurfaces with subsequent analytical representation.

In terms of the characteristics of the potential hypersurface, isomerism is interpreted straightforwardly as the appearance of more than one local minimum within a single hypersurface or within a set of hypersurfaces belonging to different electronic states (usually, only nondegenerate local minima are considered in this connection). At present, very effective procedures are available for the localization of stationary points on hypersurfaces (primarily local minima and transition states) for all the important types of quantum-chemical approximations (including the *ab initio* MC SCF level), working with energy gradients constructed from analytical formulae. The characterization of types of individual stationary points is then carried out by analysis of the eigenvalues of the matrix of the second derivatives of the energy, in particular by parity structure of the eigenvalues. The force constant matrix is most often constructed for this purpose by numerical differentiation of the energy gradient, even though analytical formulae are available for the second derivative of the energy, up to the *ab initio* SCF level, inclusive. The possibility of transitions between the individual stationary points, especially local minima, i.e. between isomers, is primarily determined by the heights of the energy barriers separating them; various concepts of minimum energy reaction paths can yield the minimal values of these barriers. The intrinsic reaction path is physically most valuable; this concept has a clear relationship with classical reaction trajectories. Its construction requires only the ability to generate the energy gradient in the general hypersurface point but not this hypersurface itself. It is useful for many applications to characterize the stationary points more closely, especially local minima, by passage from the force constant matrix to the vibrational **GF** problem yielding the harmonic vibrational frequencies. For simple systems, these values are further refined by corrections for anharmonic behaviour of the potential hypersurface. The highest (and final) step in the description in terms of the Born-Oppenheimer approximation represents knowledge of the whole potential hypersurface, opening a pathway to comprehensive determination of the nuclear wave function and thus to a description that would be exact within this approximation. Nonetheless, at present problems in isomeric chemistry are typically studied theoretically by representation of the hypersurface in terms of its isolated stationary points (or harmonic description of the neighbourhood of these points) and mapping of the relationship between them in terms of suitable reaction paths.

REFERENCES

(1) Born, M., and Oppenheimer, R.: 1927, Ann. Phys. 4. Volge *84* (der ganzen Reihe *389*), pp. 457—484.
(2) Woolley, R. G., and Sutcliffe, B. T.: 1977, Chem. Phys. Lett. *45*, pp. 393—398.
(3) Davydov, A. S.: 1965, Quantum Mechanics, Pergamon Press, Oxford.
(4) McWeeny, R.: 1972, Quantum Mechanics: Principles and Formalism, Pergamon Press, Oxford.
(5) Weissbluth, M.: 1978, Atoms and Molecules, Academic Press, New York.
(6) Born, M.: 1951, Nachr. Akad. Wiss. Göttingen, Math. Phys. Kl. No 6, pp. 1—3.
(7) Born, M., and Huang, K.: 1954, Dynamical Theory of Crystal Lattices, Oxford University Press, London, pp. 402—407.
(8) Kołos, W.: 1970, Advan. Quantum Chem. *5*, pp. 99—133.
(9) Sutcliffe, B. T.: 1975, in Computational Techniques in Quantum Chemistry and Molecular Physics, Ed. G. H. F. Diercksen, B. T. Sutcliffe, and A. Veillard, D. Reidel Publ. Comp., Dordrecht, pp. 1—105.
(10) Sutcliffe, B. T.: 1980, in Quantum Dynamics of Molecules. The New Experimental Challenge to Theorists, Ed. R. G. Woolley, Plenum Press, New York, pp. 1—37.
(11) Combes, J. M., Duclos, P., and Seiler, R.: 1981, in Rigorous Atomic and Molecular Physics, Ed. G. Velo, and A. S. Wightman, Plenum Press, New York, pp. 185—212.
(12) Hellmann, H.: 1937, Einführung in die Quantenchemie, F. Deuticke, Leipzig.
(13) Feynman, R. P.: 1939, Phys. Rev. *56*, pp. 340—343.
(14) Schutte, C. J. H.: 1971, Structure and Bonding *9*, pp. 213—263.
(15) Özkan, İ., and Goodman, L.: 1979, Chem. Rev. *79*, pp. 275—285.
(16) Essén, H.: 1977, Int. J. Quantum Chem. *12*, pp. 721—735.
(17) Longuet-Higgins, H. C.: 1961, Advan. Spectrosc. *2*, p. 429.
(18) Lathouwers, L., van Leuven, P., and Bouten, M.: 1977, Chem. Phys. Lett. *52*, pp. 439—441.
(19) van Leuven, P., and Lathouwers, L.: 1980, in Quantum Dynamics of Molecules. The New Experimental Challenge to Theorists, Ed. R. G. Woolley, Plenum Press, New York, pp. 197—220.
(20) Kołos, W., and Wolniewicz, L.: 1964, J. Chem. Phys. *41*, pp. 3663—3673.
(21) Kołos, W., and Wolniewicz, L.: 1964, J. Chem. Phys. *41*, pp. 3674—3678.
(22) Kołos, W., and Wolniewicz, L.: 1966, J. Chem. Phys. *45*, pp. 509—514.
(23) Wolniewicz, L.: 1966, J. Chem. Phys. *45*. pp. 515—523.
(24) Kołos, W., and Wolniewicz, L.: 1968, J. Chem. Phys. *49*, pp. 404—410.

(25) Czub, J., and Wolniewicz, L.: 1978, Mol. Phys. *36*, pp. 1301—1308.
(26) Bunker, P. R.: 1977, J. Mol. Spectrosc. *68*, pp. 367—371.
(27) Brickmann, J., and Bitto, H.: 1978, Chem. Phys. *32*, pp. 143—152.
(28) Goscinski, O., and Palma, A.: 1979, Int. J. Quantum Chem. *15*, pp. 197—205.
(29) Hutson, J. M., and Howard, B. J.: 1980, Mol. Phys. *41*, pp. 1113—1122.
(30) Fulton, R. L., and Gouterman, M.: 1964, J. Chem. Phys. *41*, pp. 2280—2286.
(31) Brickmann, J.: 1978. Mol. Phys. *35*, pp. 155—176.
(32) Goodman, L., and Özkan, İ.: 1979, Chem. Phys. Lett. *61*, pp. 216—222.
(33) Özkan, İ., and Goodman, L.: 1979, Chem. Phys. Lett. *64*, pp. 32—38.
(34) Gustav, K., and Colditz, R.: 1981, Z. Chem. *21*, pp. 417—418.
(35) Gustav, K., and Colditz, R.: 1982, Int. J. Quantum Chem. *22*, pp. 31—36.
(36) Levine, R. D., and Bernstein, R. B.: 1974, Molecular Reaction Dynamics, Oxford University Press, New York.
(37) Nikitin, E. E., and Zülicke, L.: 1978, Selected Topics of the Theory of Chemical Elementary Processes, Springer-Verlag, Berlin.
(38) Christov, S. G.: 1980, Collision Theory and Statistical Theory of Chemical Reactions, Springer-Verlag, Berlin.
(39) Hobey, W. D., and McLachlan, A. D.: 1960, J. Chem. Phys. *33*, pp. 1695—1703.
(40) Englman, R.: 1972, The Jahn-Teller Effect in Molecules and Crystals, Wiley-Interscience, New York.
(41) Bersuker, I. B.: 1976, Electronic Structure and Properties of Coordination Compounds, Khimiya, Leningrad, (in Russian).
(42) Hougen, J. T.: 1970, in Physical Chemistry. An Advanced Treatise, Vol. IV, Ed. D. Henderson, Academic Press, New York, pp. 307—348.
(43) Renner, R.: 1934, Z. Phys. *92*, pp. 172—193.
(44) Jahn, H. A., and Teller, E.: 1937, Proc. Roy. Soc. *A 161*, pp. 220—235.
(45) Longuet-Higgins, H. C.: 1975, Proc. Roy. Soc. *A 344*, pp. 147—156.
(46) Stone, A. J.: 1976, Proc. Roy. Soc. *A 351*, pp. 141—150.
(47) Davidson, E. R.: 1977, J. Am. Chem. Soc. *99*, pp. 397—402.
(48) Aronowitz, S.: 1978, Int. J. Quantum Chem. *14*, pp. 253—269.
(49) Gerber, W. H., and Schumacher, E.: 1978, J. Chem. Phys. *69*, pp. 1692—1703.
(50) Meyer, R., Graf, F., Ha, T.-K., and Günthard, Hs. H.: 1979, Chem. Phys. Lett. *66*, pp. 65—71.
(51) Kellman, M. E.: 1982, Chem. Phys. Lett. *87*, pp. 171—176.
(52) King, G. W.: 1970, in Physical Chemistry. An Advanced Treatise, Vol. IV, Ed. D. Henderson, Academic Press, New York, pp. 67—121.
(53) Liehr, A. D.: 1961, Progr. Inorg. Chem. *3*, pp. 281—314f.
(54) Liehr, A. D.: 1962, Progr. Inorg. Chem. *4*, pp. 455—540.
(55) Liehr, A. D.: 1962, Annu. Rev. Phys. Chem. *13*, pp. 41—76.
(56) Liehr, A. D.: 1963, Progr. Inorg. Chem. *5*, pp. 385—430.
(57) Liehr, A. D.: 1963, J. Phys. Chem. *67*, pp. 389—471.
(58) Liehr, A. D.: 1963, J. Phys. Chem. *67*, pp. 471—494.
(59) Bauer, S. H.: 1970, in Physical Chemistry. An Advanced Treatise, Vol. IV, Ed. D. Henderson, Academic Press, New York, pp. 1—17.
(60) Muetterties, E. L.: 1965, Inorg. Chem. *4*, pp. 769—771.
(61) Muetterties, E. L.: 1970, Accounts Chem. Res. *3*, pp. 266—273.
(62) von Doering, W. E., and Roth, W. R.: 1963, Angew. Chem. *75*, pp. 27—35.
(63) Cotton, F. A.: 1968, Accounts Chem. Res. *1*, pp. 257—265.
(64) Ugi, I., Marquarding, D., Klusacek, H., and Gillespie, P.: 1971, Accounts Chem. Res. *4*, pp. 288—296.

(65) Gillespie, P., Hoffman, P., Klusacek, H., Marquarding, D., Pfohl, S., Ramirez, F. Tsolis, E. A., and Ugi, I.: 1971, Angew. Chem., Int. Ed. Engl. *10*, pp. 687—715.
(66) Sergeev, N. M.: 1973, Usp. Khim. *42*, pp. 769—798.
(67) Spirodonov, V. P., Ischenko, A. A., and Zasorin, E. Z.: 1978, Usp. Khim. *47*, pp. 101 to 126.
(68) Ref.[41], pp. 279—280.
(69) Boldyrev, A. I., Charkin, O. P., Rambidi, N. G., and Avdeev, V. I.: 1976, Chem. Phys. Lett. *44*, pp. 20—24.
(70) Boldyrev, A. I., Charkin, O. P., Rambidi, N. G., and Avdeev, V. I.: 1977, Chem. Phys. Lett. *50*, pp. 239—242.
(71) Slanina, Z., Beran, S., and Knížek, P.: 1979, in Heterogeneous Catalysis, Part 2, Ed. D. Shopov, A. Andreev, A. Palazov, and L. Petrov, Publ. House Bulg. Acad. Sci., Sofia, pp. 69—74.
(72) Charkin, O. P., and Boldyrev, A. I.: 1980, Potential Surfaces and Structural Nonrigidity of Inorganic Molecules, VINITI, Moscow, (in Russian).
(73) Boldyrev, A. I., Sukhanov, L. P., Zakhzevskii, V. G., and Charkin, O. P.: 1981, Chem. Phys. Lett. *79*, pp. 421—426.
(74) Yamada, K., and Winnewisser, M.: 1976, Z. Naturforsch. *a 31*, pp. 139—144.
(75) Amar, F., Kellman, M. E., and Berry, R. S.: 1979, J. Chem. Phys. *70*, pp. 1973—1985.
(76) Berry, R. S.: 1979, in The Permutation Group in Physics and Chemistry, Ed. J. Hinze, Springer-Verlag, Berlin, pp. 92—120.
(77) Kellman, M. E., Amar, F., and Berry, R. S.: 1980, J. Chem. Phys. *73*, pp. 2387—2404.
(78) Berry, R. S.: 1980, in Quantum Dynamics of Molecules. The New Experimental Challenge to Theorists, Ed. R. G. Woolley, Plenum Press, New York, pp. 143—195.
(79) Baker Jr., G. A.: 1956, Phys. Rev. *103*, pp. 1119—1120.
(80) Liskow, D. H., Bender, C. F., and Schaefer III, H. F.: 1972, J. Chem. Phys. *56*, pp. 5075—5080.
(81) Diercksen, G. H. F., Kraemer, W. G., and Roos, B. O.: 1975, Theor. Chim. Acta *36*, pp. 249—274.
(82) Weber, J., Yoshimine, M., and McLean, A. D.: 1976, J. Chem. Phys. *64*, pp. 4159—4164.
(83) Poppinger, D., Radom, L., and Pople, J. A.: 1977, J. Am. Chem. Soc. *99*, pp. 7806—7816.
(84) Bouma, W. J., and Radom, L.: 1978, Aust. J. Chem. *31*, pp. 1649—1660.
(85) Almlöf, J., and Ischenko, A. A.: 1979, Chem. Phys. Lett. *61*, pp. 79—82.
(86) Demuynck, J., Fox, D. J., Yamaguchi, Y., and Schaefer III, H. F.: 1980, J. Am. Chem. Soc. *102*, pp. 6204—6207.
(87) Osamura, Y., Schaefer III, H. F., Gray, S. K., and Miller, W. H.: 1981, J. Am. Chem. Soc. *103*, pp. 1904—1907.
(88) Krishnan, R., Frisch, M. J., and Pople, J. A.: 1981, Chem. Phys. Lett. *79*, pp. 408—411.
(89) Osamura, Y., and Schaefer III, H. F.: 1981, Chem. Phys. Lett. *79*, pp. 412—415.
(90) Andrade, J. G., Chandrasekhar, J., and von Schleyer, P. R.: 1981, J. Comput. Chem. *2*, pp. 207—211.
(91) Minkin, V. L., Olekhnovich, L. P., and Zhdanov, Yu. A.: 1981, Accounts Chem. Res. *14*, pp. 210—217.
(92) Smeyers, Y. G., and Bellido, M. N.: 1981, Int. J. Quantum Chem. *19*, pp. 553—565.
(93) Walsh, A. D.: 1953, J. Chem. Soc., pp. 2260—2331.
(94) Brown, H. C. (with comments by P. v. R. Schleyer): 1977, The Nonclassical Ion Problem, Plenum Press, New York.
(95) Nemukhin, A. V., Almlöf, J. E., and Heiberg, A.: 1981, Theor. Chim. Acta *59*, pp. 9—16.
(96) le Roy, R. J., and Carley, J. S.: 1980, Advan. Chem. Phys. *42*, pp. 353—420.
(97) Berns, R. M., and van der Avoird, A.: 1980, J. Chem. Phys. *72*, pp. 6107—6116.

(98) van der Avoird, A.: 1981, in Intermolecular Forces, Ed. B. Pullman, D. Reidel Publ. Comp. Dordrecht, pp. 1—14.
(99) Slanina, Z.: 1981, Chem. Phys. Lett. *82*, pp. 33—35.
(100) Slanina, Z.: 1981, Chem. Phys. Lett. *83*, pp. 418—422.
(101) Tennyson, J., and van der Avoird, A.: 1982, J. Chem. Phys. *77*, pp. 5664—5681.
(102) Brigot, N., Odiot, S., and Walmsley, S. H.: 1982, Chem. Phys. Lett. *88*, pp. 543—546.
(103) Slanina, Z.: 1983, Theochem, *11*, pp. 401—405.
(104) Levitt, M., and Warshel, A.: 1978, J. Am. Chem. Soc. *100*, pp. 2607—2613.
(105) Sundaralingam, M., and Westhof, E.: 1979, Int. J. Quantum Chem., Quantum Biol. Symp. *6*, pp. 115—130.
(106) Venkatachalam, C. M., and Urry, D. W.: 1979, Int. J. Quantum Chem., Quantum Biol. Symp. *6*, pp. 343—351.
(107) Northrup, S. H., and McCammon, J. A.: 1980, Biopolymers *19*, pp. 1001—1016.
(108) Kumar, N. V., and Govil, G.: 1980, Int. J. Quantum Chem., Quantum Biol. Symp. *7*, pp. 101—113.
(109) Karplus, M.: 1981, in Biomolecular Stereodynamics, Vol. II, Ed. R. H. Sarma, Adenine Press, New York.
(110) Langlet, J., Claverie, P., Caron, F., and Boeuve, J. C.: 1981, Int. J. Quantum Chem. *19*, pp. 299—338.
(111) Gupta, V. D.: 1981, Int. J. Quantum Chem. *20*, pp. 9—21.
(112) Paterson, Y., Rumsey, S. M., Benedetti, E., Némethy, G., and Scheraga, H. A.: 1981, J. Am. Chem. Soc. *103*, pp. 2947—2955.
(113) Tosi, C., and Saenger, W.: 1982, Chem. Phys. Lett. *90*, pp. 277—281.
(114) 1976, Landolt-Börnstein Zahlenwerte und Funktionen aus Naturwissenschaften und Technik. Neue Serie, Gruppe II, Band 7, Ed. J. H. Callomon, E. Hirota, K. Kuchitsu, W. J. Lafferty, A. G. Maki, and C. S. Pote, Springer-Verlag, Berlin.
(115) Hargittai, M., and Hargittai, I.: 1977, The Molecular Geometries of Coordination Compounds in the Vapour Phase, Akadémiai Kiadó, Budapest.
(116) Allen, F. H., Bellard, S., Brice, M. D., Cartwright, B. A., Doubleday, A., Higgs, H., Hummelink, T., Hummelink-Peters, B. G., Kennard, O., Motherwell, W. D. S., Rodgers, J. R., and Watson, D. G.: 1979, Acta Cryst. *B 35*, pp. 2331—2339.
(117) Langer, V., and Bláha, K.: 1981, Chem. Listy *75*, pp. 1070—1074.
(118) Laurie, V. W.: 1974, in Critical Evaluation of Chemical and Physical Structural Information, Ed. D. R. Lide, Jr., and M. A. Paul, National Academy of Sciences, Washington, pp. 67—76.
(119) Hedberg, K.: 1974, in Critical Evaluation of Chemical and Physical Structural Information, Ed. D. R. Lide, Jr., and M. A. Paul, National Academy of Sciences, Washington, pp. 77—93.
(120) Schwendeman, R. H.: 1974, in Critical Evaluation of Chemical and Physical Structural Information, Ed. D. R. Lide, Jr., and M. A. Paul, National Academy of Sciences, Washington, pp. 94—115.
(121) Kuchitsu, K.: 1974, in Critical Evaluation of Chemical and Physical Structural Information, Ed. D. R. Lide, Jr., and M. A. Paul, National Academy of Sciences, Washington, pp. 132—139.
(122) Kuchitsu, K., and Oyanagi, K.: 1977, Faraday Discuss. Chem. Soc. *62*, pp. 20—28.
(123) Papoušek, D., and Aliev, M. R.: 1982, Molecular Vibrational/Rotational Spectra, Elsevier, Amsterdam & Academia, Prague.
(124) Watson, J. K. G.: 1973, J. Mol. Spectrosc. *48*, pp. 479—502.
(125) Zeil, W., and Christen, D.: 1980, J. Phys. Chem. *84*, pp. 1790—1792.
(126) Boggs, J. E., von Carlowitz, M., and von Carlowitz, S.: 1982, J. Phys. Chem. *86*, pp. 157—159.

(127) HEDBERG, K., and IWASAKI, M.: 1962, J. Chem. Phys. *36*, pp. 589—594.
(128) RYAN, R. R., and HEDBERG, K.: 1969, J. Chem. Phys. *50*, pp. 4986—4995.
(129) DANIELSON, D. D., PATTON, J. V., and HEDBERG, K.: 1977, J. Am. Chem. Soc. *99*, pp. 6484—6487.
(130) CYVIN, S. J.: 1968, Molecular Vibrations and Mean Square Amplitudes, Universitetsforlaget, Oslo.
(131) SCHOEN, P. E., PRIEST, R. G., SHERIDAN, J. P., and SCHNUR, J. M.: 1977, Nature *270*, pp. 412—413.
(132) LEWIS, G. J., and WHALLEY, E.: 1978, J. Chem. Phys. *68*, pp. 1119—1127.
(133) WHALLEY, E.: 1980, Rev. Phys. Chem. Jap. *50*, pp. 119—131.
(134) TANIGUCHI, Y., TAKAYA, H., WONG, P. T. T., and WHALLEY, E.: 1981, J. Chem. Phys. *75*, pp. 4815—4822.
(135) TAKAYA, H., TANIGUCHI, Y., WONG, P. T. T., and WHALLEY, E.: 1981, J. Chem. Phys. *75*, pp. 4823—4828.
(136) CARDILLO, M. J., and BAUER, S. H.: 1969, Inorg. Chem. *8*, pp. 2086—2092.
(137) Ref.[59], pp. 796—798.
(138) GILBERT, M. M., GUNDERSEN, G., and HEDBERG, K.: 1972, J. Chem. Phys. *56*, pp. 1691—1697.
(139) BROWN, D. E., and BEAGLEY, B.: 1977, J. Mol. Struct. *38*, pp. 167—176.
(140) FUJIWARA, F. G., PAINTER, J. L., and KIM, H.: 1977, J. Mol. Struct. *41*, pp. 169—175.
(141) FUJIWARA, F. G., CHANG, J. C., and KIM, H.: 1977, J. Mol. Struct. *41*, pp. 177—182.
(142) MARSTOKK, K.-M., and MØLLENDAL, H.: 1978, J. Mol. Struct. *49*, pp. 221—237.
(143) STEIGEL, A.: 1978, in Dynamic NMR Spectroscopy, Ed. P. Diehl, E. Fluck, and R. Kosfeld, Springer-Verlag, Berlin, pp. 1—53.
(144) JARDETZKY, O.: 1980, Biochim. Biophys. Acta *621*, pp. 227—232.
(145) BUTT, G., WALTER, R., RENUGOPALAKRISHNAN, V., and DRUYAN, M. E.: 1979, Int. J. Quantum Chem., Quantum Biol. Symp. *6*, pp. 453—458.
(146) WILSON, E. B.: 1976, Pure Appl. Chem. *47*, pp. 41—47.
(147) POPLE, J. A.: 1976, Bull. Soc. Chim. Belg. *85*, pp. 347—361.
(148) HA, T.-K.: 1976, Chimia *30*, pp. 297—305.
(149) DANNENBERG, J. J.: 1976, Angew. Chem., Int. Ed. Engl. *15*, pp. 519—525.
(150) SCHAEFER III, H. F.: 1976, Annu. Rev. Phys. Chem. *27*, pp. 261—290.
(151) BAIRD, N. C.: 1977, Pure Appl. Chem. *49*, pp. 223—236.
(152) MULLIKEN, R. S.: 1978, Annu. Rev. Phys. Chem. *29*, pp. 1—30.
(153) JUG, K.: 1980, Theor. Chim. Acta *54*, pp. 263—300.
(154) LESTER JR., W. A.: 1980, in Computational Methods in Chemistry, Ed. J. Bargon, Plenum, New York, pp. 301—319.
(155) LÖWDIN, P.-O.: 1980, Advan. Quantum Chem. *12*, pp. 263—316.
(156) ČÁRSKY, P., and URBAN, M.: 1980, Ab Initio Calculations, Springer-Verlag, Berlin.
(157) ROOS, B: 1975, in Computational Techniques in Quantum Chemistry and Molecular Physics, Ed. G. H. F. Diercksen, B. T. Sutcliffe, and A. Veillard, D. Reidel Publ. Comp., Dordrecht, pp. 251—297.
(158) HURLEY, A. C.: 1976, Electron Correlation in Small Molecules, Academic Press, London.
(159) MCWEENY, R.: 1976, in The New World of Quantum Chemistry, Ed. B. Pullman, and R. Parr, D. Reidel Publ. Comp., Dordrecht, pp. 3—31.
(160) KVASNIČKA, V.: 1977, Advan. Chem. Phys. *36*, pp. 345—412.
(161) BRANDOW, B. H.: 1977, Advan. Quantum Chem. *10*, pp. 187—249.
(162) POPLE, J. A., and BEVERIDGE, D. L.: 1970, Approximate Molecular Orbital Theory, McGraw-Hill, New York.
(163) DEWAR, M. J. S.: 1969, The Molecular Orbital Theory of Organic Chemistry, McGraw-Hill, New York.

(164) COFFEY, P., and JUG, K.: 1973, J. Am. Chem. Soc. *95*, pp. 7575—7580.
(165) BÖHM, M. C., and GLEITER, R.: 1981, Theor. Chim. Acta *59*, pp. 127—151.
(166) BÖHM, M. C., and GLEITER, R.: 1981, Theor. Chim. Acta *59*, pp. 153—179.
(167) LIFSON, S., and WARSHEL, A.: 1968, J. Chem. Phys. *49*, pp. 5116—5129.
(168) 1971, Conformational Analysis. Scope and Present Limitations, Ed. G. Chiurdoglu, Academic Press, New York.
(169) DASHEVSKY, V. G.: 1974, Conformations of Organic Molecules, Khimija, Moscow, (in Russian).
(170) NIKETIĆ, S. R., and RASMUSSEN, K.: 1977, The Consistent Force Field: A Documentation, Springer-Verlag, Berlin.
(171) MCIVER JR., J. W., and KOMORNICKI, A.: 1972, J. Am. Chem. Soc. *94*, pp. 2625—2633.
(172) ERMER, O.: 1976, Structure and Bonding *27*, pp. 161—217.
(173) MURRELL, J. N.: 1977, Structure and Bonding *32*, pp. 93—146.
(174) MEZEY, P. G.: 1977, Progr. Theor. Org. Chem. *2*, pp. 127—161.
(175) MEZEY, P. G.: 1981, in Computational Theoretical Organic Chemistry, Ed. I. G. Csizmadia, and R. Daudel, D. Reidel Publ. Comp., Dordrecht, pp. 101—128.
(176) SLANINA, Z.: 1981, Advan. Quantum Chem. *13*, pp. 89—153.
(177) DAUDEL, R.: 1973, Quantum Theory of Chemical Reactivity, D. Reidel Publ. Comp., Dordrecht.
(178) BERAN, S., ČÁRSKY, P., HOBZA, P., PANCÍŘ, J., POLÁK, R., SLANINA, Z., and ZAHRADNÍK, R.: 1978, Russ. Chem. Rev. *47*, pp. 1025—1041.
(179) ZHIDOMIROV, G. M., BAGATURJANC, A. A., and ABRONIN, I. A.: 1979, Applied Quantum Chemistry, Khimija, Moscow, (in Russian).
(180) NELDER, J. A., and MEAD, R.: 1965, Comput. J. *7*, pp. 308—313; *8*, p. 27.
(181) BROWN, A., DEWAR, M. J. S., and SCHOELLER, W.: 1970, J. Am. Chem. Soc. *92*, pp. 5516 to 5517.
(182) DEWAR, M. J. S., and KOHN, M. C.: 1972, J. Am. Chem. Soc. *94*, pp. 2704—2706.
(183) PULAY, P.: 1969, Mol. Phys. *17*, pp. 197—204.
(184) PULAY, P.: 1970, Mol. Phys. *18*, pp. 473—480.
(185) PULAY, P.: 1971, Mol. Phys. *21*, pp. 329—339.
(186) MCIVER JR., J. W., and KOMORNICKI, A.: 1971, Chem. Phys. Lett. *10*, pp. 303—306.
(187) GARTON, D., and SUTCLIFFE, B. T.: 1974, in Theoretical Chemistry, Specialist Periodical Reports, Vol. 1, Ed. R. N. Dixon, The Chemical Society, London, pp. 34—59.
(188) OLSEN, J., YEAGER, D. L., and JØRGENSEN, P.: 1983, Advan. Chem. Phys. *54*, pp. 1—176.
(189) MOCCIA, R.: 1967, Theor. Chim. Acta *8*, pp. 8—17.
(190) GERRATT, J., and MILLS, I. M.: 1968, J. Chem. Phys. *49*, pp. 1719—1729.
(191) HURLEY, A. C.: 1954, Proc. Roy. Soc. *A 226*, pp. 170—178.
(192) BRATOŽ, S.: 1958, Colloq. Int. Cent. Nat. Rech. Sci. *82*, pp. 287—301.
(193) PULAY, P: 1977, in Modern Theoretical Chemistry, Vol. 4., Ed. H. F. Schaefer III, Plenum Press, New York, pp. 153—185.
(194) EPSTEIN, S. T.: 1980, Theor. Chim. Acta *55*, pp. 251—253.
(195) FLETCHER, R.: 1970, Mol. Phys. *19*, pp. 55—63.
(196) KUMANOVA, M. D.: 1972, Mol. Phys. *23*, pp. 407—412.
(197) BLOEMER, W. L., and BRUNER, B. L.: 1973, J. Chem. Phys. *58*, pp. 3735—3744.
(198) KOMORNICKI, A., IHSIDA, K., MOROKUMA, K., DITCHFIELD, R., and CONRAD, M.: 1977, Chem. Phys. Lett. *45*, pp. 595—602.
(199) BOYS, S. F.: 1950, Proc. Roy. Soc. *A 200*, pp. 542—554.
(200) GERRATT, J., and MILLS, I. M.: 1968, J. Chem. Phys. *49*, pp. 1730—1739.
(201) BERAN, S., SLANINA, Z., and ZIDAROV, D. C.: 1978, Int. J. Quantum Chem. *13*, pp. 227—233.
(202) HALL, G. G.: 1961, Philos. Mag. *6*, pp. 249—258.

(203) COHEN, M., and DALGARNO, A.: 1961, Proc. Phys. Soc. London *77*, pp. 748—750.
(204) STANTON, R. E.: 1962, J. Chem. Phys. *36*, pp. 1298—1300.
(205) TUAN, D. F.-T.: 1969, J. Chem. Phys. *51*, pp. 607—611.
(206) COULSON, C. A.: 1971, Mol. Phys. *20*, pp. 687—694.
(207) NAKATSUJI, H., KANDA, K., and YONEZAWA, T.: 1980, Chem. Phys. Lett. *75*, pp. 340—346.
(208) NAKATSUJI, H., HAYAKAWA, T., and HADA, M.: 1981, Chem. Phys. Lett. *80*, pp. 94—100.
(209) NAKATSUJI, H., KANDA, K., and YONEZAWA, T.: 1982, J. Chem. Phys. *77*, pp. 1961—1968.
(210) HUBER, H.: 1979, Chem. Phys. Lett. *62*, pp. 95—99.
(211) TACHIBANA, A., YAMASHITA, K., YAMABE, T., and FUKUI, K.: 1978, Chem. Phys. Lett. *59*, pp. 255—260.
(212) FRAGA, S., SAXENA, K. M. S., and TORRES, M.: 1978, Biomolecular Information Theory, Elsevier, Amsterdam.
(213) WOLIŃSKI, K., and SADLEJ, A. J.: 1978, Acta Univ. Wratisl. *425*, pp. 19—33.
(214) WOLIŃSKI, K., and SADLEJ, A. J.: 1978, Acta Univ. Wratisl. *425*, pp. 35—57.
(215) DAVIDON, W. C.: 1959, A. E. C. Res. Develop. Rept. ANL 5990 (quoted according to Refs[170,198]).
(216) FLETCHER, R., and POWELL, M. J. D.: 1963, Comput. J. *6*, pp. 163—168.
(217) MURTAGH, B. A., and SARGENT, R. W. H.: 1970, Comput. J. *13*, pp. 185—194.
(218) MCIVER JR., J. W., and STANTON, R. E.: 1972, J. Am. Chem. Soc. *94*, pp. 8618—8620.
(219) STANTON, R. E., and MCIVER JR., J. W.: 1975, J. Am. Chem. Soc. *97*, pp. 3632—3646.
(220) ERMER, O.: 1975, Tetrahedron *31*, pp. 1849—1854.
(221) BEREMERMANN, H.: 1970, Math. Biosci. *9*, pp. 1—15.
(222) HILDERBRANDT, R. L.: 1977, Comput. Chem. *1*, pp. 179—186.
(223) GILL, P. E., MURRAY, W., and PICKEN, S. M.: 1972, Natl. Phys. Lab. Report NAC *24* (quoted according to Ref.[170]).
(224) SCHARFENBERG, P.: 1979, Theor. Chim. Acta *53*, pp. 279—292.
(225) FLANIGAN, M. C., KOMORNICKI, A., and MCIVER JR., J. W.: 1977, in Semiempirical Methods of Electronic Structure Calculation, Part B: Applications, Ed. G. A. Segal, Plenum Press, New York, p. 1.
(226) ILLEK, B., GRIMMER, M., and HEIDRICH, D.: 1979, Wissen. Z. KMU Leipzig, Math.-Naturwissen. R. *28*, pp. 633—639.
(227) POWELL, M. J. D.: 1965, Comput. J. *7*, pp. 303—307.
(228) ROTHMAN, M. J., and LOHR JR., L. L.: 1980, Chem. Phys. Lett. *70*, pp. 405—409.
(229) CERJAN, C. J., and MILLER, W. H.: 1981, J. Chem. Phys. *75*, pp. 2800—2806.
(230) HALGREN, T. A., PEPPERBERG, I. M., and LIPSCOMB, W. N.: 1975, J. Am. Chem. Soc. *97*, pp. 1248—1250.
(231) HALGREN, T. A., and LIPSCOMB, W. N.: 1977, Chem. Phys. Lett. *49*, pp. 225—232.
(232) MEZEY, P. G., PETERSON, M. R., and CSIZMADIA, I. G.: 1977, Can. J. Chem. *55*, pp. 2941 to 2945.
(233) MÜLLER, K., and BROWN, L. D.: 1979, Theor. Chim. Acta *53*, pp. 75—93.
(234) MÜLLER, K.: 1980, Angew. Chem., Int. Ed. Engl. *19*, pp. 1—13.
(235) SCHARFENBERG, P.: 1980, Theor. Chim. Acta *58*, pp. 73—79.
(236) POPPINGER, D.: 1975, Chem. Phys. Lett. *34*, pp. 332—336.
(237) COLLINS, J. B., VON SCHLEYER, P. R., BINKLEY, J. S., and POPLE, J. A.: 1976, J. Chem. Phys. *64*, pp. 5142—5151.
(238) PAYNE, P. W.: 1976, J. Chem. Phys. *65*, pp. 1920—1926.
(239) DETAR, D. F.: 1977, Comput. Chem. *1*, pp. 141—144.
(240) LAUER, G., SCHULTE, K.-W., SCHWEIG, A., and THIEL, W.: 1979, Theor. Chim. Acta *52*, pp. 319—328.
(241) HAVLAS, Z., and MALOŇ, P.: 1980, Collect. Czech. Chem. Commun. *45*, pp. 321—329.

(242) Sellers, H. L., Klimkowski, V. J., and Schäfer, L.: 1978, Chem. Phys. Lett. *58*, pp. 541—544.
(243) Komornicki, A., and McIver Jr., J. W.: 1974, J. Am. Chem. Soc. *96*, pp. 5798—5800.
(244) Lewis, P. N., Momany, F. A., and Scheraga, H. A.: 1973, Isr. J. Chem. *11*, pp. 121—152.
(245) Gō, N., and Scheraga, H. A.: 1973, Macromolecules *6*, pp. 525—535.
(246) Rao, G. S., Tyagi, R. S., and Mishra, R. K.: 1981, Int. J. Quantum Chem. *20*, pp. 273—279.
(247) Jorgensen, W. L.: 1979, J. Am. Chem. Soc. *101*, pp. 2011—2015.
(248) Slanina, Z.: 1980, Collect. Czech. Chem. Commun. *45*, pp. 3417—3435.
(249) Slanina, Z.: 1974, Collect. Czech. Chem. Commun. *39*, pp. 3187—3191.
(250) Weintraub, H. J. R., and Hopfinger, A. J.: 1975, Int. J. Quantum Chem., Quantum Biol. Symp. *2*, pp. 203—208.
(251) Faber, D. H., and Altona, C.: 1977, Comput. Chem. *1*, pp. 203—213.
(252) Ferguson, T. R., and Beckel, C. L.: 1973, J. Chem. Phys. *59*, pp. 1905—1913.
(253) Slanina, Z., Beran, S., and Zidarov, D. C.: 1978, Proc. Int. Conf. Coord. Chem. 19th, p. S4.
(254) Beran, S., Slanina, Z., and Zidarov, D. C.: 1981, Int. J. Quantum Chem. *19*, pp. 585—591.
(255) Zeiss, G. D., and Whitehead, M. A.: 1971, J. Chem. Soc. *A*, pp. 1727—1738.
(256) Rinaldi, D., and Rivail, J.-L.: 1972, C. R. Hebd. Seances Acad. Sci. *C 274*, pp. 1664—1667.
(257) Pulay, P., and Török, F.: 1973, Mol. Phys. *25*, pp. 1153—1161.
(258) Pancíř, J.: 1973, Theor. Chim. Acta *29*, pp. 21—28.
(259) Török, F., Hegedüs, Á., and Pulay, P.: 1973, Theor. Chim. Acta *32*, pp. 145—150.
(260) Grimmer, M., and Heidrich, D.: 1973, Z. Chem. *13*, pp. 356—358.
(261) Rinaldi, D.: 1976, Comput. Chem. *1*, pp. 109—114.
(262) Dewar, M. J. S., and Komornicki, A.: 1977, J. Am. Chem. Soc. *99*, pp. 6174—6179.
(263) Khalil, S. M., and Shanshal, M.: 1977, Theor. Chim. Acta *46*, pp. 23—37.
(264) Scharfenberg, P.: 1978, Theor. Chim. Acta *49*, pp. 115—122.
(265) Dewar, M. J. S., and Yamaguchi, Y.: 1978, Comput. Chem. *2*, pp. 25—29.
(266) Leška, J., Zahradník, P., and Ondrejičková, D.: 1979, Theor. Chim. Acta *53*, pp. 253 to 268.
(267) Schlegel, H. B., Wolfe, S., and Bernardi, F.: 1975, J. Chem. Phys. *63*, pp. 3632—3638.
(268) Huber, H., Čársky, P., and Zahradník, R.: 1976, Theor. Chim. Acta *41*, pp. 217—221.
(269) Dupuis, M., and King, H. F.: 1978, J. Chem. Phys. *68*, pp. 3998—4004.
(270) Pople, J. A., Krishnan, R., Schlegel, H. B., and Binkley, J. S.: 1978, Int. J. Quantum Chem. *14*, pp. 545—560.
(271) Pulay, P.: 1979, Theor. Chim. Acta *50*, pp. 299—312.
(272) Yamashita, K., Tachibana, A., Yamabe, T., and Fukui, K.: 1980, Chem. Phys. Lett. *69*, pp. 413—416.
(273) Brooks, B. R., Laidig, W. D., Saxe, P., Goddard, J. D., Yamaguchi, Y., and Schaefer III, H. F.: 1980, J. Chem. Phys. *72*, pp. 4652—4653.
(274) Krishnan, R., Schlegel, H. B., and Pople, J. A.: 1980, J. Chem. Phys. *72*, pp. 4654—4655.
(275) Dupuis, M.: 1981, J. Chem. Phys. *74*, pp. 5758—5765.
(276) Osamura, Y., Yamaguchi, Y., and Schaefer III, H. F.: 1981, J. Chem. Phys. *75*, pp. 2919—2922.
(277) Goddard, J. D.: 1981, in Computational Theoretical Organic Chemistry, Ed. I. G. Csizmadia, and R. Daudel, D. Reidel Publ. Comp., Dordrecht, pp. 161—174.
(278) Kato, S., and Morokuma, K.: 1979, Chem. Phys. Lett. *65*, pp. 19—25.
(279) Goddard, J. D., Handy, N. C., and Schaefer III, H. F.: 1979, J. Chem. Phys. *71*, pp. 1525—1530.
(280) Pople, J. A., Krishnan, R., Schlegel, H. B., and Binkley, J. S.: 1979, Int. J. Quantum Chem., Quantum Chem. Symp. *13*, pp. 225—241.

(281) Dashevsky, V. G.: 1973, Usp. Khim. *42*, pp. 2097–2129.
(282) Dashevsky, V. G.: 1974, Usp. Khim. *43*, pp. 491–518.
(283) Pople, J. A.: 1973, in Energy, Structure and Reactivity, Ed. D. W. Smith, and W. B. McRae, Wiley, New York, pp. 51–61.
(284) Pople, J. A.: 1974, Tetrahedron *30*, pp. 1605–1615.
(285) Bell, S.: 1978, J. Chem. Phys. *69*, pp. 3879–3880.
(286) Bock, C. W., George, P., Mains, G. J., and Trachtman, M.: 1978, J. Mol. Struct. *49*, pp. 215–219.
(287) Almlöf, J., and Helgaker, T.: 1981, Chem. Phys. Lett. *83*, pp. 125–128.
(288) Mezey, P. G., and Haas, E.-C.: 1982, J. Chem. Phys. *77*, pp. 870–876.
(289) Cremer, D.: 1978, J. Chem. Phys. *69*, pp. 4440–4455.
(290) Cremer, D.: 1978, J. Chem. Phys. *69*, pp. 4456–4471.
(291) Davidson, E. R.: 1981, Int. J. Quantum Chem., Quantum Chem. Symp. *15*, pp. 65–68.
(292) Lüthi, H. P., Ammeter, J. H., Almlöf, J., and Faegri Jr., K.: 1982, J. Chem. Phys. *77*, pp. 2002–2009.
(293) Stillinger, F. H.: 1978, Int. J. Quantum Chem. *14*, pp. 649–657.
(294) Marcus, R. A.: 1966, J. Chem. Phys. *45*, pp. 4493–4499.
(295) Polanyi, J. C., and Schreiber, J. L.: 1974, in Physical Chemistry. An Advanced Treatise, Vol. VIa, Ed. W. Jost, Academic Press, New York, pp. 383–487.
(296) Fukui, K.: 1970, J. Phys. Chem. *74*, pp. 4161–4163.
(297) Fukui, K., Kato, S., and Fujimoto, H.: 1975, J. Am. Chem. Soc. *97*, pp. 1–7.
(298) Ishida, K., Morokuma, K., and Komornicki, A.: 1977, J. Chem. Phys. *66*, pp. 2153–2156.
(299) Basilevsky, M. V.: 1977, Chem. Phys. *24*, pp. 81–89.
(300) Tachibana, A., and Fukui, K.: 1978, Theor. Chim. Acta *49*, pp. 321–347.
(301) Fukui, K.: 1981, Accounts Chem. Res. *14*, pp. 363–368.
(302) Murrell, J. N., and Laidler, K. J.: 1968, Trans. Faraday Soc. *64*, pp. 371–377.
(303) Schaeffer III, H. F.: 1975, Chem. Brit. *11*, pp. 227–230.
(304) Pearson, P. K., Schaefer III, H. F., and Wahlgren, U.: 1975, J. Chem. Phys. *62*, pp. 350–354.
(305) Kato, S., and Fukui, K.: 1976, J. Am. Chem. Soc. *98*, pp. 6395–6397.
(306) Joshi, B. D., and Morokuma, K.: 1977, J. Chem. Phys. *67*, pp. 4880–4883.
(307) Morokuma, K., Kato, S., and Hirao, K.: 1980, J. Chem. Phys. *72*, pp. 6800–6802.
(308) Kato, S., and Morokuma, K.: 1980, J. Chem. Phys. *73*, pp. 3900–3914.
(309) Morokuma, K., Kato, S., Kitaura, K., Obara, S., Ohta, K., and Hanamura, M.: 1983, in New Horizons of Quantum Chemistry, Ed. P.-O. Löwdin, and B. Pullman, D. Reidel Publ. Comp., Dordrecht, pp. 221–241.
(310) Mezey, P. G.: 1980, Theor. Chim. Acta *54*, pp. 95–111.
(311) Tachibana, A., and Fukui, K.: 1979, Theor. Chim. Acta *51*, pp. 189–206.
(312) Tachibana, A., and Fukui, K.: 1979, Theor. Chim. Acta *51*, pp. 275–296.
(313) Yamashita, K., Yamabe, T., and Fukui, K.: 1982, Theor. Chim. Acta *60*, pp. 523–533.
(314) Fukui, K., Tachibana, A., and Yamashita, K.: 1981, Int. J. Quantum Chem.,Quantum Chem. Symp. *15*, pp. 621–632.
(315) Fukui, K.: 1981, Int. J. Quantum Chem., Quantum Chem. Symp. *15*, pp. 633–642.
(316) Pancíř, J., 1975, Collect. Czech. Chem. Commun. *40*, pp. 1112–1118.
(317) Basilevsky, M. V., and Shamov, A. G.: 1981, Chem. Phys. *60*, pp. 347–358.
(318) Empedocles, P.: 1969, Int. J. Quantum Chem., Quantum Chem. Symp. *3*, pp. 47–62.
(319) Dewar, M. J. S., and Kirschner, S.: 1971, J. Am. Chem. Soc. *93*, pp. 4290–4291.
(320) Dewar, M. J. S., and Kirschner, S.: 1971, J. Am. Chem. Soc. *93*, pp. 4291–4292.
(321) Dewar, M. J. S., and Kirschner, S.: 1971, J. Am. Chem. Soc. *93*, pp. 4292–4294.
(322) McCullough Jr., E. A., and Silver, D. M.: 1975, J. Chem. Phys. *62*, pp. 4050–4052.

(323) BASKIN, C. P., BENDER, C. F., BAUSCHLICHER Jr., C. W., and SCHAEFER III, H. F.: 1974, J. Am. Chem. Soc. *96*, pp. 2709—2713.

(324) BAUSCHLICHER JR., C. W., SCHAEFER III, H. F., and BENDER, C. F.: 1976, J. Am. Chem. Soc. *98*, pp. 1653—1658.

(325) EHRENSON, S.: 1974, J. Am. Chem. Soc. *96*, pp. 3778—3784.

(326) EHRENSON, S.: 1974, J. Am. Chem. Soc. *96*, pp. 3784—3793.

(327) RICE, F. O., and TELLER, E.: 1938, J. Chem. Phys. *6*, pp. 489—498.

(328) RICE, F. O., and TELLER, E.: 1939, J. Chem. Phys. *7*, p. 199.

(329) FUKUI, K.: 1974, in The World of Quantum Chemistry, Ed. R. Daudel, and B. Pullman, D. Reidel Publ. Comp., Dordrecht, pp. 113—141.

(330) METIU, H., ROSS, J., SILBEY, R., and GEORGE, T. F.: 1974, J. Chem. Phys. 61, pp. 3200—3209.

(331) SWANSON, B. I.: 1976, J. Am. Chem. Soc. *98*, pp. 3067—3071.

(332) PECHUKAS, P.: 1976, J. Chem. Phys. *64*, pp. 1516—1521.

(333) NALEWAJSKI, R. F.: 1978, Int. J. Quantum Chem., Quantum Chem. Symp. *12*, pp. 87—102.

(334) NALEWAJSKI, R. F., and CARLTON, T. S.: 1978, Acta Phys. Polon. *A 53*, pp. 321—338.

(335) SCHARFENBERG, P.: 1981, Chem. Phys. Lett. *79*, pp. 115—117.

(336) TACHIBANA, A.: 1981, Theor. Chim. Acta *58*, pp. 301—308.

(337) NAUTS, A., and CHAPUISAT, X.: 1982, Chem. Phys. Lett. *85*, pp. 212—219.

(338) BASILEVSKY, M. V.: 1982, Chem. Phys. *67*, pp. 337—346.

(339) HASE, W. L., MROWKA, G., BRUDZYNSKI, R. J., and SLOANE, C. S.: 1978, J. Chem. Phys. *69*, pp. 3548—3562.

(340) EYRING, H., and LIN, S. H.: 1978, in Physical Chemistry. An Advanced Treatise, Vol. VIa, Ed. W. Jost, Academic Press, New York, pp. 121—186.

(341) JOHNSTON, H. S., and PARR, C.: 1963, J. Am. Chem. Soc. *85*, pp. 2544—2551.

(342) JOHNSTON, H. S.: 1966, Gas Phase Reaction Rate Theory, The Ronald Press Comp., New York.

(343) BADER, R. F. W., and GANGI, R. A.: 1975, in Theoretical Chemistry. Specialist Periodical Reports, Vol. 2, Ed. R. N. Dixon, and C. Thomson, The Chemical Society, London, pp. 1—65.

(344) ELLISON, F. O.: 1963, J. Am. Chem. Soc. *85*, pp. 3540—3544.

(345) MILLS, I. M.: 1977, Faraday Discuss. Chem. Soc. *62*, pp. 7—19.

(346) SIMONS, G.: 1974, J. Chem. Phys. *61*, pp. 369—374.

(347) DIAB, S., and SIMONS, G.: 1977, Mol. Phys. *33*, pp. 843—847.

(348) SORBIE, K. S., and MURRELL, J. N.: 1975, Mol. Phys. *29*, pp. 1387—1407.

(349) MURRELL, J. N., SORBIE, K. S., and VARANDAS, A. J. C.: 1976, Mol. Phys. *32*, pp. 1359 to 1372.

(350) FARANTOS, S., LEISEGANG, E. C., MURRELL, J. N., SORBIE, K., TEXEIRA-DIAS, J. J. C., and VARANDAS, A. J. C.: 1977, Mol. Phys. *34*, pp. 947—962.

(351) VARANDAS, A. J. C., and MURRELL, J. N.: 1977, Faraday Discuss. Chem. Soc. *62*, pp. 92—109.

(352) MURRELL, J. N., and FARANTOS, S.: 1977, Mol. Phys. *34*, pp. 1185—1188.

(353) CARTER, S., MILLS, I. M., and MURRELL, J. N.: 1980, Mol. Phys. *39*, pp. 455—469.

(354) GERBER, R. B., ROTH, R. M., and RATNER, M. A.: 1981, Mol. Phys. *44*, pp. 1335—1353.

(355) REES, A. L. G.: 1947, Proc. Phys. Soc. London *59*, p. 998.

(356) MCGUIRE, R. F., VANDERKOVI, G., MOMANY, F. A., INGWALL, R. T., CRIPPEN, G. M., LOTAN, N., TUTTLE, R. W., KASHUBA, K. L., and SCHERAGA, H. A.: 1971, Macromolecules *4*, pp. 112—124.

(357) MOMANY, F. A., CARRUTHERS, L. M., MCGUIRE, R. F., and SCHERAGA, H. A.: 1974, J. Phys. Chem. *78*, pp. 1595—1620.

(358) MOMANY, F. A., CARRUTHERS, L. M., and SCHERAGA, H. A.: 1974, J. Phys. Chem. *78*, pp. 1621—1630.
(359) DUNFIELD, L. G., BURGESS, A. W., and SCHERAGA, H. A.: 1978, J. Phys. Chem. *82*, pp. 2609—2616.
(360) SCOTT, R. L.: 1971, in Physical Chemistry. An Advanced Treatise, Vol. VIIIA, Ed. D. Henderson, Academic Press, New York, pp. 1—83.
(361) DYMOND, J. H., and SMITH, E. B.: 1969, The Virial Coefficients of Gases. A Critical Compilation, Clarendon Press, Oxford.
(362) HUTSON, J. M., and HOWARD, B. J.: 1981, Mol. Phys. *43*, pp. 493—516.
(363) BUCK, U., and SCHLEUSENER, J.: 1981, J. Chem. Phys. *75*, pp. 2470—2472.
(364) POPKIE, H., KISTENMACHER, H., and CLEMENTI, E.: 1973, J. Chem. Phys. *59*, pp. 1325—1336.
(365) KISTENMACHER, H., POPKIE, H., CLEMENTI, E., and WATTS, R. O.: 1974, J. Chem. Phys. *60*, pp. 4455—4465.
(366) KISTENMACHER, H., LIE, G. C., POPKIE, H., and CLEMENTI, E.: 1974, J. Chem. Phys. *61*, pp. 546—561.
(367) MATSUOKA, O., CLEMENTI, E., and YOSHIMINE, M.: 1976, J. Chem. Phys. *64*, pp. 1351—1361.
(368) CLEMENTI, E., BARSOTTI, R., FROMM, J., and WATTS, R. O.: 1976, Theor. Chim. Acta *43*, pp. 101—120.
(369) CLEMENTI, E.: 1976, Determination of Liquid Water Structure, Coordination Numbers for Ions, and Solvation for Biological Molecules, Springer-Verlag, Berlin.
(370) CLEMENTI, E., CAVALLONE, F., and SCORDAMAGLIA, R.: 1977, J. Am. Chem. Soc. *99*, pp. 5531—5545.
(371) SCORDAMAGLIA, R., CAVALLONE, F., and CLEMENTI, E.: 1977, J. Am. Chem. Soc. *99*, pp. 5545—5550.
(372) BOLIS, G., and CLEMENTI, E.: 1977, J. Am. Chem. Soc. *99*, pp. 5550—5557.
(373) CAROZZO, L., CORONGIU, G., PETRONGOLO, C., and CLEMENTI, E.: 1978, J. Chem. Phys. *68*, pp. 787—793.
(374) RAGAZZI, M., FERRO, D. R., and CLEMENTI, E.: 1979, J. Chem. Phys. *70*, pp. 1040—1050.
(375) CLEMENTI, E.: 1980, Computational Aspects for Large Chemical Systems, Springer-Verlag, Berlin.
(376) CLEMENTI, E.: 1983, in New Horizons of Quantum Chemistry, Ed. P.-O. Löwdin, and B. Pullman, D. Reidel Publ. Comp., Dordrecht.
(377) BERNAL, J. D., and FOWLER, R. H.: 1933, J. Chem. Phys. *1*, pp. 515—548.
(378) NOVARO, O., and NIEVES, F.: 1976, J. Chem. Phys. *65*, pp. 1109—1113.
(379) KOŁOS, W., NIEVES, F., and NOVARO, O.: 1976, Chem. Phys. Lett. *41*, pp. 431—434.
(380) EVANS, D. J., and WATTS, R. O.: 1974, Mol. Phys. *28*, pp. 1233—1240.
(381) LIE, G. C., and CLEMENTI, E.: 1976, J. Chem. Phys. *64*, pp. 5308—5309.
(382) SLANINA, Z.: 1980, J. Chem. Phys. *73*, pp. 2519—2521.
(383) SLANINA, Z.: 1980, Collect. Czech. Chem. Commun. *45*, pp. 3417—3435.
(384) SLANINA, Z.: 1981, Advan. Mol. Relax. Interact. Process. *19*, pp. 117—128.
(385) HOLMGREN, S. L., WALDMAN, M., and KLEMPERER, W.: 1978, J. Chem. Phys. *69*, pp. 1661—1669.
(386) LE ROY, R. J., and VAN KRANENDONK, J.: 1974, J. Chem. Phys. *61*, pp. 4750—4769.
(387) VAN DER AVOIRD, A., WORMER, P. E. S., MULDER, F., and BERNS, R. M.: 1980, Top. Curr. Chem. *93*, pp. 1—51.
(388) DOWNING, J. W., MICHL, J., ČÍŽEK, J., and PALDUS, J.: 1979, Chem. Phys. Lett. *67*, pp. 377—380.
(389) DOWNING, J. W., and MICHL, J.: 1981, in Potential Energy Surfaces and Dynamics Calculations for Chemical Reactions and Molecular Energy Transfer, Ed. D. G. Truhlar, Plenum Press, New York, pp. 199—212.

(390) BAIRD, N. C.: 1977, Pure Appl. Chem. *49*, pp. 223—236.
(391) DEVAQUET, A.: 1975, Fortschr. Chem. Forsch. *54*, pp. 1—73.
(392) POPPINGER, D., RADOM, L., and POPLE, J. A.: 1977, J. Am. Chem. Soc. *99*, pp. 7806—7816.
(393) FOWLER, P. W.: 1982, Chem. Phys. Lett. *85*, pp. 313—316.
(394) HENRY, L., and AMAT, G.: 1960, J. Mol. Spectrosc. *5*, pp. 319—325.
(395) HENRY, L., and AMAT, G.: 1965, J. Mol. Spectrosc. *15*, pp. 168—179.
(396) PAPOUŠEK, D., and SARKA, K.: 1968, J. Mol. Spectrosc. *28*, pp. 125—143.
(397) HOUGEN, J. T.: 1970, in Physical Chemistry. An Advanced Treatise, Vol. IV, Ed. D. Henderson, Academic Press, New York, pp. 307—348.
(398) PAPOUŠEK, D., and ŠPIRKO, V.: 1976, Top. Curr. Chem. *68*, pp. 59—102.
(399) BUNKER, P. R.: 1979, Molecular Symmetry and Spectroscopy, Academic Press, New York.
(400) CARREIRA, L. A., LORD, R. C., and MALLOY JR., T. B.: 1979, Top. Curr. Chem. *82*, pp. 1—95.
(401) SØRENSEN, G. O.: 1979, Top. Curr. Chem. *82*, pp. 97—175.
(402) FORTUNE, P. J., ROSENBERG, B. J., and WAHL, A. C.: 1976, J. Chem. Phys. *65*, pp. 2201 to 2205.
(403) PERIĆ, M., RUNAU, R., RÖMELT, J., and PEYERIMHOFF, S. D.: 1979, J. Mol. Spectrosc. *78*, pp. 309—332.
(404) CARNEY, G. D.: 1980, Mol. Phys. *39*, pp. 923—933.
(405) RAO, C. V. S. R.: 1981, J. Mol. Spectrosc. *89*, pp. 197—213.
(406) ŠPIRKO, V., and BERAN, S.: J. Mol. Spectrosc., in press.
(407) ŠPIRKO, V., and BERAN, S.: J. Mol. Spectrosc., in press.
(408) LEECH, J. W.: 1958, Classical Mechanics, Methuen, London.
(409) WILSON JR., E. B.: 1939, J. Chem. Phys. *7*, pp. 1047—1052.
(410) WILSON JR., E. B.: 1941, J. Chem. Phys. *9*, pp. 76—84.
(411) ELYASHEVICH, M. A.: 1940, Dokl. Akad. Nauk SSSR *28*, pp. 605—609.
(412) WILSON JR., E. B., DECIUS, J. C., and CROSS, P. C.: 1955, Molecular Vibrations, McGraw-Hill, New York.
(413) CRAWFORD JR., B., and OVEREND, J.: 1964, J. Mol. Spectrosc. *12*, pp. 307—318.
(414) VAN ZANDT, L. L., and LU, K.-C.: 1977, J. Chem. Phys. *67*, pp. 2636—2641.
(415) WILLIAMS, I. H.: 1982, Chem. Phys. Lett. *88*, pp. 462—466.
(416) SHIMANOUCHI, T.: 1970, in Physical Chemistry. An Advanced Treatise, Vol. IV, Ed. D. Henderson, Academic Press, New York, pp. 233—306.
(417) BARAN, J., and KOŁOS, W.: 1962, J. Mol. Spectrosc. *8*, pp. 121—125.
(418) MCLEAN, A. D.: 1964, J. Chem. Phys. *40*, pp. 243—244.
(419) PALDUS, J., and HRABĚ, P.: 1968, Theor. Chim. Acta *11*, pp. 401—410.
(420) DEWAR, M. J. S., and METIU, H.: 1972, Proc. Roy. Soc. *A 330*, pp. 173—184.
(421) NELANDER, B., and RIBBEGÅRD, G.: 1974, J. Mol. Struct. *20*, pp. 325—333.
(422) SMIT, W. M. A., and ROOS, F. A.: 1978, Mol. Phys. *36*, pp. 1017—1023.
(423) BISHOP, D. M., and RANDIĆ, M.: 1966, J. Chem. Phys. *44*, pp. 2480—2487.
(424) BISHOP, D. M., and MACIAS, A.: 1969, J. Chem. Phys. *51*, pp. 4997—5001.
(425) BISHOP, D. M., and MACIAS, A.: 1970, J. Chem. Phys. *53*, pp. 3515—3521.
(426) BISHOP, D. M., and MACIAS, A.: 1971, J. Chem. Phys. *55*, pp. 647—651.
(427) BLOEMER, W. L., and BRUNER, B. L.: 1972, J. Mol. Spectrosc. *43*, pp. 452—466.
(428) YDE, P. B., THOMSEN, K., and SWANSTRØM, P.: 1972, Mol. Phys. *23*, pp. 691—697.
(429) THOMSEN, K., and SWANSTRØM, P.: 1973, Mol. Phys. *26*, pp. 735—750.
(430) THOMSEN, K., and SWANSTRØM, P.: 1973, Mol. Phys. *26*, pp. 751—763.
(431) POPLE, J. A., SCHLEGEL, H. B., KRISHNAN, R., DEFREES, D. J., BINKLEY, J. S., FRISCH, M. J., WHITESIDE, R. A., HOUT, R. F., and HEHRE, W. J.: 1981, Int. J. Quantum Chem., Quantum Chem. Symp. *15*, pp. 269—278.

(432) KANAKAVEL, M., CHANDRASEKHAR, J., SUBRAMANIAN, S., and SINGH, S.: 1976, Theor. Chim. Acta *43*, pp. 185—196.
(433) SCHLEGEL, H. B., WOLFE, S., and BERNARDI, F.: 1977, J. Chem. Phys. *67*, pp. 4181—4193.
(434) AL-JIBURI, A.-L. K., AL-NIAMI, K. H., and SHANSHAL, M.: 1979, Theor. Chim. Acta *53*, pp. 327—335.
(435) SANA, M.: 1982, Theor. Chim. Acta *60*, pp. 543—559.
(436) KERN, C. W., and KARPLUS, M.: 1964, J. Chem. Phys. *40*, pp. 1374—1389.
(437) SWANSTRØM, P., THOMSEN, K., and YDE, P. B.: 1971, Mol. Phys. *20*, pp. 1135—1146.
(438) DEB, B. M.: 1973, Rev. Mod. Phys. *45*, pp. 22—43.
(439) PULAY, P., and MEYER, W.: 1971, J. Mol. Spectrosc. *40*, pp. 59—70.
(440) MEYER, W., and PULAY, P.: 1972, J. Chem. Phys. *56*, pp. 2109—2116.
(441) MEYER, W., and PULAY, P.: 1974, Theor. Chim. Acta *32*, pp. 253—264.
(442) TÖRÖK, F., HEGEDÜS, Á., KÓSA, K., and PULAY, P.: 1976, J. Mol. Struct. *32*, pp. 93—99.
(443) PANCHENKO, Yu. N., PULAY, P., and TÖRÖK, F.: 1976, J. Mol. Struct. *34*, pp. 283—289.
(444) FOGARASI, G., and PULAY, P.: 1977, J. Mol. Struct. *39*, pp. 275—280.
(445) MEYER, W.: 1973, J. Chem. Phys. *58*, pp. 1017—1035.
(446) SLANINA, Z., BERÁK, P., and ZAHRADNÍK, R.: 1977, Collect. Czech. Chem. Commun. *42*, pp. 1—15.
(447) DEWAR, M. J. S., and FORD, G. P.: 1977, J. Am. Chem. Soc. *99*, pp. 1685—1691.
(448) BLOM, C. E., and ALTONA, C.: 1976, Mol. Phys. *31*, pp. 1377—1391.
(449) BLOM, C. E., OTTO, L. P., and ALTONA, C.: 1976, Mol. Phys. *32*, pp. 1137—1149.
(450) BLOM, C. E., ALTONA, C., and OSKAM, A.: 1977, Mol. Phys. *34*, pp. 557—571.
(451) LAKDAR, T. B., SUARD, M., TAILLANDIER, E., and BERTHIER, G.: 1978, Mol. Phys. *36*, pp. 509—518.
(452) YAMAGUCHI, Y., and SCHAEFER III, H. F.: 1980, J. Chem. Phys. *73*, pp. 2310—2318.
(453) RAGHAVACHARI, K.: 1982, J. Chem. Phys. *76*, pp. 3668—3672.
(454) GODDARD, J. D., and CLOUTHIER, D. J.: 1982, J. Chem. Phys. *76*, pp. 5039—5042.
(455) TACHIBANA, A.: 1981, Theor. Chim. Acta *58*, pp. 301—308.
(456) SAWODNY, W., and PULAY, P.: 1974, J. Mol. Spectrosc. *51*, pp. 135—141.
(457) MOLT, K., SAWODNY, W., PULAY, P., and FOGARASI, G.: 1976, Mol. Phys. *32*, pp. 169—176.
(458) HA, T.-K., MEYER, R., and GÜNTHARD, Hs. H.: 1978, Chem. Phys. Lett. *59*, pp. 17—20.
(459) MIYAZAKI, T., SHIGETANI, T., and YAMAMOTO, H.: 1972, Bull. Chem. Soc. Jap. *45*, pp. 678—682.
(460) BRUNS, R. E., and PERSON, W. B.: 1972, J. Chem. Phys. *57*, pp. 324—331.
(461) BROWNLEE, R. T. C., MUNDAY, J., TOPSOM, R. D., and KATRITZKY, A. R.: 1973, J. Chem. Soc., Faraday Trans. *II 69*, pp. 349—354.
(462) BRUNS, R. E., and PERSON, W. B.: 1973, J. Chem. Phys. *58*, pp. 2585—2592.
(463) VUČELIĆ, M., ÖHRN, Y., and SABIN, J. R.: 1973, J. Chem. Phys. *59*, pp. 3003—3007.
(464) PERKAMPUS, H.-H., KLAMPFER, M., and KNOP, J. V.: 1974, Theor. Chim. Acta *34*, pp. 293 to 299.
(465) BLECKMANN, P.: 1974, Z. Naturforsch. *29a*, pp. 1485—1488.
(466) KAMENSKII, JU. B., KOVALEV, I. F., and MOLODENKOVA, I. D.: 1974, Opt. Spektrosk. *37*, pp. 668—671.
(467) POUCHAN, C., DARGELOS, A., CHAILLET, M., FORD, G., TOPSOM, R. D., and KATRITZKY, A. R.: 1974, J. Chim. Phys. *71*, pp. 934—942.
(468) MIYAZAKI, T., IKEDA, M., and SHIBATA, M.: 1975, Bull. Chem. Soc. Jap. *48*, pp. 1138—1145.
(469) BLOM, C. E., and ALTONA, C.: 1977, Mol. Phys. *34*, pp. 177—192.
(470) KOMORNICKI, A., and MCIVER Jr., J. W.: 1979, J. Chem. Phys. *70*, pp. 2014—2016.
(471) KOMORNICKI, A., and JAFFE, R. L.: 1979, J. Chem. Phys. *71*, pp. 2150—2155.
(472) PULAY, P., FOGARASI, G., and BOGGS, J. E.: 1981, J. Chem. Phys. *74*, pp. 3999—4014.

(473) Nafie, L. A., and Freedman, T. B.: 1981, J. Chem. Phys. *75*, pp. 4847—4851.
(474) Schaad, L. J., Hess Jr., B. A., and Ewig, C. S.: 1982, J. Org. Chem. *47*, pp. 2904—2906.
(475) Spiekermann, M., Bougeard, D., and Schrader, B.: 1982, J. Comput. Chem. *3*, pp. 354—362.
(476) Ermler, W. C., and Kern, C. W.: 1971, J. Chem. Phys. *55*, pp. 4851—4860.
(477) Almlöf, J.: 1972, Chem. Phys. Lett. *17*, pp. 49—52.
(478) Janoschek, R., Weidemann, E. G., and Zundel, G.: 1973, J. Chem. Soc., Faraday Trans. *II 69*, pp. 505—520.
(479) Pecul, K., and Janoschek, R.: 1974, Theor. Chim. Acta *36*, pp. 25—36.
(480) Botschwina, P.: 1974, Chem. Phys. Lett. *29*, pp. 98—101.
(481) Støgård, A., Strich, A., Almlöf, J., and Roos, B.: 1975, Chem. Phys. *8*, pp. 405—411.
(482) Janoschek, R.: 1976, in The Hydrogen Bond, Vol. I, Ed. P. Schuster, G. Zundel, and C. Sandorfy, North-Holland, Amsterdam, pp. 165—216.
(483) Sandorfy, C.: 1976, in The Hydrogen Bond, Vol. II, Ed. P. Schuster, G. Zundel, and C. Sandorfy, North-Holland, Amsterdam, pp. 613—654.
(484) Blom, C. E., Slingerland, P. J., and Altona, C.: 1976, Mol. Phys. *31*, pp. 1359—1376.
(485) Botschwina, P.: 1976, Mol. Phys. *32*, pp. 729—733.
(486) Jönsson, B., and Nelander, B.: 1977, Chem. Phys. *25*, pp. 263—269.
(487) Botschwina, P.: 1979, Chem. Phys. *40*, pp. 33—44.
(488) Sellers, H. L., Pinegar, J. F., and Schäfer, L.: 1979, Chem. Phys. Lett. *61*, pp. 499—502.
(489) Bouteiller, Y., Allavena, M., and Leclercq, J. M.: 1980, Chem. Phys. Lett. *69*, pp. 521—524.
(490) Meyer, W., Botschwina, P., Rosmus, P., and Werner, H.-J.: 1980, in Computational Methods in Chemistry, Ed. J. Bargon, Plenum, New York, pp. 157—174.
(491) Botschwina, P.: 1982, Chem. Phys. *68*, pp. 41—63.
(492) Carney, G. D., and Porter, R. N.: 1976, J. Chem. Phys. *65*, pp. 3547—3565.
(493) Zahradník, R.: 1982, Chem. Listy *76*, pp. 1009—1027 (esp. p. 1026).
(494) Wolfram, T., and Asgharian, A.: 1981, J. Chem. Phys. *74*, pp. 1661—1675.
(495) Crowder, C. D.: 1982, J. Chem. Phys. *76*, pp. 5655—5656.

4 Quantum Mechanics and Isomerism

In recent years, interest has once again turned to the use and limitations of the Born-Oppenheimer approximation in molecular physics. This is undoubtedly connected with the great progress made in experimental techniques, especially in spectroscopy, permitting high resolution experiments on atoms and small molecules in the gaseous phase. It appears that thorough theoretical interpretation of these experiments requires the abandonment of the Born-Oppenheimer approximation, as only the complete quantum-mechanical theory employing molecular wave functions is appropriate here[1,2]. This implies the need to overcome the traditional artificial distinction between the structural and dynamic aspects of molecular theory, which could lead to an avoidance of the present paradoxical state where quantum chemistry is, in fact, practically an autonomous theory in relation to quantum mechanics. In this connection, special attention has been paid to the re-examination of the classical concept of molecular structure and its relationship to rigorous quantum-mechanical theory. This re-evaluation of the conventional approximation framework of quantum chemistry is directly connected with and improves our understanding of isomerism.

4.1 Criticism and Generalization of the Concept of Molecular Structure

The classical concept of molecular structure is used so widely and commonly that the elementary fact that it is not consistent with the requirements of quantum mechanics is often neglected. Within the concept of the molecular structure, the position and momentum variables and energy values are simultaneously specified, which is contrary to the uncertainty principle. If an energy state is assigned to the system, then it can be described by a wave function (or density matrix) and the amplitude (generally time-dependent) corresponding to a given configuration of particles can be determined. On the other hand, postulation of a certain molecular configuration implies an uncertainty as regards energy. This is naturally true also for the ground state – some zero-point motion must necessarily exist, leading both to the quantum ground state energy $E_{0,i}^{(q)}$ of the ith minimum on the hypersurface:

$$E_{0,i}^{(q)} = E_{0,i}^{(cl)} + \Delta_{0,i}^{(q)}, \qquad (4\text{-}1)$$

where $\Delta_{0,i}^{(q)}$ designates the corresponding quantum correction, and also to 'blurring' of the equilibrium nuclear configuration. The incompatibility of the concept of the molecular structure with quantum mechanics has recently been mentioned anew in several works[1-8]. As the concept of the molecular structure cannot be derived directly from quantum mechanics, Woolley[7] has proposed that it be conceived as a postulate (in order to maintain the conventional picture of molecular physics). Essén[9] pointed out a sense in which the concept of the molecular structure might be justifiable. If states with a structure form a complete set, expansion of the time-independent state with time-independent expansion coefficients would be possible. This would lead to time-independent structural amplitudes. Simultaneously, however, it has been pointed out[6] that criticism of the concept of the molecular structure is primarily justifiable for isolated systems. The shape of molecules surrounded by an environment might be understood as a property formed by the effect of this environment (represented, for example, by the boundary conditions) on the quantum states of the isolated molecules[10]. The exact molecular states of the isolated systems are delocalized but the effect of the surroundings leads to their localization[11]. An alternative concept to that of the molecular structure could be considered[6] to include the maxima in the molecular wave function in the position representation.

Criticism of the concept of the molecular structure also leads to questions concerning the replacement of the conventional notions of bond lengths and angles. In place of the classical bond length R_e, quantum mechanics yields only the average (expectation) value $\langle R \rangle$ of operator $\hat{R}$

$$\langle R \rangle = \int \Psi^* \hat{R} \Psi \, d\tau, \qquad (4\text{-}2)$$

Table 4-1. Comparison[a] of structural parameters R_e and $\langle R \rangle$ for the ground and first excited vibrational states of H_2 isotopomers

Parameter (10^{-10} m)	Ground state			First excited vibrational state		
	H_2	D_2	T_2	H_2	D_2	T_2
R_e	0.771 (0.741)[b]	0.764	0.762	0.801	0.783	0.775
$\langle R \rangle$	0.767	0.759	0.756	0.818	0.795	0.785

[a] From Ref.[12]; values from nonadiabatic calculations.
[b] The value from the adiabatic approximation is given in parenthesis.

where Ψ is the molecular wave function from the solution of the molecular Schrödinger equation. This value cannot, of course, be measured. Table 4-1 gives a comparison of parameters R_e and $\langle R \rangle$ for the ground and first excited vibrational states of the H_2, D_2 and T_2 molecules. Calculations[12] have been carried out in the non-adiabatic approximation; here, of course, quantity R_e has only the sense of a formal variational parameter. The difference between values R_e and $\langle R \rangle$ are most marked for the first excited vibrational state. This fact can be readily understood from Fig. 4-1 which

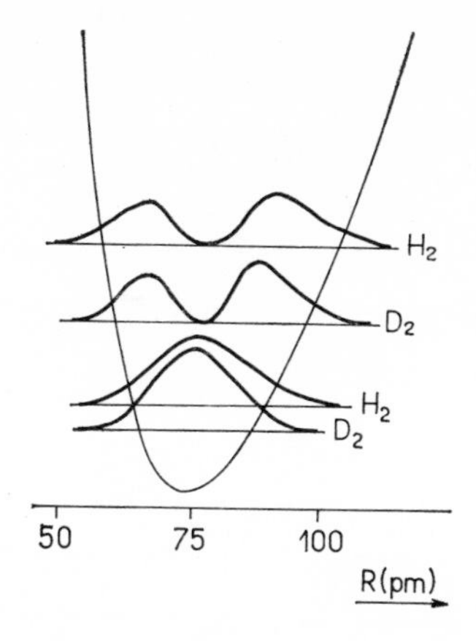

Figure 4-1. Probability distributions for intermolecular distance R for the ground and first excited vibrational states of the H_2 and D_2 molecules (based on the nonadiabatic calculations of Kołos and Wolniewicz[12] – the potential energy curve indicated has only formal meaning)

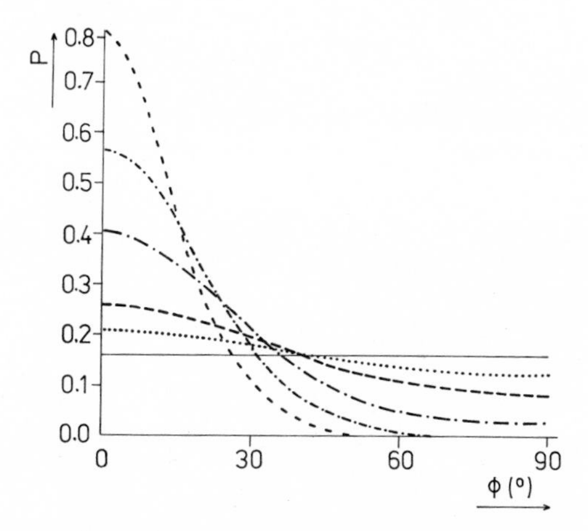

Figure 4-2. Probability P distributions for torsional angle Φ for an ensemble of quantum-mechanical rotators with reduced moments of inertia equal to 2.74×10^{-47} kg m^2 (averaged over 61 rotational energy levels at 305 K). Barrier height (in kJ mol^{-1}): 0 (———), 2 (·······), 4 (— — —), 10 (— ———), 20 (— · — · —), and 500 (-----) (from Ref.[14])

gives probability distributions for distance R for the ground and first excited vibrational states of the H_2 and D_2 molecules, obtained from the non-adiabatic approximation[12]. It should be noted that the results obtained (adiabatically) for the harmonic oscillator alone[13] qualitatively correspond well with the information in Fig. 4-1.

On transition to a less rigid system than the H_2 molecule, it can be expected that the differences between parameters of the R_e and $\langle R \rangle$ types will be greater. An example is study[14] of the expectation value of the torsional angle in molecules with hindered rotation (cf.[15,16]), carried out by ensemble averaging over the quantum states. Fig. 4-2 gives an illustration of the distribution of the relative probabilities as a function of the torsion angle. Even for a barrier of 500 kJ mol^{-1}, where only the ground state is populated under the given conditions, the expectation value of the torsion angle $\langle \Phi \rangle$ is very different from zero, and is equal[14] to 12°. Non-rigid clusters are another example of a system where large differences can be expected between the equilibrium and average values of the structural parametes[17,18]. For example, Fig. 4-3 depicts the probability distribution[19] for the stretching coordinate in the $(N_2)_2$ system.

Reconsideration[1-8] of the classical nature of the molecular structure concept – and the detailed reanalysis of the Born-Oppenheimer approximation – has lead to a principal result: in contrast to diatomic molecules, it has been demonstrated for polyatomic molecules[4,20,21] that there is no consistent manner of expansion of the true eigenfunctions of the molecular Hamiltonian in terms of adiabatic functions,

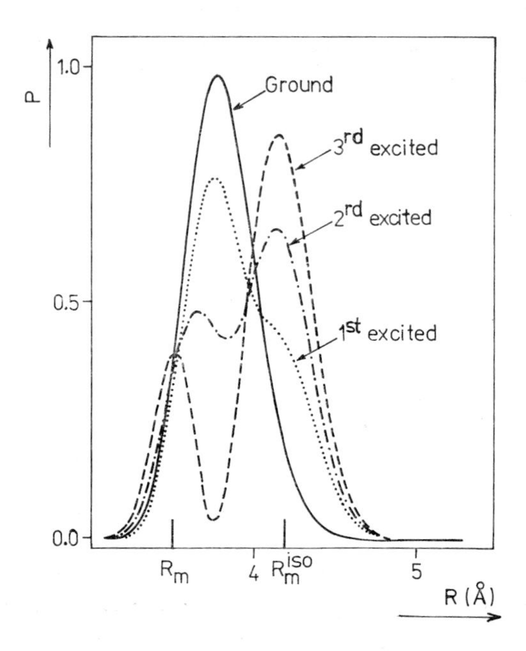

Figure 4-3. Probability P distributions for the stretching coordinate of $(N_2)_2$ for the four lowest $J = 0$ states of A_1^+ symmetry. For comparison, the equilibrium separation (R_m) and the minimum of the isotropic potential (R_m^{iso}) are given (from Ref.[19])

as these function do not lie[4] in the Hilbert space of the molecular eigenstates. Consequently, the existence of the expansion which is mostly postulated in the introduction of the adiabatic approximation is not valid[21]. This indicates that the Born-Oppenheimer approximation cannot be considered simply as an approximation introducing a certain numerical imprecision, but as a concept that is qualitatively inadequate from a physical point of view.

Another important product of the criticism of the classical concept of the molecular structure is the attempt to retain or broaden the concept of the molecular structure while abandoning the Born Oppenheimer approximation. Bader and Nguyen-Dang[22] formulated requirements foı an acceptable molecular structure theory. They pointed out that chemical observations are carried out at finite temperatures and require times such that the values obtained are averaged over the nuclear motion in the system. Thus the definition of the molecular structure must be based on time dependent quantum-mechanical formulation and must be independent of, and distinct from, the molecular geometry. Instead of strictly localized atomic positions, such a definition of the molecular structure should be associated with an open neighbourhood of the most probable molecular geometry, which would lead to a network of chemical bonds invariant with the nuclear motions of a stable system. In order to ensure the independence on particular models, including the Born-Oppenheimer approximation, the concept of the molecular structure must be formulated in terms of observable physical properties of the system. Bader et al.[23,24] utilized the observable distribution of the charge of a molecular system in real space as a basis for a new concept of the molecular structure; they developed the theory of the charge distribution previously[25-30] (it is mostly termed the theory of quantum topology). It is apparent that this approach is completely free of the concept of a potential energy hypersurface. Instead, the charge distribution of any one of N electrons is employed; this is essentially only a three-dimensional function of its position $\boldsymbol{r}_i$:

$$\varrho(\boldsymbol{r}_i, \boldsymbol{R}, t) = N \sum_{\text{spins}} \int \Psi^*(\boldsymbol{x}, \boldsymbol{R}, t)\, \Psi(\boldsymbol{x}, \boldsymbol{R}, t) \{\prod_{j \neq i} \mathrm{d}\boldsymbol{r}_j\}\,, \tag{4-3}$$

where $\boldsymbol{x}$ designates the collection of space and spin coordinates of N electrons, $\boldsymbol{R}$ characterizes the fixed nuclear configuration and Ψ is a (properly antisymmetrized) solution to the general time-dependent Schrödinger equation. Density $\varrho(\boldsymbol{r}_i, \boldsymbol{R}, t)$ represents the distribution of any one of the electrons, as determined by averaging over the motions of the remaining electrons. Apparently, transition from the wave function $\Psi(\boldsymbol{x}, \boldsymbol{R}, t)$ to charge density $\varrho(\boldsymbol{r}_i, \boldsymbol{R}, t)$ leads to a great reduction in the amount of information that can be obtained; nonetheless, it can be argued that the information lost is physically irrelevant. It is clear that the wave function must contain certain information that is redundant because of the indistinguishability of the electrons, symmetry of the interactions and two-body nature of the Coulombic interaction. The topological properties of the molecular charge density ϱ in three dimensions represent a basis for a new, rigorous approach to molecular structure. Primarily, it follows that only local maxima appear in function ϱ in the ground state of the molecular system at the positions of the nuclei. It is thus useful to replace scalar ϱ by its gradient $\nabla\varrho$ and to introduce the notions of critical points ($\nabla\varrho = \mathbf{0}$), gradient paths

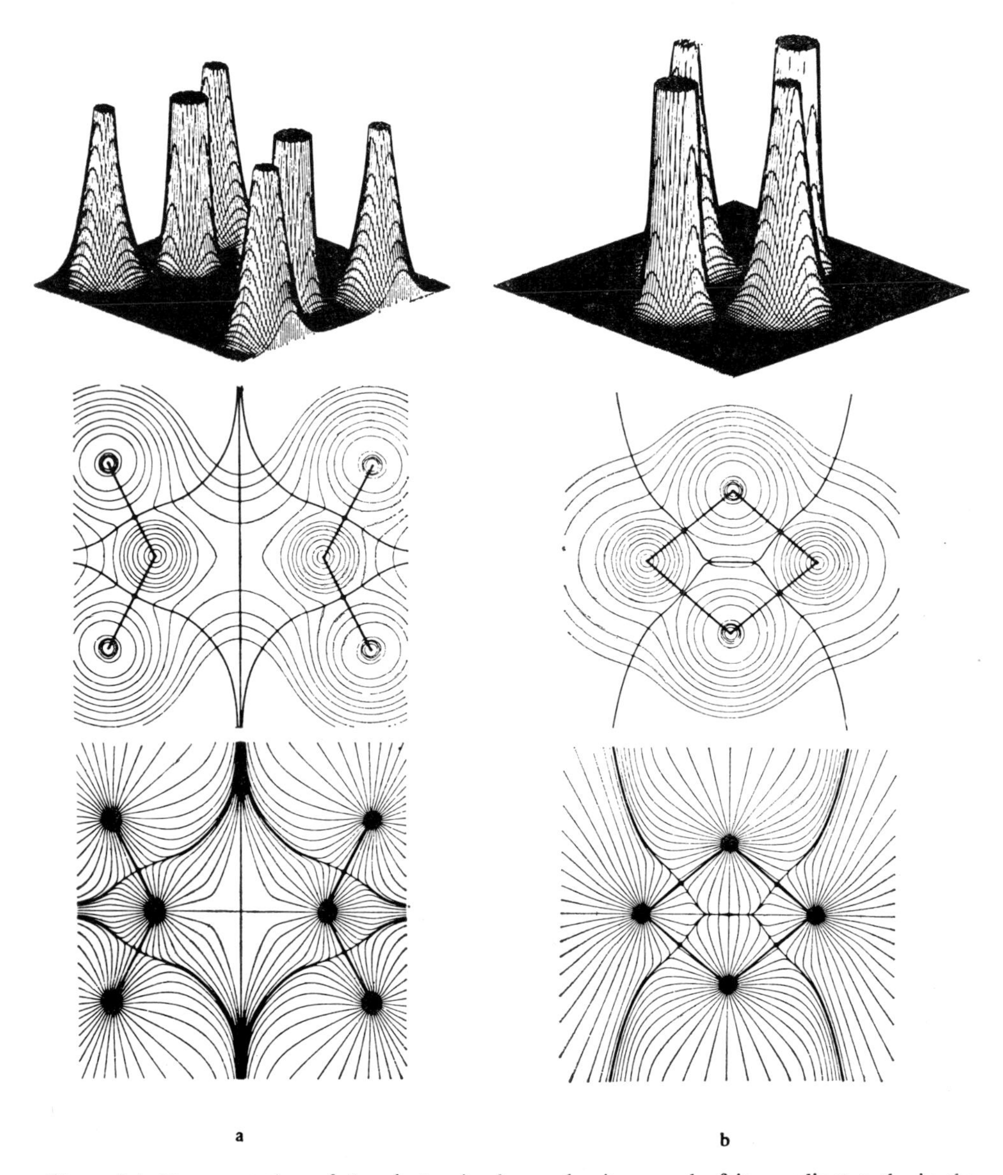

Figure 4-4. Representation of the electronic charge density ϱ and of its gradient paths in the Al_2F_4 plane (a), and in the Al_2F_2 bridging plane of the molecule Al_2F_6 (b). The upper diagram in each case is a three-dimensional plot of ϱ in the indicated plane, then contour plots and gradient maps follow (from Ref.[23])

orthogonal to the constant density ϱ contours, attractors of $\nabla\varrho$ (closed subsets of three-dimensional space exhibiting certain properties related to the course of the gradient paths, e.g. for every point of the attractor the gradient path through this point also belongs to the attractor), and basins (the maximal open invariant neighbourhood of the attractor such that every trajectory starting in the basin ends in the attractor). Critical points can be characterized[31,32] in terms of the rank (the number of non-zero eigenvalues of the Hessian matrix) and the signature (the excess number of positive over negative eigenvalues). There are on the surfaces four types of critical points with all the eigenvalues non-zero (i.e. with rank equal to three): (3,3), (3,1), (3,-1), and (3,-3). This classification of the critical points is not only formal – its topological significance is considered in ref.[31]. Thus, in this terminology, atomic nuclei are attractors of $\nabla\varrho$, where the associated basins separate molecular systems into atomic fragments. Each atom is then not considered as a point but rather as the union of an attractor and its associated basin (see Fig. 4-4). The common boundary between two atomic fragments contains a critical point where a pair of gradient paths connecting the two adjacent attractors originates. The concept of a molecular graph of the system and an equivalence relation for molecular graphs is unambiguously introduced on the basis of such pairs of gradient paths, the associated critical points and the path terminal points (the neighbouring atoms). The molecular structure is then defined as an equivalence class (see Chap. 5.1.1) of molecular graphs. The equivalence relation of molecular graphs leads to nontrivial partitioning of the nuclear configurational space into a finite number of regions, each connected with a particular structure.

In this approach, chemical reactions connected with the formation and/or disappearance of chemical bonds are then trajectories in configurational space that must intersect the boundary between structural regions. In set theory terminology, the interior and the boundary of a structural region can be readily defined; these two concepts are clearly topologically different as, while the interior is an open set, the boundary of the structural region is a closed set. The concept of structural stability[23,24] is further introduced and permits description of changes in the molecular structure by the catastrophe theory[31,32]. Summarizing, it is clear that, in the framework of the Bader *et al.* theory of quantum topology[22–30], the concept of atoms, bonds and molecular structure can be introduced rigorously and independently of the Born-Oppenheimer approximation. In addition, a predictive theory of structural stability is obtained.

The question of whether molecular wave functions can reflect at least some of the characteristics of the classical concept of molecular structure was discussed in detail by Claverie and Diner[33]. Here a key argument is the fact that there is no reason why all the nuclear configurations should be equally probable. This non-uniformity of the probability distribution permits the formation of maxima of the probability density. This indicates the existence of correlation, not only between electrons, but also between nuclei. Claverie and Diner[33] call such a structure, deduced from the complete molecular wave function, the quantum or potential structure. Recently, available nonadiabatic calculations were re-examined[34] from this point of view and it was documented, using the examples of H_2^+ and H_2, that certain features of classical molecular structure are indeed retained in the molecular eigenstates of these systems. This is not surprising, as even the atomic nuclei exhibit structuration features (cf.[35]). Claverie and Diner[33] also defend the concept of stepwise structuration – i.e. localization with respect to some degrees of freedom, but not all. For example, they explain the existence of separated optical isomers by pointing out that the localization of the wave functions for the individual stereoisomers is retained even with considerable isolation of the molecules (when delocalization already exists, for example, with respect to the rotational variables), so that there need be no contradiction with the requirement that the total molecular (fully symmetric) wave function exhibits the same probability for both optical isomers. In addition, the fact[10,33,36] is pointed out that a molecule is never completely isolated (during the measurement) because of the presence of interactions with the electromagnetic field; this represents a special example of wave function localization and formation of a structure through interactions of a molecule with its environment.

Later, García-Sucre and Bunge[37] formalized the concept of geometry and shape of a quantum system; their definitions can be used (in the nonrelativistic approximation) for both stationary and nonstationary states. For an introduction to, and study of, chemical structure they used pairwise relationships within a set of local maxima of the nuclear density functions. They demonstrated for such a concept of the chemical structure that this type of property is exhibited by every molecule in the ground state. It especially follows from their rather abstract analysis that the environment undoubtedly affects the structure that they introduced, but that its determination is internal rather than external. In this formalized concept, the geometry appears as a primary consequence of the composition of the system and the relationships between the components, and only secondarily as a consequence of the effect of the medium surrounding the system.

In order to obtain molecular wave functions, the generator coordinate method was recently proposed[38–40], originally developed for description of the collective motion of nucleons in the nucleus. Although the nuclear configuration appears during the process of construction of the wave function in this method, it is integrated out of the final result[38]. However, this is not contradictory to the previously mentioned possibility of the presence of quantum or potential structures in the complete molecular wave function. It can be obtained by a suitable analysis of the nuclear probability density. The above-mentioned integrating out process naturally concerns only the classical molecular structure.

A further way of overcoming problems with the concept of molecular structure in the quantum-mechanical approach involves a transition[33,41] to stochastic electrodynamics. In the stochastic approach to the description of the micro-world, various interpretation problems connected with the quantum theory itself disappear. However, in the light of existing[33,41] quantitative failures, stochastic electrodynamics cannot be understood, at least at present, as a genuine alternative to the quantum theory.

Recently, Wilson[42] described a versatile generalization of the classical concept of molecular structure including non-adiabatic wave functions (although still essentially within the Born-Oppenheimer framework). This extension is again based on probability interpretation of the exact molecular function $\Psi(r, \boldsymbol{R})$ introduced in terms of the electronic and nuclear coordinates $\boldsymbol{r}$ and $\boldsymbol{R}$. The probability $P(\boldsymbol{r}, \boldsymbol{R})\,\mathrm{d}\boldsymbol{r}\,\mathrm{d}\boldsymbol{R}$ of finding electrons in $\mathrm{d}\boldsymbol{r}$ at point $\boldsymbol{r}$ and of finding the nuclei in $\mathrm{d}\boldsymbol{R}$ at point $\boldsymbol{R}$ is given by the relationship

$$P(\boldsymbol{r}, \boldsymbol{R})\,\mathrm{d}\boldsymbol{r}\,\mathrm{d}\boldsymbol{R} = |\Psi(\boldsymbol{r}, \boldsymbol{R})|^2\,\mathrm{d}\boldsymbol{r}\,\mathrm{d}\boldsymbol{R}\,. \tag{4-4}$$

The probability $P(\boldsymbol{R})\,\mathrm{d}\boldsymbol{R}$ of finding the nuclei in $\mathrm{d}\boldsymbol{R}$ at $\boldsymbol{R}$ regardless of the positions of the electrons is clearly given as

$$P(\boldsymbol{R})\,\mathrm{d}\boldsymbol{R} = \mathrm{d}\boldsymbol{R}\int|\Psi(\boldsymbol{r}, \boldsymbol{R})|^2\,\mathrm{d}\boldsymbol{r}\,. \tag{4-5}$$

Furthermore, introducing the probability $P_{\boldsymbol{R}}(\boldsymbol{r})\,\mathrm{d}\boldsymbol{r}$ of finding the electrons in $\mathrm{d}\boldsymbol{r}$ at $\boldsymbol{r}$, when the nuclei are localized at $\boldsymbol{R}$, it clearly holds that

$$P(\boldsymbol{r}, \boldsymbol{R})\,\mathrm{d}\boldsymbol{r}\,\mathrm{d}\boldsymbol{R} = P(\boldsymbol{R})\,\mathrm{d}\boldsymbol{R}\,P_{\boldsymbol{R}}(\boldsymbol{r})\,\mathrm{d}\boldsymbol{r}\,, \tag{4-6}$$

leading directly to

$$P_{\boldsymbol{R}}(\boldsymbol{r}) = \frac{|\Psi(\boldsymbol{r}, \boldsymbol{R})|^2}{\int|\Psi(\boldsymbol{r}, \boldsymbol{R})|^2\,\mathrm{d}\boldsymbol{r}}\,. \tag{4-7}$$

Relationship (4-7) yields the probability distribution of the electrons for a given configuration of atomic nuclei in terms of the molecular wave function (where the denominator only ensures renormalization of the distribution). Knowledge of this distribution permits calculation of the expectation value $E_{\mathrm{elec}}(\boldsymbol{R})$ of the electronic energy (this explanation of quantity $E_{\mathrm{elec}}(\boldsymbol{R})$, however, assumes treatment of the nuclei as classical particles; otherwise, it is only an operator):

$$E_{\mathrm{elec}}(\boldsymbol{R}) = \frac{\int\Psi^*(\boldsymbol{r}, \boldsymbol{R})\,\hat{H}_{\mathrm{elec}}\,\Psi(\boldsymbol{r}, \boldsymbol{R})\,\mathrm{d}\boldsymbol{r}}{\int|\Psi(\boldsymbol{r}, \boldsymbol{R})|^2\,\mathrm{d}\boldsymbol{r}}\,, \tag{4-8}$$

where the electronic Hamiltonian $\hat{H}_{\mathrm{elec}}$ was introduced during discussion of the Born-Oppenheimer approximation (Chap. 3.1). It can now readily be demonstrated that, provided that the exact wave function of the system $\Psi(r, \boldsymbol{R})$ is replaced by the usual Born-Oppenheimer term, Eq. (4-8) is reduced to the conventional Born-Oppenheimer potential energy hypersurface. Thus function (4-8) represents an extension of the concept of an energy hypersurface beyond the framework of the Born-Oppenheimer approximation and thus permits extension of the conventional concept of the molecular structure to a non-adiabatic situation (however, with the above condition of treating the nuclei as classical particles) through the concept of the non-adiabatic energy hypersurface. Here, the existence of a sufficiently deep local minimum on the hypersurface (4-8) will be sufficient. For the sake of completeness, an interesting relationship with the uncertainty principle will be mentioned. Operator $\hat{H}_{elec}$ commutes with the nuclear coordinates $\boldsymbol{R}$ and thus it is possible in Born-Oppenheimer situations to measure the electronic energy and nuclear coordinates simultaneously.

Although Wilson's extension of the concept of molecular structure to non-adiabatic cases is not quantum-mechanically rigorous, it represents an alternative to the above-mentioned approach utilizing[6,33,37] the nuclear probability distribution, e.g. through maximization of probability $P(\boldsymbol{R})$ (Eq. (4-5)). It should be noted that both approaches have a common characteristic (in nonadiabatic situations): in contrast to the classical equilibrium molecular structure, they are dependent on the vibrational and rotational states. This is because generalization (4-8) of the definition of the energy hypersurface in non-adiabatic situations generally depends on all the quantum numbers. This may be a strong dependence, as demonstrated, for example, for the probability density $P(\boldsymbol{R})$ by comparing its shape for the two lowest vibrational states of H_2 and D_2 (Fig. 4-1). In addition to changes in the position of the probability maxima with a change, for example, of the vibrational state, the existence of a greater number (compared with the classical concept) of configurations with the same, or almost the same, probability must be considered (cf. the two local maxima for the first excited vibrational state in Fig. 4-1).

It is apparent that, rather than a single definition of the molecular structure, a greater number of alternatives is possible. From this point of view, the molecular structure is a concept involving a convention, an agreed upon definition. In certain situations the ambiguity can be removed by the ease with which the individual types of structural information can be extracted from the experimental data.

At present it is apparent that the extensive critical discussion[1-8] of the concept of the molecular structure from a quantum-mechanical point of view involves negative conclusions only on the classical concept of structure. Extension of this concept or an alternative definition not dependent on the particular manner of approximation of the wave function are clearly possible. It is true also of the exact molecular wave function that (certain) structural information can be obtained from it. At present, however, there is no satisfactory derivation of the classical molecular structure based on a strict quantum approach. This can be understood as a special case of the general, and far from clear[33], problem of the relationship between quantum theory and classical physics. Incidentally, clarification of these questions forms an essential component of development in the theory of measurements.

4.2 The Molecular Wave Function and Isomerism

Every member of a set of chemical species with the same number of atomic nuclei of given types, as well as the same number of electrons, can naturally be described[21,42–44] by the same total molecular Hamiltonian. This molecular Hamiltonian does not describe a particular chemical species. The formal definition of a molecule as a quantum system does not differentiate between isomers. However, experimental chemistry has provided a huge number of very well defined distinct pairs (or even series) of isomers, each of which exhibits clear uniqueness, including isolability as a defined species. This fact can be rationalized in terms of the molecular Hamiltonian either by transition to localized Hamiltonians[45] acting only in certain regions of the configurational space or by directly studying the localization of the molecular wave function[21,42–49]. In an attempt to interpret systems in which the isomers were clearly unique, Woolley[3] suggested a conjecture that the stationary states of various isomeric structures correspond to different subsets of states of the Hilbert space spanned by a given molecular Hamiltonian, where selection rules prevent transitions between different subsets of states. It was, however, clear that this approach cannot be exact; strong arguments for this fact are given in the subsequent more detailed discussions[4,6].

Recently, this question was discussed in greater detail by Aronowitz[21], who concluded that the molecular Hamiltonian spans a single coherent Hilbert space rather than a series of coherent subspaces. In addition, he pointed out that no expansion in terms of the product of electronic and nuclear wave functions for only a single chemical species can yield true eigenfunctions of the molecular Hamiltonian and can span the full Hilbert space. Thus, the assignment of various eigenfunctions of the molecular Hamiltonian to distinct chemical species must generally be considered to be only an approximation. Various situations can arise here, e.g. the lowest molecular wave function may exhibit negligible amplitude everywhere outside a certain region in the nuclear configurational space or, on the other hand, it may have non--negligible amplitude in several domains. In any case, a certain unity is retained between the isomeric compounds, even between those that are apparently entirely distinct. In short, the more closely isomers are examined, the less unique they appear as species. Nonetheless, the uniqueness of isomers in certain situations is a good, realistic approximation and simplification. This depends on the relative isolation of one stability domain from another. An assumption has been formulated[21] that direct transformation between isomers can occur only when two species are in states that are mutually degenerate. In this assumption, intreconversion of isomers is controlled[21] by symmetry rules and orthogonality requirements. In addition, however, it has been demonstrated that, provided that two stability regions have no degeneracies in common, there must exist pathways leading to a common region (otherwise the space spanned by the molecular Hamiltonian would be divided into coherent subspaces). Consequently, all the stability regions should, in principle, be accessible to each other. It is apparent that the interpretation of isomerism in terms of stationary molecular wave functions admits the existence of isomeric species that are distinguishable only with difficulty as well as those distinguishable clearly, where the distinct character of an isomer is only a relative concept. Moreover, similar to the quantum concepts of molecular structure, the understanding of isomers includes an element of convention connected with the particular means of analysis of the molecular wave function. Abandoning the conventional interpretation in terms of the Born-Oppenheimer approximation also leads to differentiation of the formerly uniform picture according to the vibrational and rotational quantum numbers, including the possibility of differing quality for various choices of these numbers.

So far, this problem has been treated in terms of stationary state functions, i.e. time-independent functions. However, as was pointed out in the classical work of Berry[50], observations are never carried out on true stationary states of the system (cf.[51,52]). Indeed, it is impossible to prepare a totally isolated system. In addition, chemical preparation yields systems in a certain geometry rather than in an exact energy state. This is a result of the fact that[42] the properties important in separation processes depend largely on the molecular geometry. A chemically prepared substance is thus initially localized in a certain geometry, so that it cannot be in a stationary state. Consequently, observations are carried out on time-dependent states; however, under certain situations, changes with time can be small or even negligible. This is particularly true of tunnelling times for transitions to other isomeric forms that may amount to years or even longer[53]. Under such conditions, precise energy measurements can be reliably carried out without disturbing the geometry. It is apparent that, as in the discussion connected with stationary states, interpretation in terms of non-stationary states admits the existence of isomers with very different degrees of distinctness, including both limiting situations. It should, however, be pointed out that this whole qualitative discussion in terms of time-dependent wave functions (tacitly) incorporates the classical assumption of molecular structure, or at least the structure itself is not sufficiently specified. In addition, an *a priori* theory of long-lived time-dependent states has not yet been developed (cf. also[81–83]).

Quantum-mechanical tunnelling, enabling passage through the potential barrier that is inpenetrable in terms of classical mechanics, has recently been the subject of a more quantitative study[54–70]. The double-well potential has mostly been used as a model, constructed either phenomenologically or on the basis of quantum-chemical calculations. Although the study of tunnelling depends on the adiabatic approximation, this is a typical quantum effect and thus discussion of its consequence for the understanding of isomerism is included in this chapter. Traditionally, the greatest attention has been paid to proton tunnelling (see for example[71]), especially in the hydrogen bond. It should be noted, however, that tunnelling is certainly not limited to the lightest particles. It has, for example, recently been demonstrated[72,73] that, at low temperatures, the rebonding of carbon monoxide to heme proteins after photodissociation may occur via tunnelling. For our purposes, comparative studies of tunnelling in a symmetric and asymmetric double minimum potential are especially important; in this respect de la Vega *et al.*[64–70] and Cribb *et al.*[63] demonstrated that there is a principal difference in the tunnelling possibilities in both situations.

Provided that the potential profile is symmetric, both potential wells are indistinguishable and the wave function must be distributed over both regions to reflect the symmetry of the Hamiltonian. The first system for which the tunnelling effect was widely utilized to explain anomalies, the inversion of ammonia and the splitting of the lines in its vibrational spectrum, is a useful illustrative example[74,75] (see Fig. 4-5). The energy levels in both wells have the same values and occur in pairs with symmetric and antisymmetric components. The splitting separating the two levels of the pair increases with increasing energy. The eigenfunctions of the Hamiltonian (see Fig. 4-6a) have two branches with the same amplitude in both regions. Tunnelling exists for all finite barriers and all finite interminimal distances, as a result of the indistinguishability of the two wells between which the system oscillates. The tunnelling frequency, splitting of the energy levels, and degree of delocalization of the vibrational wave function will, however, clearly depend on the height and shape of the barrier.

However, the situation is qualitatively different[63–70] for asymmetric potential barriers, where the symmetry may be disturbed even by small changes in the structure in either region. This is illustrated in Fig.4-6b for intramolecular proton exchange in asymmetrically substituted 1,3 malonaldehyde[67]. In contrast with the inversion

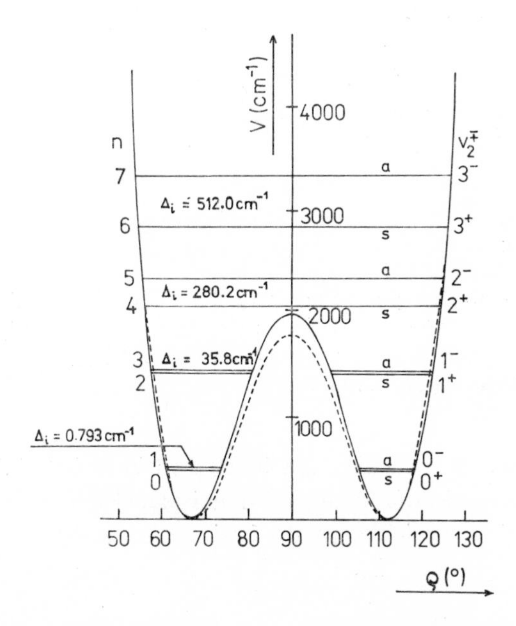

Figure 4-5. The double-minimum symmetric potential function for NH_3 inversion (inversion coordinate ϱ) with indication of the inversion splittings of the energy levels (a—s pairs). Full line — effective potential function, dashed line — "true" potential function; Δ_i — inversion splitting; n and $v_2^{\mp}$ — two sets of quantum vibrational numbers used in the literature (from Refs.[74,75])

of ammonia, energy levels in pairs do not appear here; the positions of the energy levels are very different in the two potential wells (except for accidental degeneracy). For low lying states, localization of the wave function in its respective well is typical. The possibility to extend through the classically forbidden region into the other well appears only for states close to the top of the barrier. It has been found[65] that even a small defect in the symmetry may practically eliminate tunnelling except for very low and narrow potential barriers.

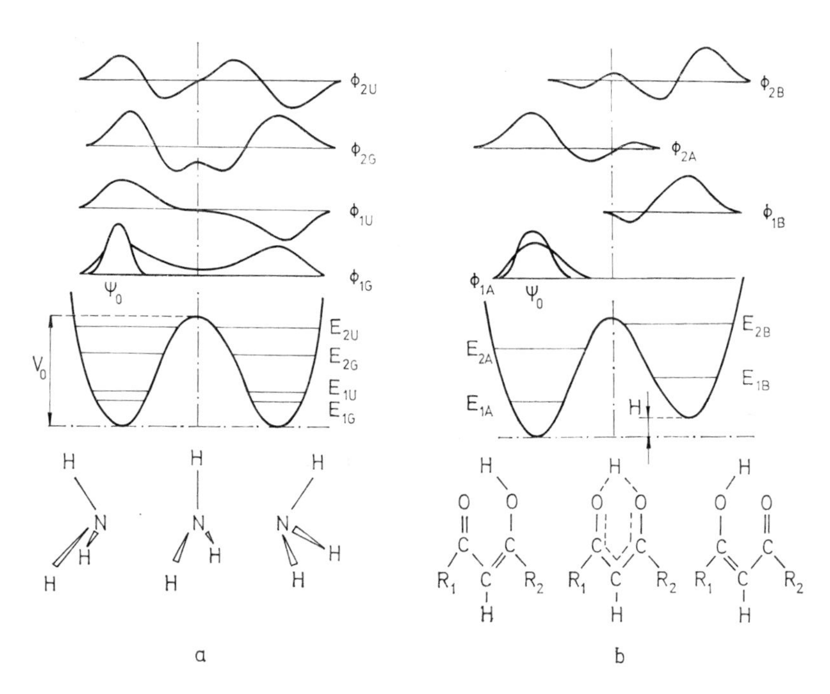

Figure 4-6. Tunnelling in bonded systems: (a) energy levels and eigenfunctions for a symmetric potential (the inversion of ammonia); (b) energy levels and eigenfunctions for an asymmetric potential (the intramolecular proton exchange in 1,3 asymmetrically substituted malonaldehyde). (From Ref.[70])

The time dependence of tunnelling can be studied in terms of the state function $\Psi(x, t)$ that can be obtained variationally[65] for an isolated system whose Hamiltonian $\hat{H}$ does not contain time explicitly, on the basis of knowledge of stationary eigenfunctions $\Psi_n(x)$ and eigenvalues E_n,

$$\Psi(x, t) = \sum_n c_n \Psi_n(x) \exp\left(-\frac{i}{\hbar} E_n t\right). \tag{4-9}$$

The time evolution of the position of the tunnelling particle cannot be given with certainty for a system whose energy is known exactly. The motion of this particle can be usefully studied in terms of the expectation value of the position

$$\langle x \rangle = \frac{\int \Psi^*(x, t)\, \hat{x} \Psi(x, t)\, \mathrm{d}x}{\int |\Psi(x, t)|^2\, \mathrm{d}x}, \tag{4-10}$$

and further described by the indeterminacy of the particle position by means of the expectation value of the square of the position. Both the expectation values are used[65,67,70] to study the time evolution of the state function representing the proton motion in the one-dimensional potential barriers describing the inversion of ammonia and intramolecular proton exchange (Fig. 4-7). Because of the delocalized character of the eigenfunctions for a symmetric barrier, the initial state $\Psi_0(x, 0)$ cannot coincide with just one stationary state, but rather several terms in expansion (4-9) are required. The initial location of the proton on one of the two symmetric potential wells always leads to oscillation of the expectation value of the proton position between the two wells, connected with proton passage through a region that is classically forbidden. The indeterminacy of the proton position increases as it approaches this classically forbidden region (Fig. 4-7a).

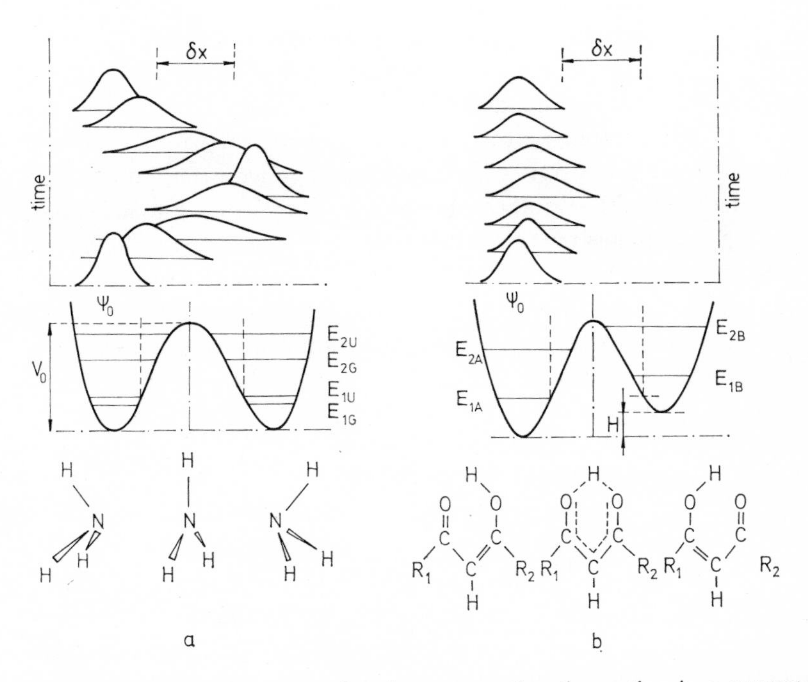

Figure 4-7. Time evolution of the state function representing the motion in a symmetric (a) and in an asymmetric (b) double minimum potential (cf. Figure 4-6; from Ref.[70])

Because of the localized character of the low-lying stationary states with an asymmetric barrier, the initial state $\Psi(x, 0)$ may coincide (or almost coincide) with one of the lowest eigenfunctions $\Psi_n(x)$. Then only a single term appears in expansion (4-9), as a result of which the time dependence of the position of the proton in the asymmetric case is qualitatively different from the symmetric barrier shape, in spite of the fact that its height is basically the same in both cases (see Fig. 4-7b). In the asymmetric case, the state function remains confined to the well in which it was initially located.

This dramatic reduction in tunnelling on transition from a symmetric to an asymmetric situation is especially important in understanding isomerism, as it leads to the completely different role of tunnelling in autoisomerizations (in which the reactant and product are identical) and during isomerizations between different structures. Except for very low barriers, tunnelling in an asymmetric potential profile cannot represent a substantial contribution to the realization of the isomerization process, that can thus occur only through the classical pathway over the barrier. At the same time, however, tunnelling may be significant for symmetric barriers of similar heights. These results were utilized[65], for example, in the rationalization of experimental information on proton exchange between the methyloxonium ion and methanol, methanol and the methoxide ion, or the hydronium ion and water:

$$H_2O.H_2OH^+ + OH_2.OH_2 \rightleftarrows H_2O.H_2O + {}^+HOH_2.OH_2 . \qquad (4\text{-}11)$$

Provided that the arrangement of species in the latter process is that of ice, the corresponding potential barrier is symmetric. If the possibility of internal rotation is, however, considered, corresponding to conditions in the liquid phase, the symmetry is lost and calculations for the same barrier height indicate loss of the possibility of tunnelling for this asymmetric case. This is in agreement with the finding that the experimentally determined rate of proton transfer in ice is ten times higher than in water. Similarly, in both the other two processes mentioned, internal rotation of the methyl group leads to loss of symmetry and so suppresses the possiblility of tunnelling. Thus, the rate of proton exchange in these cases is determined by the rate of rotation of the methyl group. The effect of the introduction of a methyl group into β-hydroxyacrolein[68] or naphthazarin[69] has been explained in a similar manner.

It is apparent that, in addition to clearly asymmetric potential profiles, there may be intermediate situations where the deviation from symmetry is small. It thus becomes interesting to determine how great the energy difference H between the energies of the two minima must be (see Fig. 4–6b or 4–7b) in order for the given profile to be clearly definable as asymmetric. Similarly, it is of interest to determine how close the energy level can lie to the top of the barrier before it becomes uncertain whether it lies above or below this barrier. It is apparent that these differences will be controlled[65] by the energy indeterminacy following from the Heisenberg principle. If these differences are to be significant, it is necessary that they be larger than the energy indeterminacy δE. Numerical results have demonstrated that, if difference ΔE between the top of the barrier and the ground state energy is less than δE, the proton always emerges in the second well; however, the degree of tunnelling decreases with increasing H value. Provided that ΔE was slightly larger than δE, in the particular case studied in ref.[65] the possibility of tunnelling disappeared at $H > 2.6$ kJ mol^{-1}. When $\Delta E \gg \delta E$, tunneling was suppressed at $H > 1.8$ kJ mol^{-1}. In contrast to the asymmetric situation, in the symmetric case tunnelling occurred even when $\Delta E \gg \delta E$. These results indicate that it is really necessary to consider tunnelling differently for symmetric and asymmetric barriers. In the latter case, with a barrier of less than about 20 kJ mol^{-1}, tunnelling must be considered as a direct result of the Heisenberg principle. In the symmetric case, tunnelling always exists (except when the barrier and/or the interminimal distance is infinite) and is a result of the indistinguishability of the two potential wells. Tunnelling appears in a nearly symmetric barrier higher than about 20 kJ mol^{-1} only for a very small H (where the system cannot be described as asymmetric with certainty). For the sake of completeness, it should be noted that the existence of accidental degeneracy between the levels of the two regions of an asymmetric potential profile can lead to an increase in tunelling[63]. This can especially occur[57] for large polyatomic systems with a large difference in the sizes of the two potential wells.

The optical isomers of a given asymmetric substance represent a system in which interconversion is realized within a symmetric potential energy profile. The degenerate states Ψ_L and Ψ_R of the laevorotatory and dextrorotatory species completely localized in one or the other potential well are, however, not eigenfunctions of the Hamiltonian. The symmetry requirements are satisfied by true eigenfunctions of the Hamiltonian represented by combinations

$$\Psi_{1,2} = (\Psi_L \pm \Psi_R)/\sqrt{2}\,. \qquad (4\text{-}12)$$

Hund[51] pointed out the paradox of optical isomers, i.e. their apparent stability in spite of the fact that they are not the ground state of the system, and explained this by the very long tunnelling times for complicated (biological) systems. Recently, however, Harris and Stodolsky[76,77] pointed out that Hund's explanation may be incomplete. Weak interactions (at the level of elementary particles) could lead to parity violation[76], i.e. to removal of the symmetry of the potential profile as a result of which Ψ_L and Ψ_R would become exact eigenfunctions. Tunnelling would then be suppressed, so that explanation of the paradox for isolated molecules would depend on weak interactions. Moreover, the lifetime of a single form would be increased considerably, even in very small isolated systems. The discovery of the retention of optical activity for times much longer than those corresponding to tunnelling would then be proof of the action of weak interactions in the molecular system. This assumes, of course, that (unperturbed) tunnelling times can be determined from an independent and reliable approach. The situation is different for molecules in media[78] where interactions with the environment may apparently be the determining factor in stabilization. This can be interpreted straightforwardly[11] as a result of the localization of the molecular eigenfunctions owing to the influence of the medium. In addition, tunnelling in both directions could[76,77] lead to observable oscillations in the optical activity, whose quality would change in the presence of the above-mentioned weak interactions. In connection with this hypothesis, a dynamic theory of relaxation of optical isomers has been developed[77] utilizing the density matrix formalism. It should be noted for the sake of completeness that this density matrix technique (widely used with remarkable success, e.g. in statistical mechanics or in the study of electronic structure[79,80]) is certainly a promising future tool for a rigorous approach to the study of isomerism as a whole.

4.3 Summary

The quantum-mechanical concept of isomerism is developing in close connection with the development of the quantum concept of molecular structure. The recent quite extensive discussion has demonstrated that the concept of molecular structure can be extended so that it becomes independent of the validity of the Born-Oppenheimer approximation. Even the exact molecular wave function contains some structural information; selection among various methods of extracting this information, however, is at present rather determined by convention (which should follow from the directness of the relationship with the experimentally obtainable information). However, a satisfactory derivation of the classical molecular structure based on a strictly quantum approach has not yet been described[81,82].

Similarly as in connection with the Born-Oppenheimer potential energy hypersurfaces, there is a certain unity among isomers also in terms of the complete molecular wave function, even between apparently distinct isomers. The uniqueness of isomers becomes a good approximation when the individual stability domains are relatively well isolated. However, because observations are never carried out on exactly stationary states it is desirable that the theoretical description be based on the time evolution of the state function. At present, this problem is mostly studied in the framework of study of tunnelling in a double-well potential profile. It has been found that the behaviour of a number of isomeric systems can be usefully explained by qualitative differences in the behaviour of this model in symmetric and asymmetric situations. The main characteristic of the contemporary quantum-mechanical interpretation of isomerism (at present developed only at a basic qualitative level) is the organic incorporation of the clear uniqueness of some isomers as well as of the representation of other species by a greater number of stability regions.

REFERENCES

(1) Woolley, R. G.: 1976, Chem. Phys. Lett. *44*, pp. 73—75.

(2) 1980, Quantum Dynamics of Molecules. The New Experimental Challenge to Theorists, Ed. R. G. Woolley, Plenum Press, New York.

(3) WOOLLEY, R. G.: 1976, Advan. Phys. *25*, pp. 27—52.
(4) WOOLLEY, R. G., and SUTCLIFFE, B. T.: 1977, Chem. Phys. Lett. *45*, pp. 393—398.
(5) WOOLLEY, R. G.: 1977, Int. J. Quantum Chem. *12*, Suppl. 1, pp. 307—313.
(6) WOOLLEY, R. G.: 1978, J. Am. Chem. Soc. *100*, pp. 1073—1079.
(7) WOOLLEY, R. G.: 1978, Chem. Phys. Lett. *55*, pp. 443—446.
(8) BERTHIER, G.: 1981, Int. J. Quantum Chem. *19*, pp. 985—989.
(9) ESSÉN, H.: 1977, Int. J. Quantum Chem. *12*, pp. 721—735.
(10) PRIMAS, H.: 1975, Theor. Chim. Acta *39*, pp. 127—148.
(11) BIXON, M.: 1982, Chem. Phys. Lett. *87*, pp. 271—273.
(12) KOŁOS, W., and WOLNIEWICZ, L.: 1964, J. Chem. Phys. *41*, pp. 3674—3678.
(13) HERZBERG, G.: 1960, Molecular Spectra and Molecular Structure, II. Infrared and Raman Spectra of Polyatomic Molecules, Van Nostrand Company, Princeton.
(14) WILDMAN, T. A.: 1980, Chem. Phys. Lett. *74*, pp. 383—387.
(15) HONEGGER, E., and HEILBRONNER, E.: 1981, Chem. Phys. Lett. *81*, pp. 615—619.
(16) STRATT, R. M., and DESJARDINS, S. G.: 1982, J. Chem. Phys. *76*, pp. 5134—5144.
(17) BERRY, R. S.: 1979, in The Permutation Group in Physics and Chemistry, Ed. J. Hinze, Springer-Verlag, Berlin, pp. 92—120.
(18) BERRY, R. S.: 1980, in Ref.[2], pp. 143—195.
(19) TENNYSON, J., and VAN DER AVOIRD, A.: 1982, J. Chem. Phys. *77*, pp. 5664—5681.
(20) SUTCLIFFE, B. T.: 1975, in Computational Techniques in Quantum Chemistry and Molecular Physics, Ed. G. H. F. Diercksen, B. T. Sutcliffe, and A. Veillard, D. Reidel Publ. Comp., Dordrecht, pp. 1—105.
(21) ARONOWITZ, S.: 1978, Int. J. Quantum Chem. *14*, pp. 253—269.
(22) BADER, R. F. W., and NGUYEN-DANG, T. T.: 1982, Advan. Quantum Chem. *14*, pp. 63—124.
(23) BADER, R. F. W., TAL, Y., ANDERSON, S. G., and NGUYEN-DANG, T. T.: 1980, Isr. J. Chem. *19*, pp. 8—29.
(24) BADER, R. F. W., NGUYEN-DANG, T. T., and TAL, Y.: 1981, Rep. Prog. Phys. *44*, pp. 893 to 948.
(25) BADER, R. F. W., and BEDDALL, P. M.: 1972, J. Chem. Phys. *56*, pp. 3320—3329.
(26) BADER, R. F. W.: 1975, Accounts Chem. Res. *8*, pp. 34—40.
(27) BADER, R. F. W., ANDERSON, S. G., and DUKE, A. J.: 1979, J. Am. Chem. Soc. *101*, pp. 1389—1395.
(28) BADER, R. F. W., NGUYEN-DANG, T. T., and TAL, Y.: 1979, J. Chem. Phys. *70*, pp. 4316 to 4329.
(29) BADER, R. F. W.: 1980, J. Chem. Phys. *73*, pp. 2871—2883.
(30) TAL, Y., BADER, R. F. W., NGUYEN-DANG, T. T., OJHA, M., and ANDERSON, S. G.: 1981, J. Chem. Phys. *74*, pp. 5162—5167.
(31) COLLARD, K., and HALL, G. G.: 1977, Int. J. Quantum Chem. *12*, pp. 623—637.
(32) POSTON, T., and STEWART, I. N.: 1976, Taylor Expansions and Catastrophes, Pitman, London.
(33) CLAVERIE, P., and DINER, S.: 1980, Isr. J. Chem. *19*, pp. 54—81.
(34) TOSTES, J. G. R.: 1981, Theor. Chim. Acta *59*, pp. 229—235.
(35) POLIKANOV, S. M.: 1977, Isomerism of Atomic Nucleus Shape, Atomizdat. Moscow, (in Russian).
(36) PRIMAS, H.: 1967, Int. J. Quantum Chem. *1*, pp. 493—519.
(37) GARCÍA-SUCRE, M., and BUNGE, M.: 1981, Int. J. Quantum Chem. *19*, pp. 83—93.
(38) LATHOUWERS, L., VAN LEUVEN, P., and BOUTEN, M.: 1977, Chem. Phys. Lett. *52*, pp. 439 to 441.
(39) VAN LEUVEN, P., and LATHOUWERS, L.: 1980, in Ref.[2], pp. 197—220.
(40) LATHOUWERS, L., and VAN LEUVEN, P.: 1980, in Ref.[2], pp. 221—237.

(41) CLAVERIE, P., and DINER, S.: 1977, Int. J. Quantum Chem. *12*, Suppl. 1, pp. 41–82.
(42) WILSON, E. B.: 1979, Int. J. Quantum Chem., Quantum Chem. Symp. *13*, pp. 5–14.
(43) LAFORGUE, A.: 1981, Int. J. Quantum Chem. *19*, p. 989.
(44) SLANINA, Z.: 1981, Advan. Quantum Chem. *13*, pp. 89–153,
(45) GILLES, J. M. F., and PHILIPPOT, J.: 1978, Int. J. Quantum Chem. *14*, pp. 299–311.
(46) WATSON, J. K. G.: 1965, Can. J. Phys. *43*, pp. 1996–2007.
(47) DALTON, B. J.: 1966, Mol. Phys. *11*, pp. 265–285.
(48) GILLES, J.-M. F., and PHILIPPOT, J.: 1972, Int. J. Quantum Chem. *6*, pp. 225–261.
(49) RIESS, J., and MÜNCH, W.: 1981, Theor. Chim. Acta *58*, pp. 295–300.
(50) BERRY, R. S.: 1960, Rev. Mod. Phys. *32*, pp. 447–454.
(51) HUND, F.: 1927, Z. Phys. *43*, pp. 805–826.
(52) WILSON JR., E. B.: 1959, Bull. Am. Phys. Soc. Ser. *II 4*, p. 164.
(53) LÖWDIN, P. O.: 1965, Advan. Quantum Chem. *2*, pp. 213–360.
(54) WEINER, J. H.: 1978, J. Chem. Phys. *68*, pp. 2492–2506.
(55) WEINER, J. H.: 1978, J. Chem. Phys. *69*, pp. 4743–4749.
(56) CRIBB, P. H., NORDHOLM, S., and HUSH, N. S.: 1979, Chem. Phys. *44*, pp. 315–335.
(57) HELLER, E. J., and DAVIS, M. J.: 1981, J. Phys. Chem. *85*, pp. 307–309.
(58) BASILEVSKY, M. V., and RYABOY, V. M.: 1981, Mol. Phys. *44*, pp. 785–798.
(59) WEINER, J. H., and TSE, S. T.: 1981, J. Chem. Phys. *74*, pp. 2419–2426.
(60) CHRISTOFFEL, K. M., and BOWMAN, J. M.: 1981, J. Chem. Phys. *74*, pp. 5057–5075.
(61) AGRESTI, A., BACCI, M., and RANFAGNI, A.: 1981, Chem. Phys. Lett. *79*, pp. 100–104.
(62) SUN, J. C., CHOI, B. H., POE, R. T., and TANG, K. T.: 1981, Chem. Phys. Lett. *82*, pp. 255 to 259.
(63) CRIBB, P. H., NORDHOLM, S., and HUSH, N. S.: 1982, Chem. Phys. *69*, pp. 259–266.
(64) FLANIGAN, M. C., and DE LA VEGA, J. R.: 1974, J. Chem. Phys. *61*, pp. 1882–1891.
(65) BUSCH, J. H., and DE LA VEGA, J. R.: 1977, J. Am. Chem. Soc. *99*, pp. 2397–2406.
(66) FLUDER JR., E. M., and DE LA VEGA, J. R.: 1978, Chem. Phys. Lett. *59*, pp. 454–456.
(67) FLUDER, E. M., and DE LA VEGA, J. R.: 1978, J. Am. Chem. Soc. *100*, pp. 5265–5267.
(68) BUSCH, J. H., FLUDER, E. M., and DE LA VEGA, J. R.: 1980, J. Am. Chem. Soc. *102*, pp. 4000–4007.
(69) DE LA VEGA, J. R., BUSCH, J. H., SCHAUBLE, J. H., KUNZE, K. L., and HAGGERT, B. E.: 1982, J. Am. Chem. Soc. *104*, pp. 3295–3299.
(70) DE LA VEGA, J. R.: 1982, Accounts Chem. Res. *15*, pp. 185–191.
(71) BELL, R. P.: 1973, The Proton in Chemistry, Chapman and Hall, London.
(72) ALBERDING, N., AUSTIN, R. H., CHAN, S. S., EISENSTEIN, L., FRAUENFELDER, H., GUNSALUS, I. C., and NORDLUND, T. M.: 1976, J. Chem. Phys. *65*, pp. 4701–4711.
(73) EISENSTEIN, L.: 1976, Int. J. Quantum Chem., Quantum Biol. Symp. *3*, pp. 21–27.
(74) PAPOUŠEK, D., and ŠPIRKO, V.: 1976, Top. Curr. Chem. *68*, pp. 59–102.
(75) PAPOUŠEK, D., and ALIEV, M. R.: 1982, Molecular Vibrational/Rotational Spectra, Elsevier, Amsterdam & Academia, Prague.
(76) HARRIS, R. A., and STODOLSKY, L.: 1978, Phys. Lett. *78B*, pp. 313–317.
(77) HARRIS, R. A., and STODOLSKY, L.: 1981, J. Chem. Phys. *74*, pp. 2145–2155.
(78) SIMONIUS, M.: 1978, Phys. Rev. Lett. *40*, pp. 980–983.
(79) DAVIDSON, E. R.: 1976, Reduced Density Matrices in Quantum Chemistry, Academic Press, New York.
(80) BLUM, K.: 1981, Density Matrix. Theory and Applications, Plenum Press, New York.
(81) WOOLLEY, R. G.: 1980, Isr. J. Chem. *19*, pp. 30–46.
(82) WOOLLEY, R. G.: 1981, Chem. Phys. Lett. *79*, pp. 395–398.
(83) WOOLLEY, R. G.: 1982, Personal Communication.

5 Algebraic Aspects of Isomerism

Algebraic methods represented the first theoretical approach to yield useful results in the study of isomerism. As the use of these techniques does not necessarily require the application of quantum theory, it is not surprising that the first serious works date[1] from as early as 1874, when three papers were published in the problem of isomer enumeration: by Cayley[2], Körner[3] and van't Hoff[4]. It should, however, be noted foı the sake of completeness that the very first application of algebraic methods in a chemical context involved the introduction of graphs for the representation of molecules. This key step in the history of chemistry was taken[5] in the middle of the last century and is connected with works published by Couper, Crum Brown, Franklad, Kekulé, etc. Nonetheless, Dalton should probably be considered as the first chemical theoretician[6], as the concept of molecular structure began to develop in his atomic hypothesis formulated in 1807.

The possibility of using algebra (concretely, for example, set theory[7,8], group theory[9-11], graph theory[12] and information theory[13]) for the classification and characterization of isomers was mentioned in the introductory chapter. In spite of their abstract character, these algebraic methods permit the rationalization or systemization of the relationships between isomers – see for example[14-26]. Graph theory represents a basis for the analysis of structure-activity relationships[27,28] in terms of topological indices (see for example[29-33]), carried out in the framework of the additivity principle (cf.[34,35]). In addition, algebraic procedures can lead to predictions of new reactions – automerizations proceeding through six-membered transition states (studied by Balaban[36]) are a classic example. Although, in contrast to quantum-chemical techniques discussed in Chap. 3, the algebraic approach does not lead to a description of the energetics, nonetheless it permits direct insight into the intrinsic logical structure behind chemical problems and, in addition, its use is not limited by the size of the system studied (or at least not as strictly as with quantum-chemical methods). This chapter will be concerned mainly with those algebraic aspects of isomerism that are related to the logical structure of chemistry, computer-assisted synthesis design, the reactivity of chemical substances and its theoretical interpretation in terms of stationary points on the energy hypersurfaces.

5.1 Algebraic Generalization of the Concept of Isomerism

5.1.1 The Equivalence Relation

A set of all ordered pairs (x_1, x_2), such that x_1 is an element of set X_1 and x_2 is an element of set X_2, is called (see for example[37]) a Cartesian product of sets X_1 and X_2. Symbolically,

$$X_1 \times X_2 = \{(x_1, x_2); x_1 \in X_1, x_2 \in X_2\} . \tag{5-1}$$

If, for example, both X_1 and X_2 are sets of all real numbers Q, then the Cartesian product $Q \times Q$ is of illustrative meaning: it is the set of all the Cartesian coordinates in the plane. The binary relation[37] R from X_1 to X_2 is defined as the subset of the Cartesian product of sets $X_1, X_2 : R \subset X_1 \times X_2$. In particular, the binary relation from X_1 to X_1 is termed the relation on set X_1. A special case of the relation on set X is the equivalence relation[37] E introduced by the properties:

(i) reflexivity: for every $x \in X$ it holds that $(x, x) \in E$,

(ii) symmetricity: if $(x, y) \in E$, then it also holds that $(y, x) \in E$,

(iii) transitivity: if $(x, y) \in E$ and $(y, z) \in E$, then also $(x, z) \in E$.

The equivalence class[37] P_{E_x} of an element $x \in X$ in relation E is the set of all elements y in X that are E-related to element x. Symbolically,

$$P_{E_x} = \{y; y \in X \text{ and } (x, y) \in E\} . \tag{5-2}$$

Of the many properties of equivalence relations, for us the most important is the fact[37] that every element of set X belongs in a single equivalence class of E. In other words, the equivalence relation permits division of set X into subsets fulfilling the following conditions:

(i) their union is set X,

(ii) intersection of any two different subsets is an empty set.

The equivalence relation is the basis for the partitioning of a set of objects into classes including only objects that are equal in some respects (and different in other respects). The equivalence relation subdivides the set into equivalence classes so that all the elements in a given class exhibit certain common characteristics. In chemistry, most considerations are based on analogy, which is an example of classification according to an equivalence relation. An example of a set subdivided by equivalence into equivalence classes is the periodic table of the elements[38] – each period represents an equivalence class including atoms with the same number of core electrons. The equality in the numbers of these electrons then represents the equivalence relation. Another useful equivalence relation on the set of all types of atoms is equality in the numbers of valence electrons. This equivalence relation subdivides the periodic system into equivalence classes given by its groups. A further example involving equivalence relations is thermodynamic equilibrium[8]. Now, consider the set of all chemical compounds and introduce the equivalence relation – molecular formula identity. Application of this equivalence relation leads to the subdivision of the set of chemical compounds into equivalence classes – groups of isomeric compounds. Thus the phenomenon of chemical isomerism can be interpreted in terms of equivalence relations.

5.1.2 Isomeric Ensembles of Molecules

The equivalence relation of isomerism can further be generalized by transition from the individual molecules to ensembles of molecules. Consider a particular set of atoms and form individual compounds or ensembles of compounds so that all the atoms are employed. Each of these atomic arrangements thus represents[8,38–42] a single isomeric ensemble of molecules and all of these ensembles form a family of isomeric ensembles of molecules. This concept represents a generalization of the equivalence relation of isomerism from molecules to ensembles of molecules.

The equality introduced by the equivalence relation of isomerism for the ensembles of molecules between the individual members of the family of isomeric ensembles of molecules has a clear quantum-mechanical justification. In fact, this generalized approach to isomerism was anticipated in a rather implicit form in earlier theoretical works[43,44]. All the members of a given family of isomeric ensembles of molecules correspond to the same total Hamiltonian. In the framework of the Born-Oppenheimer approximation, multi-membered ensembles of molecules are included in the potential energy hypersurface as regions of dissociation products at infinity (cf.[45]). The use of the equivalence relation in this connection, as pointed out by the authors[38], is dependent on the uncertainty principle. The possibility of obtaining detailed information concerning systems is limited by this principle regardless of further improvements in instrument precision. Thus the authors of ref.[38] suggested the generation of models of chemical systems that represent certain equivalence classes of states rather than the states themselves.

Table 5-1. Family of isomeric ensembles of molecules with empirical formula C_2H_6O[a]

$\{2\,C + 6\,H + O\}$; $\{CH_3{-}CH_2{-}OH\}$; $\{CH_3{-}O{-}CH_3\}$; $\{CH_3{-}CHO + H_2\}$;
$\{CH_4 + CH_2O\}$; $\{CH_3OH + CH_2\}$; $\{H_2C{-}CH_2 + H_2\}$ (with the two carbons bridged by O); $\{CH_2{=}CH_2 + H_2O\}$;
$\{H_2C{=}CHOH + H_2\}$; $\{CO + CH_4 + H_2\}$; $\{CH_2 + CH_2O + H_2\}$;
$\{H_2C{=}C{=}O + 2\,H_2\}$; $\{HC{\equiv}C{-}OH + 2\,H_2\}$; $\{HC{\equiv}CH + H_2O + H_2\}$;
$\{CO + CH_2 + 2\,H_2\}$; $\{2\,CH_2 + H_2O\}$; $\{2\,CH_3 + O\}$; etc.

[a] Only conceivable neutral components are considered (according to Ref.[38]).

Consider, for example, the set of atoms C_2H_6O. Table 5-1 gives the corresponding family of isomeric ensembles of molecules assuming that the generation is limited to reasonable neutral substances. The be-matrices[8,38–41] (bond and electron matrices) were introduced for computer-applicable recording of isomeric ensembles of molecules. The be-matrix is introduced for each member of the family of isomeric ensembles of molecules formed of n atoms as a symmetric $n \times n$ matrix with off-diagonal entries corresponding to the formal bond orders between individual atoms, while the diagonal contains the numbers of free valence electrons on the given atom. For example, for formaldehyde the be-matrix has the form

$$\begin{array}{c} H_1 \\ \diagdown \\ C_2{=}O_4 \\ \diagup \\ H_3 \end{array} \qquad \begin{array}{c|cccc|c} & H_1 & C_2 & H_3 & O_4 & \\ \hline & 0 & 1 & 0 & 0 & H_1 \\ & 1 & 0 & 1 & 2 & C_2 \\ & 0 & 1 & 0 & 0 & H_3 \\ & 0 & 2 & 0 & 4 & O_4 \end{array} \tag{5-3}$$

It is apparent that the be-matrix contains all the constitutional information on the given isomeric ensemble of molecules, but no stereochemical information. The cc--matrix was introduced[8] for distinguishing stereoisomerism; this matrix describes conformational and configurational differences.

Every chemical reaction can be interpreted as an interconversion between two isomeric ensembles of molecules – the left- and right-hand sides of the equation correspond to individual isomeric ensembles belonging to a single family of isomeric ensembles of molecules. A reaction matrix $\mathbf{R}_{\Psi \to \Phi}$ was introduced[8,38–41] for the description of the chemical reaction leading from ensemble Ψ (described by be-matrix $\mathbf{B}_\Psi$) to ensemble of products Φ (described by be-matrix $\mathbf{B}_\Phi$):

$$\mathbf{R}_{\Psi \to \Phi} = \mathbf{B}_\Phi - \mathbf{B}_\Psi \,. \tag{5-4}$$

For example, the addition of HCN to H_2CO:

$$\begin{array}{c} H_1{-}C_2{=}O_4 \\ | \\ H_3 \end{array} + H_5{-}C_6{\equiv}N_7 \;\rightarrow\; \begin{array}{c} C_6{\equiv}N_7 \\ | \\ H_1{-}C_2{-}O_4{-}H_5 \\ | \\ H_3 \end{array} \tag{5-5}$$

is described by matrices (5-6)–(5-8):

$$\mathbf{B}_\Psi = \begin{array}{|ccccccc|c} \multicolumn{7}{c}{H_1\; C_2\; H_3\; O_4\; H_5\; C_6\; N_7} & \\ 0 & 1 & 0 & 0 & 0 & 0 & 0 & H_1 \\ 1 & 0 & 1 & 2 & 0 & 0 & 0 & C_2 \\ 0 & 1 & 0 & 0 & 0 & 0 & 0 & H_3 \\ 0 & 2 & 0 & 4 & 0 & 0 & 0 & O_4 \\ 0 & 0 & 0 & 0 & 0 & 1 & 0 & H_5 \\ 0 & 0 & 0 & 0 & 1 & 0 & 3 & C_6 \\ 0 & 0 & 0 & 0 & 0 & 3 & 2 & N_7 \end{array} \tag{5-6}$$

$$
\mathbf{B}_{\Phi} = \begin{array}{c} \\ \end{array}
\begin{array}{ccccccc|c}
H_1 & C_2 & H_3 & O_4 & H_5 & C_6 & N_7 & \\
0 & 1 & 0 & 0 & 0 & 0 & 0 & H_1 \\
1 & 0 & 1 & 1 & 0 & 1 & 0 & C_2 \\
0 & 1 & 0 & 0 & 0 & 0 & 0 & H_3 \\
0 & 1 & 0 & 4 & 1 & 0 & 0 & O_4 \\
0 & 0 & 0 & 1 & 0 & 0 & 0 & H_5 \\
0 & 1 & 0 & 0 & 0 & 0 & 3 & C_6 \\
0 & 0 & 0 & 0 & 0 & 3 & 2 & N_7
\end{array} \tag{5-7}
$$

$$
\mathbf{R}_{\Psi \to \Phi} =
\begin{array}{ccccccc|c}
H_1 & C_2 & H_3 & O_4 & H_5 & C_6 & N_7 & \\
0 & 0 & 0 & 0 & 0 & 0 & 0 & H_1 \\
0 & 0 & 0 & -1 & 0 & +1 & 0 & C_2 \\
0 & 0 & 0 & 0 & 0 & 0 & 0 & H_3 \\
0 & -1 & 0 & 0 & +1 & 0 & 0 & O_4 \\
0 & 0 & 0 & +1 & 0 & -1 & 0 & H_5 \\
0 & +1 & 0 & 0 & -1 & 0 & 0 & C_6 \\
0 & 0 & 0 & 0 & 0 & 0 & 0 & N_7
\end{array} \tag{5-8}
$$

It is apparent from the natural condition of preservation of the number of valence electrons that the sum of all the entries of matrices $\mathbf{B}_{\Psi}$ and $\mathbf{B}_{\Phi}$ is also conserved. Thus the sum of all the entries of reaction matrix $\mathbf{R}_{\Psi \to \Phi}$ must be zero (cf. Eq. (5-4)). Alternatively, the reaction matrix can be replaced[39] by a reaction operator transforming $\mathbf{B}_{\Psi}$ to $\mathbf{B}_{\Phi}$. A topological reaction coordinate[8], introduced by the relationship

$$\mathbf{B}_{\Phi}^{(r)} = \mathbf{B}_{\Psi} + r(\mathbf{B}_{\Phi} - \mathbf{B}_{\Psi}), \quad (0 \leqq r \leqq 1) \tag{5-9}$$

has been proposed for description of the course of a given reaction.

Consider two isomeric ensembles of molecules described by be-matrices $A = \{a_{ij}\}$ and $B = \{b_{ij}\}$. Then the metric (see e.g.[46-48]) introduced in the set of matrices belonging to the given family of isomeric ensembles of molecules in terms of the chemical distance[8,38,40,41] D^*:

$$D(A, B) = \sum_{i,j} |a_{ij} - b_{ij}| \tag{5-10}$$

is called the chemical metric. The chemical metric has an obvious interpretation: it indicates how close the constitutional relationships are between isomeric ensembles of molecules A and B. It can be demonstrated[38,40,41] that the chemical metric is topologically equivalent (see for example[46-48]) to the Euclidian metric, i.e. both metrics determine identical topology.

* It should be noted for the sake of completeness that this chemical distance may depend on the atom indexing. It is even conceivable that different indexing of two identical isomeric ensembles of molecules could lead to non-zero chemical distance.

Generalized isomerism, its algebraic representation and chemical metric permit a general view of the relationships existing between chemical systems – they form a basis for study of the logical structure of chemistry[38,49]. Simultaneously, they provide useful tools for semi-empirical and non-empirical computer-assisted synthesis design.

5.2 Computer-assisted Design of Syntheses

Planning of syntheses consists of two phases[50]: first, pathways are sought that generally lead from the readily available starting materials to the target molecule T; then, from among these conceivable pathways, the optimal one is selected on the basis of certain selection criteria. In terms of generalized isomerism this implies the establishment[8] of a family of isomeric ensembles of molecules that contain the target molecule T in at least one ensemble and the initial materials of the synthesis in another ensemble. The synthesis design then implies finding a set of pathways connecting the ensemble of molecules including T with the ensemble of molecules containing the starting material. In the solution of the problem of synthesis design, three levels can be distinguished[8]: non-empirical, semi-empirical and empirical (this classification scheme has, however, no relationship to the standard division of quantum-chemical methods).

The non-empirical approach to synthesis design assumes the construction of a family of isomeric ensembles of molecules that includes all the chemical compounds that must be considered in the solution of the given problem. Then all of the pathways connecting the ensemble of the starting material with the target molecule ensemble are sought. A non-empirical (topological) program based on these concepts generates a complete set of synthetic pathways for a given target molecule and indiscriminately incorporates known and unknown chemical reactions into the synthetic pathway generated. The non-empirical approach is, of course, limited[8] to purely topological relationships and does not include the selection of optimal synthetic pathways from the topologically possible ones on the basis of thermodynamic and kinetic information. Nonetheless, selection of the reaction pathways is also possible in the framework of the non-empirical approach without abandoning its topological nature. For this purpose, the concept of the shortest reaction pathway has been proposed; this concept is introduced by making the overall reaction matrix as close as possible to the minimal chemical distance[41,51]. Alternative approaches to the problem of finding all the reaction pathways were given in the works by Sinanoğlu and Lee[52–55], providing a method for finding all the *a priori* possible mechanisms for the given type of overall reaction classified according to the number of elementary reaction steps and also in the work of Nemes *et al.*[56–58] on the construction of reaction networks for complex chemical processes.

If we wish to determine which of the proposed synthetic pathways will be connected with a reasonable yield and will proceed at a reasonable rate, it is necessary to include empirical selection rules. In this way, a transfer is made from the non-empirical level to the semi-empirical one. Selection rules may be based on empirical values[8]: bond energies, electronegativities, data on types of reactions leading to high yields at high rates, etc. Another example of the use of selection rules involves use of orbital symmetry rules. If the selection rules were based on the values of the equilibrium and rate constants generated on the basis of (non-empirical) quantum-chemical and statistical--thermodynamic methods, the computer-assisted system for the synthesis design would have non-empirical character. As the authors of work[8] optimistically suggest, the non-empirical approach combined with thermodynamic and kinetic selection rules could even lead to a 'new chemistry'.

In contrast with the two previous approaches, the empirical approach does not employ a generalized concept of isomerism and its algebraic representation. It is based on empirical information on known synthetic reactions: the role of the program system of this type is limited to the effective handling of information on synthetic pathways obtained from the literature. However, for organic chemistry, even this computational level is of great assistance.

The use of computers for the solution of complex synthetic problems, initiated by the work of Corey and Wipke[59], is documented in the abundant literature – for reviews see, for example[60–63]. The initially available program systems for computer--assisted synthesis design worked at an empirical level: the synthetic schemes were constructed from known reactions stored in the computer library. The best known of these systems are: LHASA[59,64,65] (Logic and Heuristics Applied to Synthetic Analysis), SYNCHEM[66] (Synthetic Chemistry) and SECS[67,68] (Simulation and Evaluation of Chemical Synthesis). The works by Hendrickson[69,70] involve a less empirical approach to the problems of organic synthesis; he also introduced a useful general method of numerical notation representing a molecular structure. Approaches based on a mathematical model of the constitutional chemistry represented by be--matrices are sometimes denoted by the abbreviation[71] CNPE (Complete Neglect of Prior Experience). The first realization of this approach is the CICLOPS system[72] (Computers in Chemistry, Logic Oriented Planning of Syntheses). The AHMOS computer program[73] (Automatisierte heuristische Modellierung organisch-chemischer Synthesen) and the EROS system[74,75] (Elaboration of Reactions for Organic Synthesis) are also based on the same mathematical model. This mathematical model of constitutional chemistry was recently also utilized for drawing up a program system for the analysis of relationships between molecular structure and properties (activities)[76].

Questions of coding of chemical substances for computer handling, exhaustive generation of isomers, and construction of reaction graphs are closely related to the problems of computer-assisted synthesis design. In the coding of chemical formulae (for reviews, see for example[77,78]), we will be particularly interested in topological codes employing a graph adjacency matrix (cf.[30,79]). Chemical Abstracts, for example, employ[78] the Morgan topological code[80]. Unambiguous condensation of the adjacency matrix into a binary code[81,82] broadens its versatility. Computer generation of all the isomers consistent with a given empirical formula, possibly with further constraints (see[62,83–91]), is an important procedure in the computer-assisted elucidation of molecular structures[62,92–97] or in the search for structures with particular properties (cf.[98,99]). Reaction graphs are employed for chemical process systemization, mostly of isomerization, in both organic (e.g.[100–107]) and inorganic chemistry. In the latter, for example, processes of rearrangement of pentacoordinate systems[11,14,38,108,109] through the Berry[110] and turnstile[111] mechanisms have been the subject of considerable interest; the whole field of algebraic study of organometallic chemistry is surveyed in ref.[109]. It should be noted that the one-dimensional representation of a chemical substance in terms of topological indices (see for example[29–33]) is useful for some applications. The indices recently suggested by Bertz[112] and Balaban[113] are marked by their generality and high discriminatory ability.

5.3. Enumeration of Isomeric Structures

If the structural information on a given chemical system is reduced to the form of a graph, then the tools of graph theory developed for enumeration of nonisomorphic graphs[114] can be employed for the determination of the number of possible isomers. A graph is understood here to be[115] an ordered pair* (X, R), where X is a set and $R \subset X \times X$ (see Chap. 5.1.1). To define a graph means to define its vertices (set X) and its edges (relation R); the concepts of vertices and edges permit simple visualization of the abstract concept of a graph. The basic definitions and properties of graphs important for chemical applications are given in the works[77,79,115,116]. The problem of the graphical enumeration of isomers itself is already so broad that its comprehensive treatment would require a separate book, so we will give only the essential concepts and results. The modern era in graphical enumeration dates[118] from 1935 when Pólya published his now classic theorem. Results obtained before the appearance of Pólya's theorem (Hauptsatz) are reviewed in the articles[1,117–119].

5.3.1 Pólya's Theorem

In the solution of the general enumeration problem, Pólya fruitfully combined the classical method of generating functions with basic results from the theory of permutation groups. The method of generating functions consists of replacing the se-

* A graph defined in these terms is an oriented (directed) graph. In a chemical context, an undirected graph is usually sufficient. Then R would designate a subset of the set of all two-element parts of set X.

quence $a_0, a_1, a_2, \ldots$ (determining the numbers of figures of certain types) by the function

$$S(x) = a_0 + a_1 x + a_2 x^2 + \ldots \tag{5-11}$$

and in treating this function according to rules valid for a polynomial. In fact, Cayley[120] used the method of generating functions in connection with enumeration. The usefulness of the mathematical theory of permutation groups for the consideration of the symmetry of enumeration problems was already pointed out in the works by Redfield[121] and Lunn and Senior[122].

Consider p points and the permutation group H of order h permuting these points. The structure of each permutation can be characterized in terms of cycles[123]. For the cycle of order r we understand permutation of r elements with the following structure of element assignment before and after the permutation:

$$\begin{pmatrix} 1, 2, \ldots, r-1, r \\ 2, 3, \ldots, \quad r, 1 \end{pmatrix}. \tag{5-12}$$

Arbitrary permutation of the elements of a set can be expressed in such a way that the given set decomposes into disjunctive subsets such that the considered permutation induces a cycle within each of these subsets. This permits expression of an arbitrary permutation as a product of cycles. Let permutation $\alpha \in H$ exhibit $j_r(\alpha)$ cycles of order r $(r = 1, \ldots, p)$. The cycle structure of group H can be described in terms of its cycle index $Z(H)$, defined as a polynomial in variables $s_1, s_2, \ldots, s_p$:

$$Z(H) = \frac{1}{h} \sum_{\alpha \in H} s_1^{j_1(\alpha)} s_2^{j_2(\alpha)} \ldots s_p^{j_p(\alpha)} . \tag{5-13}$$

If $H_{j_1, j_2, \ldots, j_p}$ designates the number of permutations of type $[j_1, j_2, \ldots, j_p]$, i.e. the number of permutations with identical sequence $j_1, j_2, \ldots, j_p$, then the summation over the elements of group H in relationship (5-13) can be replaced by the summation ranging over all possible choices of the types $[j_1, j_2, \ldots, j_p]$:

$$Z(H) = \frac{1}{h} \sum H_{j_1, j_2, \ldots, j_p} s_1^{j_1} s_2^{j_2} \ldots s_p^{j_p} , \tag{5-14}$$

where it apparently holds that

$$1 j_1 + 2 j_2 + \ldots + p j_p = p . \tag{5-15}$$

Using the cycle index of the permutation group of the corresponding problem, Pólya[124] demonstrated how the enumeration of all possible combinatorial configurations can be reduced to enumeration of their equivalence classes. Here the original version of the derivation of Pólya's theorem will be given briefly.

Consider different figures Φ, Φ^*, Φ^{**}, ... containing three sorts of objects, for simplicity designated by red, blue and white colouring. Introduce the power series

$$\sum a_{klm} x^k y^l z^m = f(x, y, z) \tag{5-16}$$

and the sequence of functions f_i derived from it:

$$f(x, y, z) = f_1 \, ; \quad f(x^2, y^2, z^2) = f_2 \, ; \quad f(x^3, y^3, z^3) = f_3 \, ; \; \ldots, \tag{5-17}$$

where a_{klm} designates the number of figures containing k red, l blue and m white objects. By placing p figures from amongst the figures Φ, Φ^*, Φ^{**}, ... at the p points used in introducing group H, one of the possible configurations $(\Phi_1, \Phi_2, \ldots, \Phi_p)$ is formed. It should be noted that it is also possible here that $\Phi_i = \Phi_j = \ldots (i \neq j)$. Two configurations $(\Phi_1, \Phi_2, \ldots, \Phi_p)$ and $(\Phi'_1, \Phi'_2, \ldots, \Phi'_p)$ are designated as equivalent mod H if a permutation

$$S_i = \begin{pmatrix} 1, 2, \ldots, & p \\ i_1, i_2, \ldots, & i_p \end{pmatrix} \tag{5-18}$$

exists in group H such that $\Phi_{i_1} = \Phi'_1, \Phi_{i_2} = \Phi'_2, \ldots, \Phi_{i_p} = \Phi'_p$. It is now desirable to find the numbers A_{klm} of nonequivalent configurations mod H containing k red, l blue and m white objects, i.e. to find the generating function

$$\sum A_{klm} x^k y^l z^m = F(x, y, z) \, . \tag{5-19}$$

It will be stated that permutation $(\Phi_1, \Phi_2, \ldots, \Phi_p)$ admits permutation (5-18) if it holds that $\Phi_1 = \Phi_{i_1}, \Phi_2 = \Phi_{i_2}, \ldots, \Phi_p = \Phi_{i_p}$. Let $A_{klm}(S_i)$ be the number of configurations of type k, l, m admitting S_i. Let permutation S_i be of type $[j_1, j_2, \ldots, j_p]$, then it follows from the properties of the generating functions that[124]:

$$\sum A_{klm}(S_i) \, x^k y^l z^m = f_1^{j_1} f_2^{j_2} \ldots f_p^{j_p} \, . \tag{5-20}$$

All the permutations that are permitted by a fixed configuration C form a group of order g that is a subgroup of H. This also holds for every configuration equivalent to C; it can readily be demonstrated[124] that the number of configurations different from one another, but equivalent to C, is equal to h/g. Consider the sum

$$A_{klm}(S_1) + A_{klm}(S_2) + \ldots + A_{klm}(S_h) \, . \tag{5-21}$$

Configuration C (and every one of h/g equivalent configurations) is included g times in this summation. It thus follows that summation (5-21) must be the $g \, . \, h/g = h$ multiple of value A_{klm}:

$$A_{klm}(S_1) + A_{klm}(S_2) + \ldots + A_{klm}(S_h) = h \, A_{klm} \, . \tag{5-22}$$

The combination of relationships (5-19), (5-20), and (5-22) yields the expression

$$F(x, y, z) = \frac{1}{h} \sum H_{j_1, j_2, \ldots, j_p} f_1^{j_1} f_2^{j_2} \ldots f_p^{j_p} \, , \tag{5-23}$$

where the summation ranges over the types $[j_1, j_2, \ldots, j_p]$. Relationship (5-23) expresses Pólya's enumeration theorem: the generating function for enumeration of non-equivalent configurations is obtained by replacing variables s_r in the cycle index of the relevant permutation group (Eq. (5-14)) by the corresponding functions derived from the generating function of figures, i.e. f_r (Eqs. (5-16) and (5-17)).

The enumeration theorem is considered to be a milestone, not only in graph theory but in mathematics as a whole[125]. For completeness, however, a few comments should be made. Mention was made above of Redfield's work[121] which was long neglected and which contains many of the ideas later rediscovered independently by Pólya. Redfield's work was recently analyzed[126] and it was shown that it includes, replaces and simplifies many of the results for molecular combinatorics of recent years. It is interesting to note that this was the only work that Redfield ever published (cf.[125]). Another recent and interesting re-analysis [127,128] of early original results related to chemical enumeration is concerned with Körner's proof of the homotopy of the hydrogens in benzene, published in 1869, pointing out the potential usefulness of analogous arguments in the discussion of the structure of fluxional molecules.

The enumeration theorem permits enumeration of arbitrary figures, including chemical isomers. Pólya treated this problem in later works[129,130]. Consider several illustrative examples. Begin with the determination of the number of positional isomers of cyclobutadiene completely substituted by two types of substituents (Fig. 5-1). Thus two colours of objects are considered, generating function for the set of figures is given by

$$f(x, y) = x + y, \tag{5-24}$$

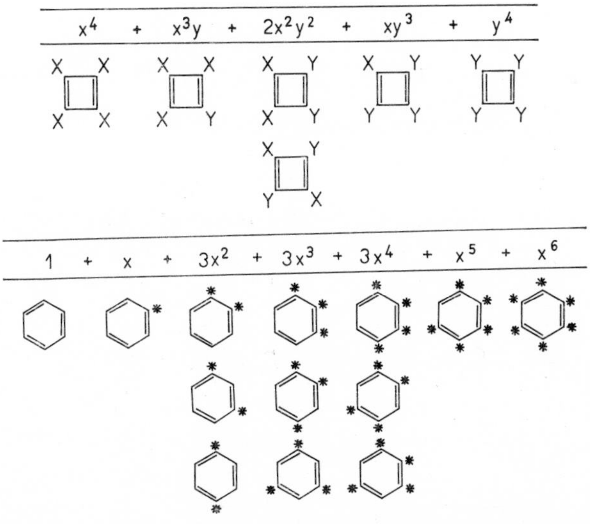

$$x^4 + x^3y + 2x^2y^2 + xy^3 + y^4$$

$$1 + x + 3x^2 + 3x^3 + 3x^4 + x^5 + x^6$$

Figure 5-1. Representation of the solution of an enumeration problem by a generating function for cyclobutadiene fully substituted by two kinds of univalent substituents (X, Y) and for benzene successively substituted by one kind of univalent group (*)

as it apparently holds for the expansion coefficients in Eq. (5-16) that $a_{10} = a_{01} = 1$, while all the others (a_{00}, a_{11}) are equal to zero. The symmetry of the fixed points on which the figures (substituents) are located is the symmetry of a square (point group D_{4h}); however, here group D_4 is sufficient. The corresponding permutation group contains these elements described in terms of cycles (the corners of the square are numbered stepwise around the sides):

$$\begin{matrix} (1)(2)(3)(4) & (1)(3)(24) & (2)(4)(13) & (13)(24) \\ (12)(34) & (14)(23) & (1234) & (1432) \end{matrix} . \tag{5-25}$$

Here the condensed notation is used, expressing permutations as products of cycles. For example, product (1) (3) (24) represents permutation $\begin{pmatrix} 1\,2\,3\,4 \\ 1\,4\,3\,2 \end{pmatrix}$, corresponding to the rotation around (two-fold) axis 13 or reflection in the plane containing this

Table 5-2. The derivation of the cycle index for each symmetry operation of the benzene ring[a]

Symmetry operation	Atom interchange	Permutation cycle groupings	Cycle index
E	$\begin{pmatrix} 1&2&3&4&5&6 \\ 1&2&3&4&5&6 \end{pmatrix}$	$\begin{pmatrix} 1 \\ 1 \end{pmatrix}\begin{pmatrix} 2 \\ 2 \end{pmatrix}\begin{pmatrix} 3 \\ 3 \end{pmatrix}\begin{pmatrix} 4 \\ 4 \end{pmatrix}\begin{pmatrix} 5 \\ 5 \end{pmatrix}\begin{pmatrix} 6 \\ 6 \end{pmatrix}$	s_1^6
C_6	$\begin{pmatrix} 1&2&3&4&5&6 \\ 2&3&4&5&6&1 \end{pmatrix}$	$\begin{pmatrix} 1&2&3&4&5&6 \\ 2&3&4&5&6&1 \end{pmatrix}$	s_6^1
C_6'	$\begin{pmatrix} 1&2&3&4&5&6 \\ 6&1&2&3&4&5 \end{pmatrix}$	$\begin{pmatrix} 6&5&4&3&2&1 \\ 5&4&3&2&1&6 \end{pmatrix}$	s_6^1
C_3	$\begin{pmatrix} 1&2&3&4&5&6 \\ 3&4&5&6&1&2 \end{pmatrix}$	$\begin{pmatrix} 1&3&5 \\ 3&5&1 \end{pmatrix}\begin{pmatrix} 2&4&6 \\ 4&6&2 \end{pmatrix}$	s_3^2
C_3'	$\begin{pmatrix} 1&2&3&4&5&6 \\ 5&6&1&2&3&4 \end{pmatrix}$	$\begin{pmatrix} 5&3&1 \\ 3&1&5 \end{pmatrix}\begin{pmatrix} 6&4&2 \\ 4&2&6 \end{pmatrix}$	s_3^2
C_2	$\begin{pmatrix} 1&2&3&4&5&6 \\ 4&5&6&1&2&3 \end{pmatrix}$	$\begin{pmatrix} 1&4 \\ 4&1 \end{pmatrix}\begin{pmatrix} 2&5 \\ 5&2 \end{pmatrix}\begin{pmatrix} 3&6 \\ 6&3 \end{pmatrix}$	s_2^3
$\sigma_{v,1}$	$\begin{pmatrix} 1&2&3&4&5&6 \\ 1&6&5&4&3&2 \end{pmatrix}$	$\begin{pmatrix} 1 \\ 1 \end{pmatrix}\begin{pmatrix} 4 \\ 4 \end{pmatrix}\begin{pmatrix} 2&6 \\ 6&2 \end{pmatrix}\begin{pmatrix} 3&5 \\ 5&3 \end{pmatrix}$	$s_1^2 s_2^2$
$\sigma_{v,2}$	$\begin{pmatrix} 1&2&3&4&5&6 \\ 5&4&3&2&1&6 \end{pmatrix}$	$\begin{pmatrix} 3 \\ 3 \end{pmatrix}\begin{pmatrix} 6 \\ 6 \end{pmatrix}\begin{pmatrix} 1&5 \\ 5&1 \end{pmatrix}\begin{pmatrix} 2&4 \\ 4&2 \end{pmatrix}$	$s_1^2 s_2^2$
$\sigma_{v,3}$	$\begin{pmatrix} 1&2&3&4&5&6 \\ 3&2&1&6&5&4 \end{pmatrix}$	$\begin{pmatrix} 2 \\ 2 \end{pmatrix}\begin{pmatrix} 5 \\ 5 \end{pmatrix}\begin{pmatrix} 1&3 \\ 3&1 \end{pmatrix}\begin{pmatrix} 4&6 \\ 6&4 \end{pmatrix}$	$s_1^2 s_2^2$
$\sigma_{v,4}$	$\begin{pmatrix} 1&2&3&4&5&6 \\ 6&5&4&3&2&1 \end{pmatrix}$	$\begin{pmatrix} 1&6 \\ 6&1 \end{pmatrix}\begin{pmatrix} 2&5 \\ 5&2 \end{pmatrix}\begin{pmatrix} 3&4 \\ 4&3 \end{pmatrix}$	s_2^3
$\sigma_{v,5}$	$\begin{pmatrix} 1&2&3&4&5&6 \\ 2&1&6&5&4&3 \end{pmatrix}$	$\begin{pmatrix} 1&2 \\ 2&1 \end{pmatrix}\begin{pmatrix} 3&6 \\ 6&3 \end{pmatrix}\begin{pmatrix} 4&5 \\ 5&4 \end{pmatrix}$	s_2^3
$\sigma_{v,6}$	$\begin{pmatrix} 1&2&3&4&5&6 \\ 4&3&2&1&6&5 \end{pmatrix}$	$\begin{pmatrix} 1&4 \\ 4&1 \end{pmatrix}\begin{pmatrix} 2&3 \\ 3&2 \end{pmatrix}\begin{pmatrix} 5&6 \\ 6&5 \end{pmatrix}$	s_2^3

[a] The numbers 1, ..., 6 stand for the ring atoms numbered successively (from Ref.[117]).

axis and perpendicular to the plane of the square. The cycle structure of elements (5-25) readily yields cycle index $Z(D_4)$ of the corresponding permutation group according to (5-13):

$$Z(D_4) = \tfrac{1}{8}(s_1^4 + 2s_1^2 s_2 + 3s_2^2 + 2s_4)\,. \tag{5-26}$$

Now, Pólya's enumeration theorem itself can be applied; substitution of (5-24) into (5-26) leads directly to

$$F(x, y) = \tfrac{1}{8}[(x + y)^4 + 2(x + y)^2 (x^2 + y^2) + + 3(x^2 + y^2)^2 + 2(x^4 + y^4)] \tag{5-27}$$

and algebraic rearrangement yields the result illustrated in Fig. 5-1. It should be noted that the same result would be obtained using the generating function of the figures $f(x) = 1 + x$ (one colour, two figures).

Now, the number of isomers involved in the substitution of benzene by a single type of substituent will be found; the generating function of the figures is used in the form

$$f(x) = 1 + x\,. \tag{5-28}$$

The symmetry of the six points on which the figures will be located is that of a regular hexagon (point group of symmetry D_{6h}). The permutation group consists of twelve elements (i.e. point group D_6), which, in terms of the cycle products, are

$$\begin{array}{llll} (1)(2)(3)(4)(5)(6) & (123456) & (654321) & (135)(246) \\ (531)(642) & (14)(25)(36) & (1)(4)(26)(35) & (3)(6)(15)(24)\,. \\ (2)(5)(13)(46) & (1)(6)(25)(34) & (12)(36)(45) & (14)(23)(56) \end{array} \tag{5-29}$$

The details of the permutation relationships are apparent from Table 5-2 and yield cycle index $Z(D_6)$ of this permutation group:

$$Z(D_6) = \tfrac{1}{12}(s_1^6 + 3s_1^2 s_2^2 + 4s_2^3 + 2s_3^2 + 2s_6)\,. \tag{5-30}$$

Substituting the generating function (5-28) in place of variables s_i converts relationship (5-30) into the final form of the generating function for nonequivalent configurations:

$$F(x) = 1 + x + 3x^2 + 3x^3 + 3x^4 + x^5 + x^6\,. \tag{5-31}$$

This result is illustrated schematically in Fig. 5-1.

The third illustrative example of the application of Pólya's theorem involves the enumeration of the number of ways in which the hydrogen atoms in toluene can be replaced by univalent substituent X. Here it is useful to first consider each rotor of toluene separately. The symmetry of the methyl group is represented by permutations (1) (2) (3), (123), (321), (1) (23), (2) (13), (3) (12). The symmetry of the second rotor is represented by permutations (4) (5) (6) (7) (8), (6) (48) (57). The cycle index of the toluene skeleton S is obtained as the product[131] of both partial cycle indices:

$$Z(S) = \tfrac{1}{6}(s_1^3 + 3s_1s_2 + 2s_3) \,.\, \tfrac{1}{2}(s_1^5 + s_1s_2^2) =$$
$$= \tfrac{1}{12}(s_1^8 + 3s_1^6s_2 + 2s_1^5s_3 + s_1^4s_2^2 + 3s_1^2s_2^3 + 2s_1s_2^2s_3)\,. \qquad (5\text{-}32)$$

The incorporation of the generating function of the figures in form (5-28) into relationship (5-32) yields the generating function of the required enumeration:

$$F(x) = 1 + 4x + 10x^2 + 16x^3 + 18x^4 + 16x^5 + 10x^6 + 4x^7 + x^8\,, \qquad (5\text{-}33)$$

indicating, for example, that there are 10 different isomers of toluene substituted by two univalent groups X and a total of 80 species resulting from successive substitution of the hydrogen atoms in toluene by this group.

Enumeration of the isomers resulting from the distribution of the given type of substituents over the fixed skeleton is a rather elementary application of Pólya's theorem. It finds more important application in enumeration in homologous series. For example, for the enumeration of alkyl radicals C_nH_{2n+1}— the generating function

$$F(x) = 1 + x + x^2 + 2x^3 + 4x^4 + 8x^5 + 17x^6 + \ldots \qquad (5\text{-}34)$$

can be obtained[124]. It should be noted that Pólya's theorem yields only the recursion formula for the expansion coefficients in the series (5-34). Function (5-34) represents a basis for the enumeration of a wide range of chemical compounds, especially acyclic compounds.

In conclusion, the application of the generating function (5-34) for alkyl radicals in the enumeration of the alkyl derivatives of benzene will be discussed. Function (5-34) will be considered in the general form

$$f(x) = A_0 + A_1x + A_2x^2 + A_3x^3 + \ldots\,. \qquad (5\text{-}35)$$

It is now apparent that the figures from the general formulation are alkyl radicals C_nH_{2n+1}— ($n = 0, 1, 2, \ldots$) located at points – the apices of a regular hexagon. The objects are now carbon atoms (only one type of object); the generating function of the figures is now series (5-35). The cycle index of the permutation group appearing in the problem was found earlier – Eq. (5-30). Now it remains to actually apply the enumeration theorem – to substitute function (5-35) into cycle index (5-30). This results in the required generating function for enumeration of the isomers of the alkyl derivatives of benzene $C_{6+n}H_{2n+6}$:

$$F(x) = \tfrac{1}{12}[f_1^6(x) + 3f_1^2(x)f_2^2(x) + 4f_2^3(x) + 2f_3^2(x) + 2f_6(x)]\,. \qquad (5\text{-}36)$$

If the expansion coefficients from (5-34) are used, then the following numbers of isomers can be obtained for the first members of the homologous series studied:

$$F(x) = 1 + x + 4x^2 + 8x^3 + 22x^4 + 51x^5 + 136x^6 + \ldots + 5749x^{10} +$$
$$+ \ldots + 690709x^{15} + \ldots + 89223734x^{20} + \ldots\,. \qquad (5\text{-}37)$$

Thus, for example, formula (5-37) indicates that there are four isomeric alkyl derivatives of benzene containing 8 C atoms. For the sake of interest, the numbers of isomers corresponding to higher homologues[131] are given in Eq. (5-37).

5.3.2 Iterative Enumeration

Before the derivation of Pólya's theorem, there was a great deal of scepticism about the possibility of a general solution, or even of obtaining an analytical formula, for an arbitrary enumeration problem. This led to the study of the possibility of the solution of a given enumeration problem iteratively, using recursion formulae. The procedure of obtaining recursion formulae will be illustrated using the first contribution of this type – iterative enumeration of the alcohols $C_nH_{2n+1}OH$ carried out by Henze and Blair[132].

Consider a general member of the enumerated homologous series in the form $R_1R_2R_3COH$, where R_1, R_2 and R_3 designate alkyl radicals or hydrogen atoms. Symbols p_n, s_n and t_n will be introduced for the numbers of primary, secondary, and tertiary alcohols containing n carbon atoms. The required number T_n of all isomers with summary formula $C_nH_{2n+1}OH$ is apparently given by

$$T_n = p_n + s_n + t_n \, . \tag{5-38}$$

If the —OH group of an arbitrary alcohol with n carbon atoms is replaced by the —CH_2OH group, a primary alcohol is always obtained, leading to the simple recursion relationship:

$$p_n = T_{n-1} \, . \tag{5-39}$$

The number s_n of secondary alcohols is based on the following scheme for their formation: alkyl radicals R_1 and R_2 are attached to the $\rangle$CHOH group so that the total number of carbon atoms equals $n - 1$. Now, it is necessary to determine all the choices of R_1 and R_2 that do not involve repetition. It is useful to distinguish two cases: (i) R_1 and R_2 have the same number of carbon atoms, and (ii) R_1 and R_2 have different numbers of carbon atoms. Let n be an even number ($n = 2k$ where k is an integer); then $n - 1$ is odd and case (i) is excluded. Now, R_1 is successively set equal to all alkyl radicals with 1, 2, ..., $n/2$-1 carbon atoms; also in this way the carbon atom content in radical R_2 is determined as $n - 2$, $n - 3$, ..., $n/2$ atoms and all possible situations are exhausted. It is thus apparent that s_n for even n is given as the sum of these $n/2$-1 terms:

$$s_n = T_1T_{n-2} + T_2T_{n-3} + \ldots + T_{n/2-1}T_{n/2} \, , \quad (n = 2k) \, . \tag{5-40}$$

If n is odd ($n = 2k + 1$), the contributions of case (i) must also be included, i.e. all the different choices of R_1 and R_2 for which each alkyl radical contains just

$(n-1)/2$ carbon atoms. Otherwise the reasoning is the same as for n even and leads to a sum of $(n-1)/2$ terms:

$$s_n = T_1T_{n-2} + T_2T_{n-3} + \dots + T_{(n-3)/2}T_{(n+1)/2} + \\ + T_{(n-1)/2}(1 + T_{(n-1)/2})/2\,, \quad (n = 2k+1)\,. \tag{5-41}$$

Finally, the recursion formula for t_n is obtained by generating tertiary alcohols by addition of alkyl radicals R_1, R_2 and R_3 to the $\geqslant$C—OH group so that the total number of carbon atoms in these radicals again equals $n - 1$. Three separate cases can be distinguished: (i) all three radicals contain different numbers of carbon atoms ($t_n^{(1)}$), (ii) two radicals have the same number of carbon atoms different from that in the third radical ($t_n^{(2)}$), and (iii) the numbers of carbon atoms in all three alkyl radicals are identical ($t_n^{(3)}$). It apparently holds that

$$t_n = t_n^{(1)} + t_n^{(2)} + t_n^{(3)}\,. \tag{5-42}$$

Case (i) leads directly to the symbolical sum

$$t_n^{(1)} = \sum T_iT_jT_k\,, \tag{5-43}$$

in which summation is carried out over all the different decompositions of number $n - 1$ into a sum of three different natural numbers i, j and k. The number of terms in sum (5-43) depends on the value of the remainder left over when dividing number n by 6 and can readily be found[132].

Condition (ii) is fulfilled for all the various decompositions of number $n - 1$ to different natural numbers i, j so that $n - 1 = 2i + j$. Summation over all these possibilities

$$t_n^{(2)} = \sum T_i(1 + T_i)\,T_j/2 \tag{5-44}$$

yields value $t_n^{(2)}$. The number of terms in relationship (5-44) is again given by the remainder on dividing n by six.

Case (iii) requires that $n - 1$ be divisible by three ($n = 3k + 1$). The relationship for $t_n^{(3)}$ can thus be summarized as:

$$t_n^{(3)} = T_{(n-1)/3}(1 + T_{(n-1)/3})\,(2 + T_{(n-1)/3})\,, \quad (n = 3k+1)\,; \\ t_n^{(3)} = 0\,, \quad (n \neq 3k+1)\,. \tag{5-45}$$

Relationships (5-39) to (5-45) represent the required recursion formulae for the determination of the number of isomeric alcohols $C_nH_{2n+1}OH$. When the number of isomers is known for 1, 2, ..., n, then the number for $n + 1$ can also be found. Recursion calculations beginning with $n = 1$ can in principle lead to the number of isomers for an arbitrarily large n. It should be noted that values T_n in relationship (5-38) naturally agree with the values of the expansion coefficients in the generating function (5-34) for alkyl radicals found on the basis of Pólya's theorem.

In contrast to Pólya's enumeration theorem, the iterative technique is not general in nature – the recursion formulae must be found separately for each type of problem. These formulae are available in the contemporary literature for a number of homologous series (see review[117]) including the classical alkane series[133]. Compared with Pólya's theorem, iterative enumerations have a mathematically simple form, as they are based only on combinatorial considerations and not on linking of combinatorial theory with group theory. On the other hand, however, recursion formulae can readily be handled by computers[117].

5.3.3 Asymptotic Enumeration Formulae

In addition to the rigorous concept of enumeration in the sense of Pólya's theorem, knowledge of asymptotic expressions for various groups of compounds with a large n is useful (cf.[131]). For example, it holds for number r_n of isomers in the series of alkanes with large numbers of carbon atoms n that

$$r_n \sim \varrho^{-n} n^{-5/2} \varrho a b^3 / 4\pi^{1/2} . \tag{5-46}$$

Similarly, it was found[131] for the series of C_nH_{2n} alkenes for asymptotic behaviour of their number s_n that

$$s_n \sim \varrho^{-n} / 4n . \tag{5-47}$$

In relationships (5-46) and (5-47), $\varrho = 0.35 \ldots$ and a and b are constants. Eqs. (5-46) and (5-47) indicate the possibility of an exponential increase in the number of isomeric structures in a homologous series with increasing content of the key atom.

5.3.4 Enumeration for Non-rigid Molecules

Suitable selection of the elements of the permutation group appearing in the enumeration theorem can lead to a description of some types of stereoisomerism (see for example[134–137]) and even molecular non-rigidity[128,137–140] and thus approach more closely the actual physical conditions existing within the given system. When considering enantiomers as different isomers, the corresponding permutations must be eliminated in the enumeration and a lower order permutation group must be used. In the same way, *cis-* and *trans-* isomers can be included in the enumeration, permitting only permutations changing the positions of all the participating substituents simultaneously. This method was used[134–137] for expanding the enumeration of structural isomers to include some types of stereoisomerism. Pólya considered[130] three types of permutation groups: a group of spatial formula leading to the number of stereoisomers, an extended group of spatial formula leading to the number of stereoisomers minus the number of pairs of optical isomers and, finally, a group of structural formula generating the number of structural isomers. For example, for cyclopropane the order of the group of spatial formula is 6, of the extended group of spatial formula is 12 and of the group of structural formula is 48. Tab. 5-3 lists[134] numerical illustrations of enumeration in acyclic series and comparison with the values from enumeration distinguishing only structural isomers.

It is well known[141] that, for non-rigid molecules, the point group of symmetry corresponding to an idealized rigid structure does not permit the correct description of the symmetry of the problem. For this purpose it must be extended by the inclusion of symmetry elements realized by intramolecular motions. Thus, in employing Pólya's theorem for enumeration of the isomers of non-rigid molecules, it is necessary to utilize[137] a permutation group including both the operations of the corresponding point group as well as every permutation feasible through the internal degrees of freedom. A permutation is considered feasible in the sense introduced by Longuet-Higgins[141] as a permutation realizable without transition over an insurmountable barrier (under the given conditions). Pólya's theorem has been extended[137] for enumeration of non-rigid compounds on the basis of a generalized wreath product method[142,143]. The generalized wreath product group represents an extension of the Longuet-Higgins results[141] for the symmetry group of non-rigid systems. This

Table 5-3. Isomer enumeration results in some acyclic series[a,b]

n (carbon atom content)	1	2	3	4	5	10	15	20	25
C_nH_{2n+1}—	1	1	2	4	8	507	48865	5622109	712566567
	1	1	2	5	11	1553	328092	82300275	22688455980
C_nH_{2n+2}	1	1	1	2	3	75	4347	366319	36797588
	1	1	1	2	3	136	18127	3396844	749329719
RCH_2OH	1	1	1	2	4	211	19241	2156010	269010485
	1	1	1	2	5	551	110500	27012286	7333282754
R_1R_2CHOH			1	1	3	194	19181	2216862	281593237
			1	2	5	768	162672	40807290	11247841224
$R_1R_2R_3COH$				1	1	102	10443	1249237	161962845
				1	1	234	54920	14480699	4107332002
$R_1C(=O)OR_2$		1	2	4	9	599	57564	6589734	832193902
		1	2	4	10	1319	273172	67819576	18581123978
$R_1R_2C{=}CR_3R_4$		1	1	3	5	377	36564	4224993	536113477
		1	1	4	6	895	185310	46244031	12704949506

[a] From Refs.[129,132–134]; the numbers of structural isomers are given in the upper line, the numbers of stereoisomers in the lower.

[b] R, R_i denote an alkyl radical or hydrogen atom; R_1, R_2, R_3, for secondary and tertiary alcohols, and R_2, for esters, are alkyl radicals.

concept has been applied[140], for example, to reactions generated by two-fold and three-fold internal rotations. While the use of conventional point groups of symmetry leads, for example, to a total of 18 isomers for $C_2H_2Br_2Cl_2$, the wreath product method yields only 6 isomers. These types of considerations are important for the rationalization and prediction[137,140] of the number of signals in NMR spectra at low and high temperatures.

A further interesting example of enumeration for a non-rigid skeleton is provided by the results obtained[138,139] for cyclohexane. The D_{3d} point group of symmetry belongs to the most stable structure of cyclohexane (Fig. 2-2). The use of the permutation group corresponding to this point group in the enumeration theorem leads to a number of isomers that is contrary[138,139] to experimental results. Simultaneously, it has been found empirically that the use of Haworth's projection formulae leads[138,139] to correct results, although they ignore the reality of the cyclohexane conformation. In the use of planar formulae, enumeration is carried out in terms of the D_{6h} point group of symmetry. Leonard *et al.*[138] provided a rationalization of these facts. The D_{3d} point group does not include operations corresponding to the non-rigidity of the cyclohexane molecular skeleton. It has been demonstrated[138] that the symmetry of cyclohexane can be adequately represented by the introduction of a new symmetry element (termed R_6), which combines a six-fold rotation with a flip motion of the ring interconverting one chair form with another. The effect of the R_6 operation is to permute all the positions on the same side of the ring. Group D_{3d} is replaced by group $D_{3d}R_6$ for construction of the permutation group; this group is obtained by expanding group D_{3d} by operation R_6. Group $D_{3d}R_6$ leads to the same cycle index as group D_{6h}, and thus also to the same enumeration results. In fact, this is a result of the circumstance that the D_{6h} and $D_{3d}R_6$ groups[144] are isomorphic. This example indicates[138,139,144] that the utilization of the symmetry groups of non--rigid molecules[137,141] should make a significant contribution towards making the results of the enumeration theorem closer to physical reality.

5.3.5 The Graph-like State of Molecules

Reduction of a real molecular system to a chemical graph or a stereograph essentially lies at the basis of any chemical enumeration. During the transfer to chemical graphs, only information on the existence of chemical bonds is retained; all information relevant to the geometrical arrangement is omitted. If it is assumed that all the bonds are flexible (which is manifested in the permutation group of the graph), the 'graph--like' concept of the state of molecules is obtained (Fig. 5-2); this concept was developed systematically by Gordon *et al.* (see e.g.[35,145–151]). This completely flexible state is hypothetical (it requires more flexibility than the real state of a molecule possesses); nonetheless, a number of molecules have a similar non-rigidity. The best known examples are the HCNO or ammonia molecule because of its inversion motion. Although this degree of abstraction of real molecules may appear unsuitable for chemical purposes, the simplicity of the graph-like state permits rigorous solution of some basic problems, especially in statistical mechanics[152,153]. A number of properties, including the entropy estimates, kinetic aspects or phase transitions, especially for polymers, can be treated in these terms, often representing a reduction of the problem to one of an enumerative nature.

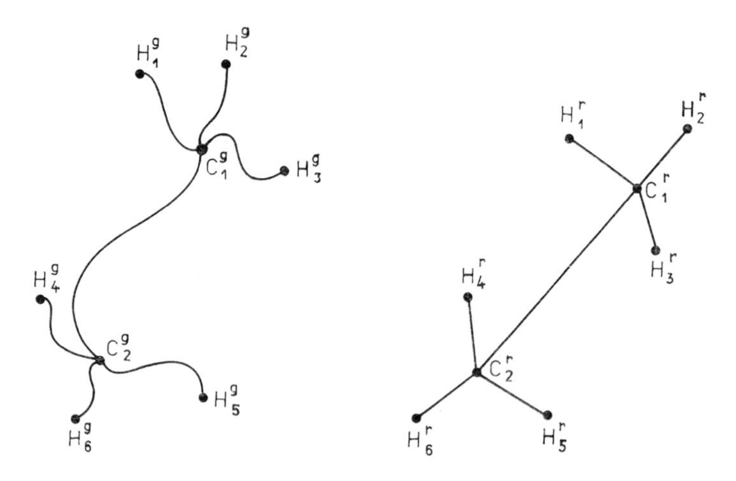

Figure 5-2. Transformation of graph-like (absolutely flexible) ethane into its three-dimensional real (rigid) state and vice versa (cf. Ref.[154])

The graph-like state of molecules corresponds[154] to the topological entropy S_t connected by a simple relationship with order $|G|$ of the symmetry group of its graph (R is the gas constant):

$$S_t = -R \ln |G| . \tag{5-48}$$

Useful rules have been derived for the calculation of order $|G|$ when the graph is of the tree type[154]. During the transition from the graph-like state of a molecule to its real state (Fig. 5-2), the integral degeneracy factor g of all the energy states of the real molecule appears, which leads to a contribution of S_m to the metric entropy with magnitude

$$S_m = R \ln g . \tag{5-49}$$

The sum of the topological and metric entropy defines the combinatorial entropy S_c:

$$S_c = S_t + S_m . \tag{5-50}$$

For suitably selected processes, this combinatorial entropy can make a decisive contribution towards the value of the overall entropy. Calculation of the combinatorial entropy does not require more information than the molecular graph and

Table 5-4. An example of the combinatorial entropy S_c with two isomeric octanes[a]

Combinatorial characteristic	*n*-octane	2,2,3,3-tetramethylbutane
g	2^8	2^8
$\|G\|$	$2^9 3^2$	$2^9 3^8$
S_c[b]	$-R \ln (2 \times 3^2)$	$-R \ln (2 \times 3^8)$

[a] From Ref.[154]
[b] $R = 8.31434 \text{ J K}^{-1} \text{ mol}^{-1}$.

the manner of its correspondence to the real state, i.e. whether, and how many, bonds of the real state of the molecule permit free rotation. Table 5-4 gives an example of the calculation of the combinatorial entropy for *n*-octane and 2,2,3,3-tetramethylbutane. It follows from the data in Table 5-4 that the contribution of the combinatorial entropy to the entropy term of the isomerization of octanes equals the value

$$\Delta S_c = 6R \ln 3 = 55.2 \text{ J mol}^{-1} \text{ K}^{-1}, \tag{5-51}$$

which represents[154] a total of 74% of the standard overall entropy change found experimentally at a temperature of 298 K. The concept of the combinatorial entropy was also utilized by Gordon[152] in discussion of the Third Law of Thermodynamics.

The graph-like state of molecules or matter might play a role in the description of polymers similar to the ideal gas state in thermodynamics and physics of gases and liquids[145]. The concept of the graph-like state of matter was used, for example, in the study of substitution effects and polymer distributions[145], the kinetics of linear polymerization reactions[147], and Rayleigh scattering[148]. An interesting relationship was found in the study of glass transitions of polymers[150]. The authors of the work[150] demonstrated that a limiting case in the statistical theory of these transitions is a famous problem in graph theory. They managed to find a connection between the number of configurations of the system and the number of Hamiltonian walks on a lattice graph. A closed Hamiltonian walk[115] on the given graph is understood to be a cycle (cf.[116]) passing through each vertex of the graph just once. Here, somewhat less stringent conditions are placed on Hamiltonian walks – their initial and final points need not be identical. In the theory of glass transitions of polymers, the value of the limit

$$\lim_{N \to \infty} (H_N)^{1/N} = n_H \tag{5-52}$$

is important, where H_N is the number of Hamiltonian walks on a lattice graph with N vertices. Graph theory has been employed to obtain characteristic (5-52) for a number of various lattices[150]. The topological approach to statistical thermodynamic theory especially for non-linear polymers (including biopolymers[155,157]) is promising for solution of a number of further problems in this field[151,155–157].

5.3.6 The Importance of Enumeration for the Theory of Chemical Reactivity

The contemporary literature contains[117] a relatively large number of enumeration results, based either on Pólya's theorem or on the iterative approach. Enumeration in acyclic series is discussed by Read[134], while the problem of enumeration of cyclic compounds was treated in the article by Balaban[158]. Enumeration of various classes of organic or inorganic substances was reviewed by Rouvray[117]; more recent results include the works[159–164]. Recently, computer-assisted enumerations have increased in number, as has the exhaustive generation of isomers (see for example[62,83,84,165,166]), especially in connection with the search for a structure consistent with experimental information, as discussed in Chap. 5.2.

As is apparent from the partial results of enumeration studies listed in Tab. 5-3 or from the asymptotic enumeration formulae – see Chap. 5.3.3 – the number of possible isomeric structures grows very rapidly with increasing size of the system. It is a key question from the point of view of the theory of chemical reactivity based on the stationary points in the corresponding energy hypersurface to determine the relationship between the numbers and types of stationary points and the corresponding results of isomer enumeration. The correspondence is, of course, not unambiguous. Every structure included in the enumeration need not necessarily correspond with a stationary point on the hypersurface. A single configuration in the enumeration may simultaneously be represented by more than one point on the hypersurface. Differentiation between the types of structures appearing in the enumeration in minima, transition states and higher types of stationary points on the corresponding hypersurface is completely beyond the capabilities of algebraic enumeration. In principle, however, the results of enumeration yield a procedure for the estimation of the order of the number of stationary points on the given hypersurface. Exhaustive generation of the enumerated structures than suggests the possible forms of stationary points.

Numerical results of enumeration clearly demonstrate that, as soon as quantum chemistry is concerned with systems with more than five atoms from the second or higher periods, the multiple appearance of stationary points on the energy hypersurface relevant for the given process is a general characteristic of the system studied. Consideration of the contributions of all possible structures in the theoretical calculation of an arbitrary structurally-dependent characteristic (and comparison with data from experiments admitting this isomerism, but not differentiating it) is then frequently necessary.

5.4 Topological Reduction and Characterization of Hypersurfaces

Basically, the potential energy hypersurface permits the detailed description of the structural and energy relationships connected with chemical changes at molecular level. Provided that the corresponding energy hypersurface is available (in an approximate form), it is possible to provide comprehensive numerical information on all the chemical processes that are possible in the framework of this hypersurface. However, the enormous number of calculations that must be carried out usually prevent such a complete numerical analysis. Considerable simplification and a better understanding of the problem can, however, be attained by the introduction of a generally applicable topological analysis of the hypersurfaces, applied in combination with a metric space, or even by replacing the latter completely. The topological approach can be understood as a generalization or abstraction of the geometric model, where the topology of the hypersurface suppresses particular geometric characteristics retaining certain general (and chemically important) properties, e.g. the continuity of functions defined on open sets. All the inherently topological properties of molecules and of the reactions between them can be analyzed in terms of intersection properties of open sets in a topological space (cf.[167–169]).

5.4.1 Reduction of Hypersurfaces to Tree-type Graphs

A basis for this non-numerical type of representation of the potential energy hypersurface was formed by the work of Krivoshey *et al.*[170–173]. Using a suitably introduced topology, a one-dimensional object can be assigned to every hypersurface (with an arbitrary number of variables) in such a manner that the differential geometrical properties of the corresponding hypersurface are completely ignored, but its topological properties (number and mutual arrangement of stationary points) are preserved. A mathematical background for this procedure is provided by the results of Kronrod[174].

For simplicity, a two-dimensional system will be considered first and a phenomenological analysis will be carried out. For example, consider a function of two variables $U = U(q_1, q_2)$ possessing two local maxima and a saddle point in a given domain (Fig. 5-3a) and set up its constant-level contours $c = U(q_1, q_2)$. These are depicted in two-dimensional space by certain lines consisting of either one or several disconnected components, i.e. curves without any common point (Fig. 5-3b). The aggregate of these components corresponding to value c is termed level set F_c. A point on a plane is assigned to each element (i.e. to each component) of level set F_c (different points to different components) .This is carried out for all values of c, where the order of values of c is considered in ordering the points. Kronrod[174] states that the point set constructed in such a way on the plane can be represented by a tree type graph for every continuous, single-valued function $U(q_1, q_2)$ (a tree is a connected graph containing no cycle as a subgraph – cf.[115,116]), as depicted in Fig. 5-3c. It is apparent that this graph reflects the overall number of stationary points of the initial surface as well as some of their interrelationships.

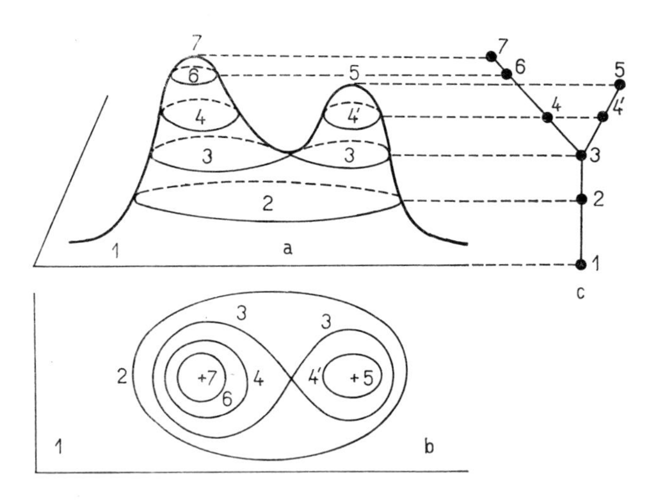

Figure 5-3. Process of mapping a two-dimensional surface (a) by tree representation **(c)**; (b) the surface contour plot (from Ref.[172])

Now consider a continuous function of n variables $U = U(q_1, q_2, \ldots, q_n)$ defined in a given n-dimensional domain D. Here level set F_c is defined[172] as a set of all points $(q_1, q_2, \ldots, q_n) \in D$, such that $U(q_1, q_2, \ldots, q_n) = c$, and its components as the corresponding maximal connected subsets. It can be demonstrated[170,172,174] using the formalism[46-48] of a topological space based on the set of components of all the level sets that, regardless of the magnitude of n, the topological space corresponding to function $U(q_1, q_2, \ldots, q_n)$ always leads to a tree type graph. This tree type graph is a geometric picture of the mapping of the given hypersurface in terms of the constant function-value contours.

The tree representing the topological characteristics of the potential energy hypersurface is the union of points and the edges connecting them. Two types of points will be considered: end points and branch points at which graph branching occurs (the branch point coincides with at least three edges). While the end points correspond[172] to local energy minima or maxima, branch points correspond to saddle points. It is apparent that every path on the given potential energy hypersurface can be interpreted as a path on the corresponding tree. The fact that the mapping of every hypersurface leads to a tree-type graph can be utilized to investigate the completeness of the localization of stationary points or to exclude some types of coincidence. For $n > 2$ it can be demonstrated[173] that there are only two principal types of saddle points that are reflected in the corresponding graph as its branch points (thus, in a sense, hypersurfaces for $n > 2$ are simpler than two-dimensional surfaces[173]).

Fig. 5-4a depicts schematically the classical LEP potential energy hypersurface[175] for the linear H_3 molecule. On the surface, domains 1 and 2 are designated, corresponding to the reactants and products, respectively, and activated complex 3. Domain 4 corresponds to a shallow minimum close to the saddle point, which is an artefact of the LEP method. Domain 5 represents a local potential energy maximum. Fig. 5-4b depicts the corresponding tree: minima 1, 2, and 4 and maximum 5 are its end points and saddle point 3 is the branch point.

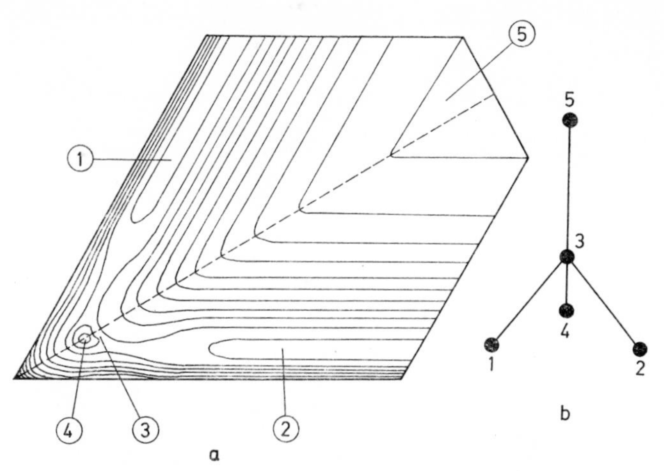

Figure 5-4. Tree representation (b) of the potential energy surface[175] of the linear H_3 molecule (a)

The usual output of a quantum-chemical study of a given hypersurface is a table of values of the internal coordinates and the corresponding energies. The following procedure has been suggested for drawing the tree of such a hypersurface[171]. Let Q_i be a point in the n-dimensional space of variables of function U, $Q_i = (q_1^i, q_2^i, \ldots, q_n^i)$, and M_c the level set of constant potential energy contours c; the number of components of this potential energy contour is to be determined. Let h designate the size of the step in the generation of points for quantum-chemical energy calculation. All these points are ordered according to their energy and sets of points corresponding to the same potential energy value c are formed, i.e. points are selected according to the correspondence with level sets M_c. The size of step h naturally determines the nicety of the mapping of the hypersurface and, in principle, even differentiation of the components may be dependent on its value (e.g. shallow minimum 4 on the surface in Fig. 5-4a – the 'Eyring Lake' – could remain undiscovered when an unsuitable step size is selected). Thus the h-components of level set M_c will be considered and the following criteria will be introduced: Two points Q_1, $Q_2 \in M_c$ lie in the same h-component if there is a sequence of points $P_1, P_2, \ldots, P_s$ in M_c such that relationships (5-53)–(5-55) are valid*. In these relationships, symbol $\| \|$ denotes the usual Euclidean distance

$$Q_1 = P_i, \quad Q_2 = P_j, \quad (i, j \in \langle 1; s \rangle); \tag{5-53}$$

$$\|P_k - P_{k+1}\| \leqq h, \quad (k = 1, \ldots, s-1); \tag{5-54}$$

$$\|P_1 - P_s\| \leqq h. \tag{5-55}$$

We will now introduce set $S_0(Q_1)$ as the set of all points in M_c that are not further than a distance of h from point $Q_1 \in M_c$, and set $S_1(Q_1)$ as the set of all points in M_c that are not further than a distance of h from at least one point in set $S_0(Q_1)$, symbolically:

$$S_1(Q_1) = \bigcup_{Q \in S_0(Q_1)} S_0(Q), \tag{5-56}$$

and recurrently for any i:

$$S_{i+1}(Q_1) = \bigcup_{Q \in S_i(Q_1)} S_0(Q). \tag{5-57}$$

It is obvious that the set relationship holds:

$$S_0(Q_1) \subset S_1(Q_1) \subset S_2(Q_1) \ldots. \tag{5-58}$$

However, in actual quantum-chemical mapping of the given potential energy hypersurface, not all the points in M_c are available, but only some of them. As soon as it is no longer possible to add a new point from M_c to set $S_i(Q_1)$, either because they are exhausted or because they are at a distance greater than h, the h-component of the constant potential energy contour c, in which point Q_1 lies, is mapped. The procedure can now be repeated for an arbitrary remaining point in M_c or for further potential levels. It is now a simple process to construct the tree[171] corresponding to the potential energy hypersurface studied. It should, however, be noted that, in addition to the above-mentioned role of the selection of step h for discovery of all stationary points, the selection of this value can, in principle, also affect the order of the stationary points.

It is apparent that this procedure readily permits the obtaining of the topological information on the given n-dimensional hypersurface that is contained in the table of potential energy values in a rather obscure form. The number of points required for this procedure, i.e. the computational demands, represent only a fraction of the requirements for obtaining an analytical description of the hypersurface. For example, in the paper[171] a tree is constructed for the hypersurface describing the reaction be-

* These formulae require the existence of a certain point sequence leading to a closed broken line. In general, however, this requirement is too strict. Sometimes only a simpler condition can be satisfied, i.e. that there exist an analogous open broken line. Then condition (5-55) should be removed.

tween ethylene and di-imide in space with five internal coordinates on the basis of the knowledge of the potential energy at only twenty points. Especially for larger systems, where the multiplicity of the occurrence of stationary points on the hyper-surface is well justified, the visualization of the relationships between them in terms of the corresponding tree can be very useful for further, more detailed studies.

The concept of topological reduction and characterization has been extended[170,172] to include systems described by a set of several potential energy hypersurfaces, namely for the purposes of description and the study of non-adiabatic reactions. Such reactions are characterized by a set of hypersurfaces and by the transition probabilities between them. The tree is constructed for each hypersurface in the usual manner. An arbitrary reaction path can then be mapped in terms of these trees and the jumps between them.

5.4.2 Topologization of Nuclear Configuration Space

Recently, Mezey[176,177] proposed two different schemes for the partitioning of nuclear configuration space R with dimension n into mutually exclusive subsets. The first scheme employs the curvature properties of the given hypersurface and leads to subsets (coordinate domains) D^i_μ characterized by the number μ of negative eigenvalues of the Hessian matrix constructed in a subspace with dimension $n - 1$. All the local minima and transition states belong to domains of the D^i_0 type. The formalism of these domains can be utilized in the discussion of the stability of the minimum energy reaction path (see Chap. 3.4). Provided that the excluded domain D_{excl} is also considered, i.e. the domain where the distance of at least two nuclei attains zero value and/or where there are no continuous second derivatives of the energy, the following complete partitioning of space R is obtained:

$$R = \bigcup_{\mu,i} D^i_\mu \cup D_{\mathrm{excl}} \,. \tag{5-59}$$

The second partitioning of R is based on the number and types of stationary (critical) points on the hypersurface and employs one of their most important properties. The extremity of every steepest descent path P_r on the hypersurface is either a critical point $r^{(l)}_c$ or a point in the excluded domain D_{excl}. Consequently, points $r \in R$ can be classified according to the critical point towards which the steepest descent path starting from these points is directed. Thus, the concept of the catchment region $C^{r_c(l)}$ of critical point $r^{(l)}_c$ is introduced. A complete partitioning of space R is then given as

$$R = \bigcup_{l} C^{r_c(l)} \bigcup C^{\bar{D}_{\mathrm{excl}}} \bigcup \bar{D}_{\mathrm{excl}} \,, \tag{5-60}$$

where $C^{\bar{D}_{\mathrm{excl}}}$ designates the catchment region of $\bar{D}_{\mathrm{excl}}$, and $\bar{D}_{\mathrm{excl}}$ represents the closure of the excluded domain D_{excl}.

Both partitionings (5-59) and (5-60) of space R consist of mutually exclusive subsets and both permit chemically significant topologization of R in a natural manner. Both the resulting topological spaces, (R, T_D) and (R, T_C), are now used for the topological analysis of selected chemical properties and concepts[167–169,178,179]; for example, molecular structure and reaction mechanism have been interpreted[179] as open sets in (R, T_C). Such an interpretation of the molecular structure is close to the concept of the structure derived from the topological properties of the molecular charge density[180–183] (cf. Chap. 3.3 and 4.1), as can be demonstrated[181] by the conjectured homeomorphism between the charge density and the potential energy.

5.4.3 Bounds for the Number of Stationary Points on Energy Hypersurfaces

A theory of critical points of functions of many variables has been developed on the basis of differential and algebraic topology leading to Morse inequalities for the lower bounds for the number of stationary or critical points[184,185]. These inequalities were introduced and adapted by Mezey[186,187] for the study of potential energy hypersurfaces. Consider subset S of the nuclear configuration space R selected to ensure the analytical behaviour of the potential energy hypersurface $U(\boldsymbol{r})$ for all $\boldsymbol{r} \in S$ ($\boldsymbol{r} = \{r_1, r_2, \ldots, r_n\}$). Stationary points will be conveniently distinguished according to the number λ of negative eigenvalues of the Hessian matrix constructed in space R. Provided there is no zero value in the spectrum of the eigenvalues of this matrix, then these are called non-degenerate critical points. The main steps in Mezey's derivation[186] of the inequalities for numbers of individual types of non-degenerate stationary points of the hypersurface, characterized according to value λ, will be discussed below. The whole procedure employs the language of set theory and leads to the enumeration of intersections of subsets of S of certain types.

Let S be provided with a coordinate system $\{e_1, e_2, \ldots, e_n\}$ and let S be an n-dimensional rectangle in terms of this coordinate system; in terms of the Cartesian product, see Chap. 5.1.1, we can formally write

$$S = S_1 \times S_2 \times \ldots \times S_n\,, \tag{5-61}$$

where the S_i are subsets of one-dimensional space ^1R described by intervals $\langle r_{i,0}; r_{i,t}\rangle$. Knowledge of the eigenvectors of the Hessian matrix of each critical point permits transfer from a coordinate system in terms of e_i to the corresponding canonical coordinates of the given critical point diagonalizing this matrix. Now, let it be possible to choose the coordinate system $\{e_i\}$ such that it holds at every non-degenerate critical point in S that $\lambda = \lambda'$, where λ' is the number of e_i coordinates along which $\partial^2 U/\partial r_i^2 < 0$. It should be noted that this condition is required only for $0 < \lambda < n$, as it is always fulfilled for local minima and maxima. Finally, the concept of faces F_0^i

and F_t^i of space S will be introduced:

$$F_\varkappa^i = \{\boldsymbol{r}; \boldsymbol{r} : r_i = r_{i,\varkappa}, \boldsymbol{r} \in S\}, \tag{5-62}$$

where $\varkappa$ designates 0 or t.

Now form sets W_0^i of all those points in S for which it holds that $\partial U/\partial r_i = 0$ and $\partial^2 U/\partial r_i^2 > 0$ and also sets W_1^i differing in terms of the second condition, viz. $\partial^2 U/$ $/\partial r_i^2 < 0$. In addition, partitioning of these two sets into disjoint subsets will be carried out:

$$W_\varepsilon^i = \bigcup_{j=1}^{\overline{m}_\varepsilon{}^i} W_\varepsilon^{i,j}, \tag{5-63}$$

$$\overline{W}_\varepsilon^{i,j} \cap \overline{W}_\varepsilon^{i,j'} \cap W_\varepsilon^i = \emptyset, \tag{5-64}$$

where ε designates 0 or 1 and the closures in relationship (5-64) are considered for all pairs j and j' (decomposition into maximum connected components). For set $W_\varepsilon^{i,j}$ the face separation property will be introduced, given by the condition that every path $P \subset S$ interconnecting faces F_0^i and F_t^i has at least one point in common with $W_\varepsilon^{i,j}$. Index arrangement is carried out in unions (5-63) such that precisely the first m_ε^i sets have the face separation property (in the subsequent discussion, it will be necessary to distinguish between indices $\overline{m}_\varepsilon^i$ and m_ε^i). Any two sets $W_\varepsilon^{i,j}$ and $W_\varepsilon^{i,j'}$ must be separated by a set $W_{1-\varepsilon}^{i,k}$, where $k \leqq m_{1-\varepsilon}^i$, as each two maxima or minima along e_i must be separated by at least one minimum or maximum, respectively. Consequently, $|m_0^i - m_1^i| \leqq 1$. As no point can be both a maximum and minimum along e_i simultaneously, it holds for each index selection that

$$W_0^{i,j} \cap W_1^{i,k} = \emptyset. \tag{5-65}$$

Consider intersections of the form

$$\bigcap_{i=1}^{n} W_{\varepsilon_i}^{i,k_i}, \tag{5-66}$$

where ε_i is 0 or 1, and $1 \leqq k_i \leqq m_{\varepsilon_i}^i$. It is apparent from the face separation property that every intersection (5-66) is non-empty. It is, however, instantly apparent from the method of introducing sets W_ε^i that every intersection (5-66) must contain at least one non-degenerate critical point $\boldsymbol{r}_c$ of hypersurface U. If various types of intersections are enumerated, lower limits are obtained for the numbers of non-degenerate critical points $\boldsymbol{r}_c^{(\lambda)}$ corresponding to the individual values of λ.

Now, introducing the union of all the intersections (5-66),

$$W = \bigcup \bigcap_{i=1}^{n} W_{\varepsilon_i}^{i,k_i}. \tag{5-67}$$

The required lower limit m_λ for the number of non-degenerate critical points of type λ on the hypersurface (which are in set W) follows from consideration of all possible choices for λ sets $W_1^{i,k}$ along λ different coordinates e_i combined with all possible choices for $n - \lambda$ sets $W_0^{i',j}$ along the remaining $n - \lambda$ different coordinates $e_{i'}$. Each such combination leads to at least one critical point. The expression for the number of various combinations is given by the relationship

$$m_\lambda = \sum_{P_{i_k}} \frac{m_1^{i_1} \ldots m_1^{i_\lambda} m_0^{i_{\lambda+1}} \ldots m_0^{i_n}}{\lambda!(n-\lambda)!}, \tag{5-68}$$

where the summation is carried out over all permutations P_{i_k} of indices $i = 1, \ldots, n$. Expression (5-68) is readily understandable: the numerator contains the products of the number of maxima along λ coordinates selected by the given permutation and of the number of minima along the remaining n-λ coordinates. The denominator considers the fact that permutation of λ (or $n - \lambda$) coordinates among themselves does not represent a contribution in the summation.

Provided that a further condition[186] is placed in hypersurface U (that should, however, usually be fulfilled), an upper limit $\overline{m}_\lambda$ to the number of non-degenerate critical points of type λ on a hypersurface can be obtained:

$$\overline{m}_\lambda = \sum_{P_{i_k}} \frac{\overline{m}_1^{i_1} \ldots \overline{m}_1^{i_\lambda} \overline{m}_0^{i_{\lambda+1}} \ldots \overline{m}_0^{i_n}}{\lambda!(n-\lambda)!}. \tag{5-69}$$

Provided that $m_\lambda = \overline{m}_\lambda$, relationships (5-68) and (5-69) would lead to determination of the precise number of non-degenerate critical points of type λ. Discussion of this, and further special cases (including periodic hypersurfaces), can be found in ref.[186].

The obtaining of limits for the number of stationary points on a hypersurface is a remarkable result of the topological approach to functions of many variables. An actual contribution to practical mapping of hypersurfaces would, however, assume that the decompositions (5-63) and (5-64) can be obtained with a sufficiently small amount of numerical information.

5.5 Symmetry Properties of Stationary Points

Symmetry has long played an essential role in the study of molecular or electronic structures. Typical uses of symmetry in molecular physics often leads to a Boolean structure of answers – 'yes – no'. Although symmetry considerations cannot replace quantum-mechanical calculations, there are broad regions where they can point out calculations useful for the given problem. Two distinct levels of utilization of the symmetry of chemical species can be distinguished (in the framework of the Born-Oppenheimer approximation). The first involves the symmetry of the electronic wave function (see for example[188–192]) of the relevant state, while the second is connected with the symmetry of the nuclear wave function, with the symmetry of the normal vibrational modes. A classic example of the utilization of symmetry rules for correlation of electronic states is contained in the results of Wigner and Witmer[193] which determine the molecular states of diatomic molecules from separate atomic states. The best known result in the field of symmetry control of chemical reactions is certainly the Woodward-Hoffmann rules[194,195], further analyzed by many other workers – see for example[196–203]. These rules are concerned with concerted processes and postulate the conservation of electronic orbital symmetry. These rules do not introduce any other limits to the size or shape of the reaction system except for the requirement of the existence of at least one symmetry element which is common to both the reactants and products. The orbital correlation diagram is then constructed considering this symmetry. If the occupied orbitals of the reactants are correlated only with the occupied orbitals of the products, the reaction is considered to be allowed. When, however, any of the occupied orbitals of the reactants is correlated with unoccupied orbitals of the products (and vice versa), the reaction is considered to be forbidden. Among other things, the numbei of isomerization reactions has been treated[194,195,204] in terms of the Woodward-Hoffmann rules. The text will now concern itself with the application of symmetry to the potential energy hypersurface or to its stationary points.

5.5.1 Symmetry Properties of Rigid and Non-rigid Molecules

The simplest approach to molecular species conceived them as rigid or quasi-rigid aggregates. It is assumed that atomic nuclei vibrate rapidly with negligibly small amplitude around a fixed equilibrium position. The symmetry properties are conventionally described using a point group reflecting the geometrical symmetry of the equilibrium nuclear configuration. It was, however, soon realized that the point group of symmetry is not the only symmetry present in a molecule and that the point group is not completely utilizable for the study of rotating molecules[205,206]. Alternatively, a transition can be made from the symmetry elements to the permutations of identical nuclei, as the complete molecular Hamiltonian is invariant under all such permutations (in contrast to the point group of symmetry). The partial results of Wilson[205,206] and Hougen[207,208] were completed by Longuet-Higgins[141], who entirely demonstrated explicitly that the permutation-inversion group is connected with the fundamental symmetries of the complete molecular Hamiltonian (molecular symmetry group), and introduced the key concept of a feasible operation as a transformation which can be achieved without passing over an insuperable barrier, i.e. within the time scale of the experimental observation. This classical analysis of Longuet-Higgins[141] provided a basis for the subsequent broad development of the study of non-rigid molecules[209–213]. The molecular symmetry group was introduced[141] as the set of all feasible (under the given conditions) permutations P of the space and spin coordinates of the identical atomic nuclei in the molecule (including the identity E) and all the feasible permutation inversions $P^* = PE^*$ (where E^* is the operation of inversion through the molecular centre of mass of the space coordinates of all the particles). The particular group of molecular symmetry of a given non-rigid molecule is a sub-group of the complete permutation-inversion group. Berry[214] interpreted (cf.[212]) the introduction of the permutation-inversion group as a process of induction from the point group to an empirically determined supergroup generated from the point group by a finite set of feasible generators.

Study of the symmetry properties of individual non-rigid molecules soon revealed that the corresponding permutation-inversion groups are quite large (typically[212] $10-10^3$ elements – cf., for example, Fig. 5-5) and also are not very common (isomorphous with the usual groups). This led to attempts to formulate groups of non-rigid molecules as semi-direct products. An example is the technique of isodynamic operations introduced by Altmann[216] or the isometric group approach developed

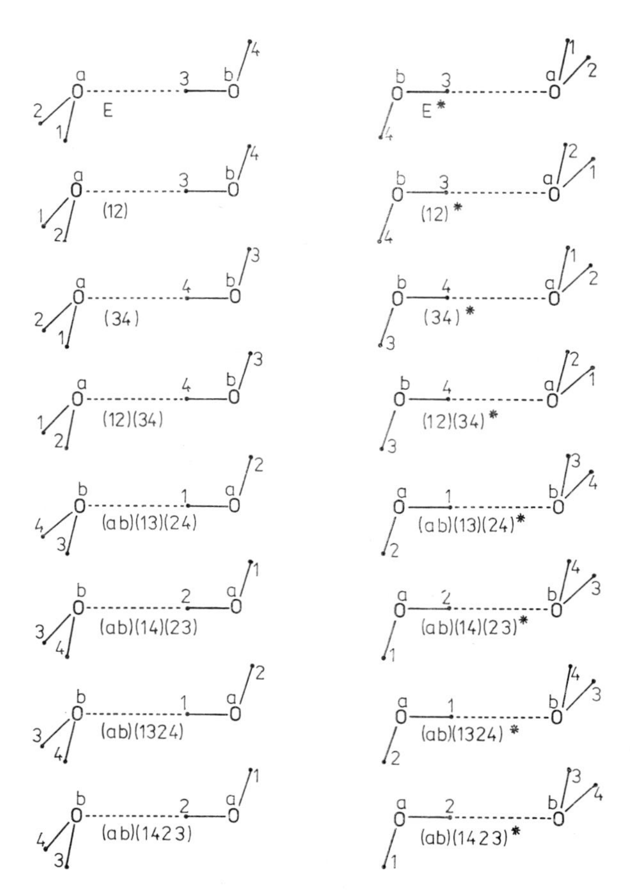

Figure 5-5. Effect of each permutation-inversion group operation on the water dimer molecule (from Ref.[215]). The group of these 16 operations is isomorphic with the D_{4h} point group. The point group C_s of the dimer has two elements (E and a plane of symmetry)

by Günthard *et al.*[217–219]. It should be noted that the relationship between permutation-inversion and the point groups is not strictly hierarchical, but rather complementary[211]. The point group of each non-linear molecule is isomorphous with the subgroup of the complete permutation-inversion group formed by the completely feasible elements[209–213]. Provided that the discussion is limited to classification of the vibronic (rather than the rovibronic) wave functions, both types of groups lead to the same classification results. This permits, for example, work with point groups of symmetry in discussion of the properties of the energy hypersurface related to the normal vibrational modes.

Another approach towards a more complete specification of molecular symmetry than that attained so far using point groups of symmetry alone involves the concept of a framework group proposed by Pople[220], which is especially useful for computer handling with structure-symmetry relations. It has been demonstrated[221] that the framework group for non-rigid molecules can readily be extended to the molecular symmetry group. The formalism of the framework group was simplified[222] in terms of the site symmetry groups[223–225].

Introduction of the concept of feasible permutations is closely connected to the understanding of the isomerism of non-rigid molecules. The existence of a larger number of geometrically non-equivalent minima on the hypersurface represents a more complicated situation[226] than that originally conceived by Longuet-Higgins[141]. Of all types of symmetry operations[141] of the complete Hamiltonian, only two (the simultaneous inversion of the positions of all the particles in the centre of mass and any permutation of the positions and spins of any set of identical nuclei) can be altered for a free chemical isomer[228]. This leads to the concept of the internal symmetry structure[227,228] of the complete Hamiltonian, that is entirely described by the group of all nuclear permutations and of the spatial inversion. However, the isomers are not associated with the irreducible representation of this group, as a particular isomer is not described by the complete Hamiltonian but by a local Hamiltonian partially localized in the corresponding domain of the configurational space. In every such domain, the complete Hamiltonian is approximated by the localized Hamiltonian invariant for permutations that are feasible without leaving that domain. Thus, operations with respect to which the complete Hamiltonian is invariant are split into operations leaving a localized Hamiltonian invariant and operations mutually transforming the localized Hamiltonians. Beginning from the complete permutation-inversion group (i.e. from the completely flexible molecule), the symmetry can be reduced by gradually forbidding some of the permutations of identical nuclei and/or the space inversion. Each step in this process yields a set of non-rigid isomers. Gradual carrying out of this forbidding process yields a set of groups embedded in one another that, at some stage of this process, become isomorphous with groups describing the symmetry of rigid species. It is apparent that every non-rigid isomer is characterized by its own Longuet-Higgins group of feasible permutations (i.e. allowed in our restriction scheme).

The symmetry properties of isomers, or of degenerate rearrangements, were studied in a number of works, see e.g.[22,38,218,228–233]. Gilles and Philippot[228] studied the structure of the Longuet-Higgins group, proposed a method for mapping all types of non-rigidity in the given system and utilized the introduction of the technique of generators for the classification of isomerization processes. Utilization of the groups of non-rigid molecules permitted direct rationalizations, for example, of knowledge concerning thermal nitroxide formation[229] or conditions during the rearrangement of $C_2H_5^+$ and $C_5H_5^+$ ions[231].

A situation described by the Jahn-Teller theorem[234], stating that the electronic wave function in an energy minimum cannot exhibit spatial degeneracy (except for linear molecular configurations) can be considered to be a special case of symmetry-conditioned non-rigidity. Thus, the potential energy hypersurface must exhibit a non-zero gradient for a degenerate electronic state, resulting in distortion of the configuration leading to this degeneracy (see for example[235–237]). The Jahn-Teller distortion leading to stabiliization of the species can lead to the production of an isomeric system[283–240] for which the name Jahn-Teller isomers has been proposed[238]. It should be noted, for the sake of completeness, that this isomerism can be considered to be a special case of orbital isomerism[241] based on alternative occupation of distinguishable orbitals by electrons.

5.5.2 Transition States and Symmetry

A number of rigorous results have been obtained concerning the symmetry of transition states. This is a result of a certain exceptional position of transition states among other types of stationary points. In simple chemical reactions, the saddle points on the hypersurface can be connected directly with the reactants and products, e.g. by the steepest descent path; the matrix of the second derivatives of the energy with respect to the coordinates in saddle points exhibits only one negative eigenvalue. It is thus intuitively clear that the symmetry of the transition state should reflect these factors in some way and should be related to the symmetry of the reactants and products (cf.[237,242,243]).

The Murrell-Laidler theorem[244] can be considered as an extremely important result in this connection; this theorem states that the conventional vibrational FG matrix analysis of the transition state can lead to only one negative* eigenvalue. To retain logical consistency, however, a transition state must now be defined in energy terms alone (the lowest energy barrier separating the reactants and products) and its differential properties should be obtained deductively. Alternatively, the Murrell-Laidler theorem can be formulated as stating that only one reaction path can lead through the transition state (provided that the possibility of its splitting in the region between

* Or imaginary, if the level of vibrational frequences is considered.

the transition state and the reactants or products is not considered). Proof through contradiction is readily carried out: assume that the given transition state contains two normal vibrational coordinates Q_r and Q_s, both with negative eigenvalues. Consider points $[Q_r^0 \pm \delta Q_r, Q_s^0 \pm \delta Q_s]$ lying in the region around transition state $[Q_r^0, Q_s^0]$ (all the other coordinates are kept constant). The Taylor expansion (regardless of the choice of signs) yields

$$U(Q_r^0 \pm \delta Q_r, Q_s^0 \pm \delta Q_s) \approx U(Q_r^0, Q_s^0) + \frac{1}{2}\left(\frac{\partial^2 U}{\partial Q_r^2}\right)_0 \delta Q_r^2 + $$

$$+ \frac{1}{2}\left(\frac{\partial^2 U}{\partial Q_s^2}\right)_0 \delta Q_s^2 < U(Q_r^0, Q_s^0) \,. \tag{5-70}$$

Naturally, points $[Q_r^0 \pm \delta Q_r, Q_s^0]$; $[Q_r^0, Q_s^0 \pm \delta Q_s]$ also lie energetically lower than the transition state. It is then apparent if the pathway passing through points $[Q_r^0 - \delta Q_r, Q_s^0]$; $[Q_r^0, Q_s^0]$; $[Q_r^0 + \delta Q_r, Q_s^0]$ is replaced by pathway $[Q_r^0 - \delta Q_r, Q_s^0]$; $[Q_r^0 - \delta Q_r, Q_s^0 - \delta Q_s]$; $[Q_r^0, Q_s^0 - \delta Q_s]$; $[Q_r^0 + \delta Q_r, Q_s^0 - \delta Q_s]$; $[Q_r^0 + \delta Q_r, Q_s^0]$ avoiding the transition state, the latter pathway is the lower of the two. This contradicts the definition of a transition state and consequently only one reaction path can actually pass through the given transition state. An elementary application of the Murrell-Laidler theorem can be demonstrated on solution of the symmetry[242] of the activated complex in the autoisomerization of ozone:

$$O_1{-}O_2{-}O_3 \rightleftarrows O_2{-}O_1{-}O_3 \rightleftarrows O_2{-}O_3{-}O_1 \,. \tag{5-71}$$

The stable form of ozone is represented by an open C_{2v} structure with a valence angle[245] close to $\frac{2}{3}\pi$. A structure with symmetry D_{3h} can be suggested as the activated complex taking part in this interconversion. The symmetry structure of the vibrational modes of this species[242] is $\Gamma_{D_{3h}} = A_1' + E'$. However, the totally symmetric mode A_1' cannot be suggested as a possible reaction coordinate, as it represents dissociation into three oxygen atoms. From this point of view, both degenerate modes with symmetry E' are suitable, but they are not compatible with the Murrell-Laidler theorem as they would represent two orthogonal directions with negative eigenvalues. Thus the structure with symmetry D_{3h} is excluded as a possible transition state but can be considered as a local minimum and/or higher type of saddle point or as a maximum. In actual fact, a quantum-chemical study has demonstrated[246] the existence of a local minimum with D_{3h} symmetry and thus isomerism of ozone. As is apparent from symmetry considerations, the search for a transition state with the latter point group is bound to be unsuccessful; the activated complex must exhibit lower symmetry. It should be noted that the transition state for reaction (5-71) has been shown[247] to be a structure with C_{2v} symmetry. However, when passing to a transition state with one or more zero eigenvalues, some peculiarities can appear in connection with the contribution of cubic or higher terms to the Taylor expansion (5-70). For example, when both eigenvalues disappear for a two-dimensional system, the corresponding hypersurface can be formed[248] by three valleys meeting at a point (called a 'monkey saddle'). For example, this situation was indicated[249] for the isomerizations in the $C_5H_5^-$ system. However, detailed re-analysis of the results[249] has demonstrated that the assumed region of a transition state with symmetry D_{3h} is formed by a maximum with this symmetry and three equivalent, but common, transition states, each corresponding to one reaction path and clearly displaced from the central maximum.

Stanton and McIver[248,250–252] further developed, formalized and illustrated the Murrell-Laidler theorem. For formulation of the symmetry rules they used the concept of a transition vector, designating the normal vibrational coordinate of the transition state with a negative eigenvalue. In terms of a transition vector, their rules are[248]: (i) the transition vector cannot belong to a degenerate representation of the transition state point group; (ii) the transition vector is antisymmetric under every transition state symmetry operation which converts the reactants to the products; and (iii) the transition vector is symmetric with respect to every operation that leaves either the reactants or the products unchanged, or changes the reactants into equivalent reactants or products into equivalent products. A special consequence of rule (ii) is, for example, the fact that a structure that would exhibit a three-fold (or higher odd) axis of rotation, converting the reactants into the products, cannot act as a transition state. For example, of the three following possible activated complexes in the four-centre reaction of $H_2 + H_2$, i.e. structures with symmetry D_{3h}, T_d and D_{4h}, the first two can be immediately excluded after analysis of the corresponding diagrams (so that the calculation need not be carried out for the species[253]). Similarly, rule (ii) demonstrates that, in narcissistic reactions[254,255] (i.e. in processes where the reactant and product are interconverted through reflection or S_n rotation), the transition vector must be antisymmetric with respect to this operation, provided it is a transition state symmetry operation. It should be noted that symmetry rules are useful not only for the selection of structures that could act as transition states, but that they can also simplify the numerical side of localization of stationary points on the hypersurface (cf.[256–258]). For example[248], it follows from rule (ii) that, if the group of the transition state includes a symmetry operation that converts the reactants to products, then the energy of the transition state will be a local minimum in the subspace of coordinates belonging to the totally symmetric representation of the group (conversion of the search for a transition state into energy minimization).

Pechukas[259] used somewhat more specific assumptions, i.e. he analyzed a situation in which the transition state is linked directly to the reactants and products by paths of steepest descent and obtained more strict rules, i.e. the rules for conservation of nuclear symmetry along the path of steepest descent. He demonstrated that the symmetry of the transition state is the symmetry of the entire path from the reactants through the transition state to the products, i.e. that the symmetry groups of the reactants and products must contain the symmetry group of the transition state linking them. Processes in which the reactant and product are physically indistinguishable, when the transition state can exhibit additional symmetries involving interchange of the reactant and product configurations, are an exception to this rule. Notably, it follows from the requirement that the symmetry of the transition state between physically distinguishable reactants and products must be included in their own symmetry that, for example[259]: (a) linear transition states must yield linear reactants and products, (b) planar transition states must yield planar reactants and products, (c) optically inactive transition states must yield optically inactive reactants and products, and (d) the point group of symmetry of the transition state can be no larger than the largest subgroup common to the point groups of both the reactants and products, etc. It should be noted that, even though the above rules for the symmetry of the activated ocmplexes were originally formulated in terms of point groups of symmetry, conversion into the terms of the Longuett-Higgins groups has also been given[231].

Formulation of the Murrell-Laidler theorem was originally inspired[244] by the once--important problem of the relationships and usefulness of symmetry numbers and statistical factors for the construction of rotational partition functions (see Chap. 6.1.2). Although it is now[260,261] apparent that the symmetry numbers are the correct objects for the calculation of the one-way rate passage through a transition state, the theory[262-264] of statistical factors is a technique that has certainly contributed considerably to the understanding of questions connected with the symmetry of transition states. In addition, these statistical factors can be used[244,265] as an alternative means for exclusion of some activated complex structures.

The Jahn-Teller theorem[234] can also be used for selection among possible transition state structures. The possible use of this theorem in connection with transition states need not be trivially apparent (it could appear that the Jahn-Teller distortion occurs along the transition vector); nonetheless, it is readily justifiable[266] and follows from the existence of a direction in which the energy decrease is linear in terms of the distortion. Use of the Jahn-Teller theorem permits (independent of the above arguments) exclusion, for example, of the T_d structure from possible participation as an activated complex in the four-centre reaction, $H_2 + H_2$, as this symmetry leads to a degenerate electronic configuration. Mixing of close but non-degenerate electronic states leads to a pseudo-Jahn-Teller effect[267], in which the change in energy is not a linear but a quadratic function of the distortion. Utilization of the pseudo-Jahn--Teller effect to predict transition states is important for structures with low-lying electronic excited states[268-273].

In conclusion, it should be noted for completeness that the use of symmetry rules assumes a certain non-trivial symmetry of the transition state, and is not useful for evaluating non-symmetric structures. As was pointed out by McIver[252], there is no *a priori* reason why activated complexes should be symmetric. Consequently, the possibility of a non-symmetric structure should always be considered. It is especially necessary for a symmetry forbidden reaction to test whether the barrier cannot be lowered by considering a path of lower symmetry. This is true of the process

$$H_2 + CH_2 \rightarrow CH_4\,, \tag{5-72}$$

which the Woodward-Hoffmann rules state to be symmetry-forbidden if the reaction path retains C_{2v} symmetry. However, the barrier can be lowered considerably[274] if, for example, only C_s symmetry is required.

5.6 Summary

The recent development of algebraic chemistry convincingly demonstrates that the use of algebraic methods is gradually becoming a useful complement to other methods for studying potential energy hypersurfaces. This is especially true of the representation of these surfaces in terms of stationary points, which is, at present, the most frequently used method for quantum-chemical description of chemical systems. The use of abstract algebraic techniques is useful because of the following main characteristics: their general approach, the possibility of obtaining direct insight into a logical structure of chemical problems, and the rigorous nature of the qualitative predictions in the framework of the (usually simple) model used. All these factors are completely valid in relation to the algebraic study of isomerism. Algebraic chemistry is gradually improving the procedures for determining the numbers of stationary points on hypersurfaces, their classification and mapping of relationships between them. Numbers of stationary points can at present be deduced primarily from enumeration studies, based either on Pólya's theorem, on iterative techniques or on other algorithms. However, this approach does not permit the classification of stationary points into individual types. This deficiency is not found in a new procedure based on set theory or graph-theoretical decompositions of the configurational space considering some of its differential properties. The individual algebraic techniques differ in their effectiveness for individual types of systems and problems – while graph theory is more universally applicable, group theory leads to useful results primarily in situations with sufficiently high symmetry. The usefulness of algebraic predictions most certainly depends on the degree of physical reality included in the model used. An example of an application that successfully retains a balance between generality of approach and retention of sufficient physical reality can be found, for example, in the enumeration of non-rigid isomers. In addition to the enumeration studies themselves, systematic algebraic analysis of complex kinetic schemes is becoming important, especially of isomerization processes: this often represents the most effective pathway to the determination of rules and to completeness of predictions. This can be exemplified, for example, by recent enumeration studies of reaction pathways (using Burnside's lemma and Pólya's theorem[92,93,275–278] as well as the double coset formalism[10,11]). Algebraic generalization of the concept of isomerism, originally proposed in connection with a non-empirical approach to computer-assisted synthesis, anticipated the quantum-mechanical or quantum-chemical view of isomerism, as all species isomeric within this generalized concept belong to the same total Hamiltonian. Consideration in terms of isomeric ensembles of molecules treats all chemical processes as generalized isomerizations and thus confirms the universality of the concept of isomerism. A number of algebraic approaches also highlight and shed new light on the problem of the distinguishability or non-distinguishability of isomeric structures: a key question in the modern theory and experimental approach to isomeric chemistry.

REFERENCES

(1) Rouvray, D. H.: 1975, Endeavour *34*, pp. 28—33.
(2) Cayley, A.: 1874, Phil. Mag. *47*, pp. 444—447.
(3) Koerner, W.: 1874, Gazz. Chim. Ital. *4*, pp. 305—446.
(4) van't Hoff, J. H.: 1874, Voorstel tot Uitbreiding der Tegenwoording in de Scheikunde gebruikte Structuur-Formules in de Ruimte, benevens een daarmeê samenhangende Opmerking omtrent het Verband tusschen Optisch Actief Vermogen en Chemische Constitutie van Organische Verbindingen, Greven, Utrecht.
(5) Rouvray, D. H.: 1974, Chem. Brit. *10*, pp. 11—15.
(6) Bader, R. F. W., and Nguyen-Dang, T. T.: 1982, Advan. Quantum Chem. *14*, 63—124.
(7) Ege, G.: 1971, Naturwissenschaften *58*, pp. 247—257.
(8) Ugi, I., Gillespie, P., and Gillespie, C.: 1972, Trans. N. Y. Acad. Sci. *34*, pp. 416—432.
(9) Ruch, E.: 1968, Theor. Chim. Acta *11*, pp. 183—192.
(10) Ruch, E., Hässelbarth, W., and Richter, B.: 1970, Theor. Chim. Acta *19*, pp. 288—300.
(11) Hässelbarth, W., and Ruch, E.: 1973, Theor. Chim. Acta *29*, pp. 259—268.
(12) Mislow, K.: 1977, Bull. Soc. Chim. Belge *86*, pp. 595—601.
(13) Bonchev, D., Mekenyan, Ov., and Trinajstić, N.: 1981, J. Comput. Chem. *2*, pp. 127 to 148.
(14) Brocas, J.: 1972, Fortschr. Chem. Forsch. *32*, pp. 43—61.
(15) Balaban, A. T.: 1973, Rev. Roum. Chim. *18*, pp. 841—854.
(16) Nourse, J. G., and Mislow K.: 1975, J. Am. Chem. Soc. *97*, pp. 4571—4578.
(17) Nourse, J. G.: 1977, J. Am. Chem. Soc. *99*, pp. 2063—2069.
(18) Randić, M.: 1977, Croat. Chem. Acta *49*, pp. 643—655.
(19) Klein, D. J., and Cowley, A. H.: 1978, J. Am. Chem. Soc. *100*, pp. 2593—2599.
(20) Randić, M.: 1979, Int. J. Quantum Chem. *15*, pp. 663—682.
(21) Lloyd, E. K.: 1979, Match *7*, pp. 255—271.
(22) Nourse, J. G.: 1979, in The Permutation Group in Physics and Chemistry, Ed. J. Hinze, pp. 28—37, Springer-Verlag, Berlin.
(23) Brocas, J., Willem, R., Buschen, J., and Fastenakel, D.: 1979, Bull. Soc. Chim. Belg. *88*, pp. 415—434.
(24) Randić, M.: 1980, Int. J. Quantum Chem., Quantum Chem. Symp. *14*, pp. 557—577.
(25) Nourse, J. G.: 1980, J. Am. Chem. Soc. *102*, pp. 4883—4889.
(26) King, R. B.: 1981, Theor. Chim. Acta *59*, pp. 25—45.
(27) Kier, L. B., and Hall, L. H.: 1976, Molecular Connectivity in Chemistry and Drug Research, Academic Press, New York.
(28) Balaban, A. T., Chiriac, A., Motoc, I., and Simon, Z.: 1980, Steric Fit in Quantitative Structure-Activity Relations, Springer-Verlag, Berlin.

(29) ROUVRAY, D. H.: 1973, Am. Scientist *61*, pp. 729—735.
(30) GRAOVAC, A., GUTMAN, I., and TRINAJSTIĆ, N.: 1977, Topological Approach to the Chemistry of Conjugated Molecules, Springer-Verlag, Berlin
(31) BONCHEV, D., and TRINAJSTIĆ, N.: 1977, J. Chem. Phys. *67*, pp. 4517—4533.
(32) BONCHEV, D.: 1979, Match *7*, pp. 65—112.
(33) MEKENYAN, O., BONCHEV, D., and TRINAJSTIĆ, N.: 1980, Int. J. Quantum Chem. *18*, pp. 369—380.
(34) ROUVRAY, D. H.: 1973, Chem. Tech. *3*, pp. 379—384.
(35) ESSAM, J. W., KENNEDY, J. W., GORDON M., and WHITTLE, P.: 1977, J. Chem. Soc., Faraday Trans. *II 73*, pp. 1289—1307.
(36) BALABAN, A. T.: 1967, Rev. Roum. Chim. *12*, pp. 875—898.
(37) MAC LANE, S., and BIRKHOFF, G.: 1968, Algebra, The Macmillan Company, New York.
(38) DUGUNDJI, J., GILLESPIE, P., MARQUARDING, D., UGI, I., and RAMIREZ, F.: 1976, in Chemical Applications of Graph Theory, Ed. A. T. Balaban, pp. 107—174, Academic Press, London.
(39) UGI, I., and GILLESPIE, P.: 1971, Angew. Chem., Int. Ed. Engl. *10*, pp. 914—915.
(40) DUGUNDJI, J., and UGI, I.: 1973, Fortschr. Chem. Forsch. *39*, pp. 19—64.
(41) UGI, I., BAUER, J., BRANDT, J., FRIEDRICH, J., GASTEIGER, J., JOCHUM, C., and SCHUBERT, W.: 1979, Angew. Chem., Int. Ed. Engl. *18*, pp. 111—123.
(42) KRATOCHVÍL, M.: 1981, Chem. Listy *75*, pp. 673—698.
(43) PREUSS, H.: 1969, Int. J. Quantum Chem. *3*, pp. 123—130.
(44) PREUSS, H.: 1969, Int. J. Quantum Chem. *3*, pp. 131—139.
(45) SLANINA, Z.: 1981, Advan. Quantum Chem. *13*, pp. 89—153.
(46) BOURBAKI, N.: 1966, Elements of Mathematics. General Topology, Addison-Wesley, Don Mills, Ontario.
(47) DUGUNDJI, J.: 1966, Topology, Allyn and Bacon, Boston.
(48) ENGELKING, R.: 1977, General Topology, PWN, Warszawa.
(49) KVASNIČKA, V.: 1983, Collect. Czech. Chem. Commun. *48*, pp. 2097—2117; pp. 2118—2129.
(50) UGI, I., and GILLESPIE, P.: 1971, Angew. Chem., Int. Ed. Engl. *10*, pp. 915—919.
(51) JOCHUM, C., GASTEIGER, J., and UGI, I.: to be published.
(52) SINANOĞLU, O.: 1975, J. Am. Chem. Soc. *97*, pp. 2309—2320.
(53) SINANOĞLU, O.: 1976, Acta Chim. Turcica *3*, p. 155—161.
(54) SINANOĞLU, O., and LEE, L.-S.: 1976, Theor. Chim. Acta *48*, pp. 287—299.
(55) SINANOĞLU, O., and LEE, L.-S.: 1979, Theoret. Chim. Acta *51*, pp. 1—9.
(56) NEMES, I., VIDÓCZY, T., BOTÁR, L., and GÁL, D.: 1977, Theor. Chim. Acta *45*, pp. 215—223.
(57) NEMES, I., VIDÓCZY, T., BOTÁR, L., and GÁL, D.: 1977, Theor. Chim. Acta *45*, pp. 225—233.
(58) NEMES, I., VIDÓCZY, T., and GÁL, D.: 1977, Theor. Chim. Acta *46*, pp. 243—250.
(59) COREY, E. J., and WIPKE, W. T.: 1969, Science *166*, pp. 178—192.
(60) BERSOHN, M., and ESACK, A.: 1976, Chem. Rev. 76, pp. 269—282.
(61) 1977, Computer-Assisted Organic Synthesis, Ed. W. T. Wipke, and W. J. Howe, ACS Symp. Ser. 61, American Chemical Society, Washington.
(62) LINDSAY, R. K., BUCHANAN, B. G., FEIGENBAUM, E. A., and LEDERBERG, J.: 1980, Applications of Artificial Intelligence for Organic Chemistry: The DENDRAL Project, McGraw-Hill, New York.
(63) BENEŠ, J., KRATOCHVÍL, M., and ŠLEZSAR, L.: 1981, Chem. Prům. *31*, pp. 379—388.
(64) COREY, E. J., ORF, H. W., and PENSAK, D. A.: 1976, J. Am. Chem. Soc. *98*, pp. 210—221.
(65) PENSAK, D. A., and COREY E. J.: 1977, in Ref.[61], pp. 1—32.
(66) GELERNTER, H., SRIDHARAN, N. S., HART, A. J., YEN, S. C., FOWLER F. W., and SHUE, H. J.: 1973, Fortschr. Chem. Forsch. *41*, pp. 113—150.
(67) WIPKE, W. T., and GUND, P.: 1976, J. Am. Chem. Soc. *98*, pp. 8107—8118.

(68) Wipke, W. T., Braun, H., Smith, G., Choplin, F., and Sieber, W.: 1977, in Ref.[61], pp. 97—127.
(69) Hendrickson, J. B.: 1971, J. Am. Chem. Soc. *93*, pp. 6847—6854.
(70) Hendrickson, J. B.: 1971, J. Am. Chem. Soc. *93*, pp. 6854—6862.
(71) Thakkar, A. J.: 1973, Fortschr. Chem. Forsch. *39*, pp. 3—18.
(72) Blair, J., Gasteiger, J., Gillespie, C., Gillespie, P. D., and Ugi, I.: 1974, Tetrahedron *30*, pp. 1845—1859.
(73) Weise, A.: 1975, Z. Chem. *15*, pp. 333—340.
(74) Brandt, J., Friedrich, J., Gasteiger, J., Jochum, C., Schubert, W., and Ugi, I.: 1977, in Ref.[61], pp. 33—59.
(75) Gasteiger, J., and Jochum, C.: 1978, Top. Curr. Chem. *74*, pp. 93—126.
(76) Friedrich, J., and Ugi, I.: 1980, J. Chem. Res. (S) *70*.
(77) Rouvray, D. H.: 1971, R. I. C. Rev. *4*, 173—195.
(78) Rouvray, D. H., and Balaban, A. T.: 1979, in Applications of Graph Theory, Ed. R. J. Wilson, and L. W. Beineke, pp. 177—221, Academic Press, London.
(79) Slanina Z.: 1978, Chem. Listy *72*, pp. 1—27.
(80) Morgan, H. L.: 1965, J. Chem. Doc. *5*, pp. 107—113.
(81) Randić, M.: 1975, J. Chem. Info. Comput. Sci. *15*, pp. 105—108.
(82) Randić, M.: 1977, J. Chem. Info. Comput. Sci. *17*, pp. 171—180.
(83) Lederberg, J., Sutherland, G. L., Buchanan, B. G., Feigenbaum, E. A., Robertson, A. V., Duffield, A. M., and Djerassi, C.: 1969, J. Am. Chem. Soc. *91*, pp. 2973—2976.
(84) Masinter, L. M., Sridharan, N. S., Lederberg, J., and Smith, D. H.: 1974, J. Am. Chem. Soc. *96*, pp. 7702—7714.
(85) Masinter, L. M., Sridharan, N. S., Carhart, R. E., and Smith, D. H.: 1974, J. Am. Chem. Soc. *96*, pp. 7714—7723.
(86) Smith, D. H.: 1975, Anal. Chem. *47*, pp. 1176—1179.
(87) Carhart, R. E., Smith, D. H., Brown, H., and Sridharan, N. S.: 1975, J. Chem. Info. Comput. Sci. *15*, pp. 124—130.
(88) Smith, D. H.: 1975, J. Chem. Info. Comput. Sci. *15*, pp. 203—207.
(89) Randić, M.: 1978, Acta Cryst. *A 34*, pp. 275—282.
(90) Nourse, J. G., and Smith, D. H.: 1979, Match *6*, pp. 259—275.
(91) Nourse, J. G.: 1979, in The Permutation Group in Physics and Chemistry, Ed. J. Hinze, pp. 19—27, Springer-Verlag, Berlin.
(92) Klemperer, W. G.: 1973, J. Am. Chem. Soc. *95*, pp. 380—396.
(93) Klemperer, W. G.: 1973, J. Am. Chem. Soc. *95*, pp. 2105—2120.
(94) Carhart, R. E., Smith, D. H., Brown, H., and Djerassi, C.: 1975, J. Am. Chem. Soc. *97*, pp. 5755—5762.
(95) Hutchings, M. G., Johnson, J. B., Klemperer, W. G., and Knight III, R. R.: 1977, J. Am. Chem. Soc. *99*, pp. 7126—7132.
(96) Cone, M. M., Venkataraghavan, R., and McLafferty, F. W.: 1977, J. Am. Chem. Soc. *99*, pp. 7668—7671.
(97) Nourse, J. G., Smith, D. H., Carhart, R. E., and Djerassi, C.: 1980, J. Am. Chem. Soc. *102*, pp. 6289—6295.
(98) Balasubramanian, K., Kaufman, J. J., Koski, W. S., and Balaban, A. T.: 1980, J. Comput. Chem. *1*, pp. 149—157.
(99) Knop, J. V., Szymanski, K., Jeričević, Ž., and Trinajstić, N.: 1983, J. Comput. Chem. *4*, pp. 23—32; Int. J. Quantum Chem. *23*, pp. 713—722.
(100) Balaban, A. T., Fărcașiu, D., and Bănică, R.: 1966, Rev. Roum. Chim. *11*, pp. 1205 to 1227.
(101) Dubois, J.-E., and Panaye, A.: 1969, Tetrahedron Lett. *19*, pp. 1501—1504.

(102) WHITLOCK Jr., H. W., and SIEFKEN, M. V.: 1968, J. Am. Chem. Soc. *90*, pp. 4929—4939.
(103) GUND, T. M., SCHLEYER, P. v. R., GUND, P. H., and WIPKE, W. T.: 1975, J. Am. Chem. Soc. *97*, pp. 743—751.
(104) JOHNSON, C. K., and COLLINS, C. J.: 1974, J. Am. Chem. Soc. *96*, pp. 2514—2523.
(105) COLLINS, C. J., JOHNSON, C. K., and RAAEN, V. F.: 1974, J. Am. Chem. Soc. *96*, pp. 2524 to 2531.
(106) MISLOW, K., GUST, D., FINOCCHIARO, P., and BOETTCHER, R. J.: 1974, Top. Curr. Chem. *47*, pp. 1—28.
(107) BALABAN, A. T.: 1977, Rev. Roum. Chim. *22*, pp. 243—255.
(108) UGI, I., MARQUARDING, D., KLUSACEK, H., GOKEL, G., and GILLESPIE, P.: 1970, Angew. Chem., Int. Ed. Engl. *9*, pp. 703—730.
(109) GIELEN, M.: 1976, in Chemical Applications of Graph Theory, Ed. A. T. BALABAN, pp. 261—298, Academic Press, London.
(110) BERRY, R. S.: 1960, J. Chem. Phys. *32*, pp. 933—938.
(111) UGI, I., MARQUARDING, D., KLUSACEK, H., GILLESPIE, P., and RAMIREZ, F.: 1971, Accounts Chem. Res. *4*, pp. 288—296.
(112) BERTZ, S. H.: 1981, J. Am. Chem. Soc. *103*, pp. 3599—3601.
(113) BALABAN, A. T.: 1982, Chem. Phys. Lett. *89*, pp. 399—404.
(114) HARARY, F., and PALMER, E. M.: 1973, Graphical Enumeration, Academic Press, New York.
(115) SEDLÁČEK, J.: 1972, Einführung in die Graphentheorie, Teubner, Leipzig.
(116) ESSAM, J. W., and FISHER, M. E.: 1970, Rev. Mod. Phys. *42*, pp. 272—288.
(117) ROUVRAY, D. H.: 1974, Chem. Soc. Rev. *3*, pp. 355—372.
(118) BALABAN, A. T., and HARARY, F.: 1976, in Chemical Applications of Graph Theory, Ed. A. T. Balaban, pp. 1—4, Academic Press, London.
(119) ROUVRAY, D. H.: 1977, Chem. Brit. *13*, pp. 52—57.
(120) CAYLEY, A.: 1857, Phil. Mag. *13*, pp. 172—176.
(121) REDFIELD, J. H.: 1927, Am. J. Math. *49*, pp. 433—455.
(122) LUNN, A. C., and SENIOR, J. K.: 1929, J. Phys. Chem. *33*, pp. 1027—1079.
(123) HALL JR., M.: 1959, The Theory of Groups, The Macmillan Company, New York.
(124) PÓLYA, G.: 1935, C. R. Acad. Sci. *201*, pp. 1167—1169.
(125) BIGGS, N. L., LLOYD, E. K., and WILSON, R, J.: 1976, Graph Theory 1736—1936, Clarendon Press, Oxford.
(126) DAVIDSON, R. A.: 1981, J. Am. Chem. Soc. *103*, pp. 312—314.
(127) MCBRIDE, J. M.: 1980, J. Am. Chem. Soc. *102*, pp. 4134—4137.
(128) BALASUBRAMANIAN, K.: 1981, Theor. Chim. Acta *59*, pp. 91—93.
(129) PÓLYA, G.: 1936, Z. Kristallogr., Kristallgeometrie, Kristallphys., Kristallchem. *93*, pp. 415—443.
(130) PÓLYA, G.: 1937, Acta Math. *68*, pp. 145—254.
(131) HARARY, F., PALMER, E. M., ROBINSON, R. W., and READ, R. C.: 1976, in Chemical Applications of Graph Theory, Ed. A. T. Balaban, pp. 11—24, Academic Press, London.
(132) HENZE, H. R., and BLAIR, C. M.: 1931, J. Am. Chem. Soc. *53*, pp. 3042—3046.
(133) HENZE, H. R., and BLAIR, C. M.: 1931, J. Am. Chem. Soc. *53*, pp. 3077—3085.
(134) READ, R. C.: 1976, in Chemical Applications of Graph Theory, Ed. A. T. Balaban, pp. 25—61, Academic Press, London.
(135) ROBINSON, R. W., HARARY, F., and BALABAN, A. T.: 1976, Tetrahedron *32*, pp. 355—361.
(136) BALASUBRAMANIAN, K.: 1978, Indian J. Chem. *16B*, pp. 1094—1096.
(137) BALASUBRAMANIAN, K.: 1979, Theor. Chim. Acta *51*, pp. 37—54.
(138) LEONARD, J. E., HAMMOND, G. S., and SIMMONS, H. E.: 1975, J. Am. Chem. Soc. *97*, pp. 5052—5054.
(139) LEONARD, J. E.: 1977, J. Phys. Chem. *81*, pp. 2212—2214.

(140) Balasubramanian, K.: 1979, Theor. Chim. Acta *53*, pp. 129—146.
(141) Longuet-Higgins, H. C.: 1963, Mol. Phys. *6*, pp. 445—460.
(142) Balasubramanian, K.: 1980, J. Chem. Phys. *72*, pp. 665—677.
(143) Balasubramanian, K.: 1980, J. Chem. Phys. *73*, pp. 3321—3337.
(144) Flurry Jr., R. L.: 1976, J. Phys. Chem. *80*, pp. 777—778.
(145) Gordon, M., and Parker, T. G.: 1971, Proc. Roy. Soc. A, Edinburgh *69*, pp. 181—197.
(146) Gordon, M., and Kennedy, J. W.: 1973, J. Chem. Soc., Faraday Trans. *II 69*, pp. 484 to 504.
(147) Gordon, M., and Temple, W. B.: 1972, Makromol. Chem. *160*, pp. 263—276.
(148) Kajiwara, K., and Gordon, M.: 1973, J. Chem. Phys. *59*, pp. 3623—3632.
(149) Gordon, M., Ross-Murphy, S. B., and Suzuki, H.: 1976, Eur. Polym. J. *12*, pp. 733—740.
(150) Gordon, M., Kapadia, P., and Malakis, A.: 1976, J. Phys. *A 9*, pp. 751—769.
(151) Gordon, M., and Temple, W. B.: 1976, in Chemical Applications of Graph Theory, Ed. A. T. Balaban, pp. 300—332, Academic Press, London.
(152) Gordon, M.: 1970, J. Chem. Soc. *A*, pp. 737—740.
(153) Gordon, M., and Temple, W. B.: 1973, J. Chem. Soc., Faraday Trans. *II 69*, pp. 282—297.
(154) Gordon, M., and Temple, W. B.: 1970, J. Chem. Soc. *A*, pp. 729—737.
(155) Frank-Kamenetskii, M. D., Lukashin, A. V., and Vologodskii, A. V.: 1975, Nature *258*, pp. 398—402.
(156) Forsman, W. C.: 1976, J. Chem. Phys. *65*, pp. 4111—4115.
(157) Frank-Kamenetskii, M. D., and Vologodskii, A. V.: 1981, Usp. Fiz. Nauk *134*, pp. 641—673.
(158) Balaban, A. T.: 1976, in Chemical Applications of Graph Theory, Ed. A. T. Balaban, pp. 64—105, Academic Press, London.
(159) Alspach, B., and Aronoff, S.: 1977, Can. J. Chem. *55*, pp. 2773—2777.
(160) King, R. B., and Rouvray, D. H.: 1978, Theor. Chim. Acta *48*, pp. 207—214.
(161) Balaban, A. T., and Baciu, V.: 1978, Match 4, pp. 131—159.
(162) Randić, M.: 1979, Match *7*, pp. 5-64.
(163) Chung, C.-S.: 1979, J. Chem. Educ. *56*, p. 398.
(164) Randić, M.: 1980, Int. J. Quantum Chem., Quantum Biol. Symp. *7*, pp. 187—197.
(165) Davis, C. C., Cross K., and Ebel, M.: 1971, J. Chem. Educ. *48*, p. 675.
(166) Bennett, W. E.: 1969, Inorg. Chem. *8*, pp. 1325—1328.
(167) Mezey, P. G.: 1981, Theor. Chim. Acta *60*, pp. 97—110.
(168) Mezey, P. G.: 1981, Int. J. Quantum Chem., Quantum Biol. Symp. *8*, pp. 185—196.
(169) Mezey, P. G.: 1982, Int. J. Quantum Chem. *22*, pp. 101—114.
(170) Krivoshey, I. V.: 1969, Zh. Strukt. Khim. *10*, pp. 316—319.
(171) Zhuravlev, V. I., Krivoshey, I. V., and Sleta, L. A.: 1975, Zh. Strukt. Khim. *16*, pp. 951—955.
(172) Krivoshey, I. V., and Sleta, L. A.: 1976, Theor. Chim. Acta *43*, pp. 165—174.
(173) Krivoshey, I. V.: 1976, Zh. Strukt. Khim. *17*, pp. 227—229.
(174) Kronrod, A. S.: 1950, Usp. Matem. Nauk *5*, pp. 24—134.
(175) Eyring, H., Gershinowitz, H., and Sun, C. E.: 1935, J. Chem. Phys. *3*, pp. 786—796.
(176) Mezey, P. G.: 1980, Theor. Chim. Acta *54*, pp. 95—111.
(177) Mezey, P. G.: 1981, Theor. Chim. Acta *58*, pp. 309—330.
(178) Mezey, P. G.: 1982, Theor. Chim. Acta *60*, pp. 409—428.
(179) Mezey, P. G.:1983, Int. J. Quantum Chem., Quantum Chem. Symp. *17*, pp. 137—152.
(180) Bader, R. F. W., Nguyen-Dang, T. T., and Yoram, T.: 1979, J. Chem. Phys. *70*, pp. 4316—4329.
(181) Tal, Y., Bader, R. F. W., and Erkku, J.: 1980, Phys. Rev. *A 21*, pp. 1—11.
(182) Bader, R. F. W.: 1980, J. Chem. Phys. *73*, pp. 2871—2883.

(183) TAL, Y., BADER, R. F. W., NGUYEN-DANG, T. T., OJHA, M., and ANDERSON, S. G.: 1981, J. Chem. Phys. *74*, pp. 5162–5167.
(184) MORSE, M., and CAIRNS, S. S.: 1969, Critical Point Theory in Global Analysis and Differential Topology, Academic Press, New York.
(185) MILNOR, J.: 1963, Morse Theory, Princeton University Press, Princeton.
(186) MEZEY, P. G.: 1981, Chem. Phys. Lett. *82*, pp. 100–104.
(187) MEZEY, P. G.: 1982, Chem. Phys. Lett. *86*, p. 562.
(188) SERRE, J.: 1964, in Molecular Orbitals in Chemistry, Physics, and Biology, Ed. B. Pullman, and P.-O. Löwdin, pp. 133–149, Academic Press, New York.
(189) LÖWDIN, P.-O.: 1967, Rev. Mod. Phys. *39*, pp. 259–287.
(190) JAFFÉ, H. H., and ELLIS, R. L.: 1974, J. Comput. Phys. *16*, pp. 20–31.
(191) ORCHIN, M., and JAFFÉ, H. H.: 1971, Symmetry, Orbitals, and Spectra, Wiley-Interscience, New York.
(192) KAPLAN, I. G.: 1979, Usp. Khim. *48*, pp. 1027–1053.
(193) WIGNER, E., and WITMER, E .E.: 1928, Z. Phys. *51*, pp. 859–886.
(194) WOODWARD, R. B., and HOFFMANN, R.: 1969, Angew. Chem. *81*, pp. 797–869.
(195) WOODWARD, R. B., and HOFFMANN, R.: 1970, Die Erhaltung der Orbitalsymmetrie, Verlag Chemie, Weinheim.
(196) LONGUET-HIGGINS, H. C., and ABRAHAMSON, E. W.: 1965, J. Am. Chem. Soc. *87*, pp. 2045–2046.
(197) GEORGE, T. F., and ROOS, J.: 1971, J. Chem. Phys. *55*, pp. 3851–3866.
(198) FUKUI, K.: 1971, Accounts Chem. Res. *4*, pp. 57–64.
(199) PEARSON, R. G.: 1971, Pure Appl. Chem. *27*, pp. 145–160.
(200) BERSON, J. A., and SALEM, L.: 1972, J. Am. Chem. Soc. *94*, pp. 8917–8918.
(201) BORDEN, W. T., and SALEM, L.: 1973, J. Am. Chem. Soc. *95*, pp. 932–933.
(202) SILVER, D. M.: 1974, J. Am. Chem. Soc. *96*, pp. 5959–5967.
(203) RASSAT, A.: 1975, Tetrahedron Lett. *46*, pp. 4081–4084.
(204) GAJEWSKI, J. J.: 1981, Hydrocarbon Thermal Isomerizations, Academic Press, New York.
(205) WILSON JR., E. B.: 1935, J. Chem. Phys. *3*, pp. 276–285.
(206) WILSON JR., E. B.: 1935, J. Chem. Phys. *3*, pp. 818–821.
(207) HOUGEN, J. T.: 1962, J. Chem. Phys. *37*, pp. 1433–1441.
(208) HOUGEN, J. T.: 1963, J. Chem. Phys. *39*, pp. 358–365.
(209) JONES, R. G.: 1969, in NMR Basic Principles and Progress, Vol. 1, Ed. P. Diehl, E. Fluck, and R. Kosfeld, pp. 97–174, Springer-Verlag, Berlin.
(210) SERRE, J.: 1974, Advan. Quantum Chem. *8*, pp. 1–36.
(211) BUNKER, P. R.: 1979, Molecular Symmetry and Spectroscopy, Academic Press, New York.
(212) EZRA, G. S.: 1982, Symmetry Properties of Molecules, Springer-Verlag, Berlin.
(213) PAPOUŠEK, D., and ALIEV, M. R.: 1982, Molecular Vibrational/Rotational Spectra, Academia, Prague.
(214) BERRY, R. S.: 1980, in Qunatum Dynamics of Molecules. The New Experimental Challenge to Theorists, Ed. R. G. Woolley, pp. 143–195, Plenum Press, New York.
(215) DYKE, T. R.: 1977, J. Chem. Phys. *66*, pp. 492–497.
(216) ALTMANN, S. L.: 1967, Proc. Roy. Soc. *A 298*, pp. 184–203.
(217) FREI, H., GRONER, P., BAUDER, A., and GÜNTHARD, Hs. H.: 1978, Mol. Phys. *36*, pp. 1469–1495.
(218) FREI, H., BAUDER, A., and GÜNTHARD, Hs. H.: 1979, Top. Curr. Chem. *81*, pp. 1–97.
(219) FREI, H., BAUDER, A., and GÜNTHARD, Hs. H.: 1981, Mol. Phys. *43*, pp. 785–797.
(220) POPLE, J. A.: 1980, J. Am. Chem. Soc. *102*, pp. 4615–4622.
(221) MCDANIEL, D. H.: 1981, J. Phys. Chem. *85*, pp. 479–481.
(222) FLURRY JR., R. L.: 1981, J. Am. Chem. Soc. *103*, pp. 2901–2902.

(223) FLURRY JR., R. L.: 1972, Int. J. Quantum Chem., Quantum Chem. Symp. *6*, pp. 455—458.
(224) FLURRY JR., R. L.: 1973, Theor. Chim. Acta *31*, pp. 221—230.
(225) FLURRY JR., R. L.: 1980, Symmetry Groups. Theory and Chemical Applications, Prentice--Hall, Englewood Cliffs.
(226) DALTON, B. J.: 1966, Mol. Phys. *11*, pp. 265—285.
(227) GILLES, J.-M. F., and PHILIPPOT, J.: 1972, Int. J. Quantum Chem. *6*, pp. 225—261.
(228) GILLES, J. M. F., and PHILIPPOT, J.: 1978, Int. J. Quantum Chem. *14*, pp. 299—311.
(229) ELLINGER, Y., and SERRE, J.: 1973, Int. J. Quantum Chem., Quantum Chem. Symp. *7*, pp. 217—221.
(230) RANDIĆ, M.: 1976, Chem. Phys. Lett. *42*, pp. 283—287.
(231) BOUMAN, T. D., DUNCAN, C. D., and TRINDLE, C.: 1977, Int. J. Quantum Chem. *11*, pp. 399—413.
(232) BERRY, R. S.: 1979, in The Permutation Group in Physics and Chemistry, Ed. J. Hinze, pp. 92—120, Springer-Verlag, Berlin.
(233) BÜRGI, H. B.: 1980, Match *9*, pp. 13—14.
(234) JAHN, H. A., and TELLER, E.: 1937, Proc. Roy. Soc. *A 161*, pp. 220—235.
(235) ENGLMAN, R.: 1972, The Jahn-Teller Effect in Molecules and Crystals, Wiley-Interscience, London.
(236) BERSUKER, I. B.: 1976, Electronic Structure and Properties of Coordination Compounds, Khimija, Leningrad, (in Russian).
(237) MURRELL, J. N.: 1977, Structure and Bonding *32*, pp. 93—146.
(238) DEWAR, M. J. S., KIRSCHNER, S., KOLLMAR, H. W., and WADE, L. E.: 1974, J. Am. Chem. Soc. *96*, pp. 5242—5244.
(239) KOMORNICKI, A., and MCIVER JR., J. W.: 1976, J. Am. Chem. Soc. *98*, pp. 4553—4561.
(240) DEWAR, M. J. S., FORD, G. P., MCKEE, M. L., RZEPA, H. S., and WADE, L. E.: 1977, J. Am. Chem. Soc. *99*, pp. 5069—5073.
(241) DEWAR, M. J. S., KIRSCHNER, S., and KOLLMAR, H. W.: 1974, J. Am. Chem. Soc. *96*, pp. 5240—5242.
(242) MURRELL, J. N.: 1980, in Quantum Theory of Chemical Reactions, Vol. 1, Ed. R. Daudel, A. Pullman, L. Salem, and A. Veillard, pp. 161—176, D. Reidel Publ. Comp., Dordrecht.
(243) PECHUKAS, P.: 1981, Annu. Rev. Phys. Chem. *32*, pp. 159—177.
(244) MURRELL. J. N., and LAIDLER, K. J.: 1968, Trans. Faraday Soc. *64*, pp. 371—377.
(245) TANAKA, T., and MORINO, Y.: 1970, J. Mol. Spectrosc. *33*, pp. 538—551.
(246) WRIGHT, J. S.: 1973, Can. J. Chem. *51*, pp. 139—146.
(247) MURRELL, J. N., SORBIE, K. S., and VARANDAS, A. J. C.: 1976, Mol. Phys. *32*, pp. 1359 to 1372.
(248) STANTON, R. E., and MCIVER JR., J. W.: 1975, J. Am. Chem. Soc. *97*, pp. 3632—3646.
(249) STOHRER, W.-D., and HOFFMANN, R.: 1972, J. Am. Chem. Soc. *94*, pp. 1661—1668.
(250) MCIVER JR., J. W.: 1972, J. Am. Chem. Soc. *94*, pp. 4782—4783.
(251) MCIVER JR., J. W., and STANTON, R. E.: 1972, J. Am. Chem. Soc. *94*, pp. 8618—8620.
(252) MCIVER JR., J. W.: 1974, Accounts Chem. Res. *7*, pp. 72—77.
(253) WILSON JR., C. W., and GODDARD III, W. A.: 1969, J. Chem. Phys. *51*, pp. 716—731.
(254) SALEM, L., DURUP, J., BERGERON, C., CAZES, D., CHAPUISAT, X., and KAGAN, H.: 1970, J. Am. Chem. Soc. *92*, pp. 4472—4474.
(255) SALEM, L.: 1971, Accounts Chem. Res. *4*, pp. 322—328.
(256) MCIVER JR., J. W., and KOMORNICKI, A.: 1971, Chem. Phys. Lett. *10*, pp. 303—306.
(257) METIU, H., ROSS, J., SILBEY, R., and GEORGE, T. F.: 1974, J. Chem. Phys. *61*, pp. 3200 to 3209.
(258) SCHARFENBERG, P.: 1979, Theor. Chim. Acta *53*, pp. 279—292.
(259) PECHUKAS, P.: 1976, J. Chem. Phys. *64*, pp. 1516—1521.

(260) Pollak, E., and Pechukas, P.: 1978, J. Am. Chem. Soc. *100*, pp. 2984—2991.
(261) Coulson, D. R.: 1978, J. Am. Chem. Soc. *100*, pp. 2992—2996.
(262) Schlag, E. W.: 1963, J. Chem. Phys. *38*, pp. 2480—2482.
(263) Schlag, E. W., and Haller, G. L.: 1965, J. Chem. Phys. *42*, pp. 584—587.
(264) Bishop, D. M., and Laidler, K. J.: 1965, J. Chem. Phys. *42*, pp. 1688—1691.
(265) Murrell, J. N., and Pratt, G. L.: 1970, Trans. Faraday Soc. *66*, pp. 1680—1684.
(266) Murrell, J. N.: 1972, Chem. Commun., pp. 1044—1045.
(267) Öpik, U., and Pryce, M. H. L.: 1957, Proc. Roy. Soc. *A 238*, pp. 425—447.
(268) Bader, R. F. W.: 1962, Can. J. Chem. *40*, pp. 1164—1175.
(269) Bartell, L. S.: 1968, J. Chem. Educ. *45*, pp. 754—767.
(270) Pearson, R. G.: 1969, J. Am. Chem. Soc. *91*, pp. 1252—1254.
(271) Pearson, R. G.: 1969, J. Am. Chem. Soc. *91*, pp. 4947—4955.
(272) Salem, L.: 1969, Chem. Phys. Lett. *3*, pp. 99—101.
(273) Salem, L., and Wright, J. S.: 1969, J. Am. Chem. Soc. *91*, pp. 5947—5955.
(274) Murrell, J. N., Pedley, J. B., and Durmaz, S.: 1973, J. Chem. Soc., Faraday Trans. *II 69*, pp. 1370—1380.
(275) Klemperer, W. G.: 1972, J. Chem. Phys. *56*, pp. 5478—5489.
(276) Klemperer, W. G.: 1972, Inorg. Chem. *11*, pp. 2668—2678.
(277) Klemperer, W. G.: 1972, J. Am. Chem. Soc. *94*, pp. 6940—6944.
(278) Klemperer, W. G.: 1972, J. Am. Chem. Soc. *94*, pp. 8360—8371.

6 Isomerism and the Theory of Chemical Reactivity

The preceding chapters dealt with the description of chemical isomerism in several ways: phenomenologically, in terms of the usual quantum-chemical or the rigorous quantum-mechanical approach, or by considering the picture resulting from the more abstract algebraic approach highlighting the intrinsic mathematical structure at the expense of the physical or chemical content. All these approaches will be utilized to a certain degree in this chapter and, moreover, their combination with statistical thermodynamics in the solution of problems in isomeric chemistry will be demonstrated. Modern chemical research can be characterized as a search for a relationship between the structure and the chemical properties of substances, especially their reactivity. If the role of isomerism is considered in this connection, then two basic considerations can be made. Under certain conditions it is possible to apply methods that, under the given observational conditions, permit one to distinguish among all the structures considered, and thus all the isomers. Chemistry naturally attempts to be as selective and specific as possible. However, it frequently happens that only the major structures are detected, especially under complex reaction conditions. Under some such situations it happens that the observed data can be equally well interpreted using various working models. Selection between these models can be made on the basis of well-designed experiments and, ever increasingly, using theoretical and especially quantum-chemical approaches. These considerations are especially valid when isomers of a given species are present but cannot be distinguished using the observation methods employed under the given conditions. During the interpretation of the results, these isomers than act as a single species under a common summary formula. The presence of several isomers can often be revealed only by a systematic theoretical search on the corresponding potential energy hypersurface, leading to the discovery of a number of various minima that can clearly be assigned to a single chemical compound and/or several different transition states clearly assignable to a single observed rate process. Energy, or other criteria, can often eliminate all localized structures except one that plays a key role under the given observation conditions. Explicit consideration of isomerism in such a situation would be of only formal or marginal value. It can, however, also happen that two or more theoretically distinguishable structures are of comparable stability, so that they coexist under the observation conditions but cannot be distinguished. Then every structure-

-dependent quantity observed can be considered to be an average value resulting from the contributions of all the isomers involved. Theoretical study then indicates that the observed quantities are intrinsically of a more complex nature, and thus that the relationships involved in interpretation of the structure-reactivity scheme are more complicated. Such situations and problems will be the main subject of this chapter and the greatest attention will be paid to the consequences of isomerism of the reaction components on the convolution and deconvolution of the characteristics of various types of equilibrium and rate processes.

6.1 Calculation of the Characteristics of Equilibrium and Rate Processes

It is generally known that quantum mechanics can be considered to be the meta--theory of theoretical chemistry[1] and that, in its general form, the theory of chemical reactivity can, in principle, be rigorously derived from quantum mechanics (and quantum statistics, provided that it is not considered to be part of quantum mechanics but of statistical physics). This is most generally carried out by solution of the time--dependent Schrödinger equation (6-1) with Hamiltonian $\hat{H}$ of the system (it will be assumed that $\hat{H}$ is not explicitly dependent on time):

$$\hat{H}\Psi(r, \boldsymbol{R}, t) = i\hbar \frac{\partial \Psi(r, \boldsymbol{R}, t)}{\partial t} . \tag{6-1}$$

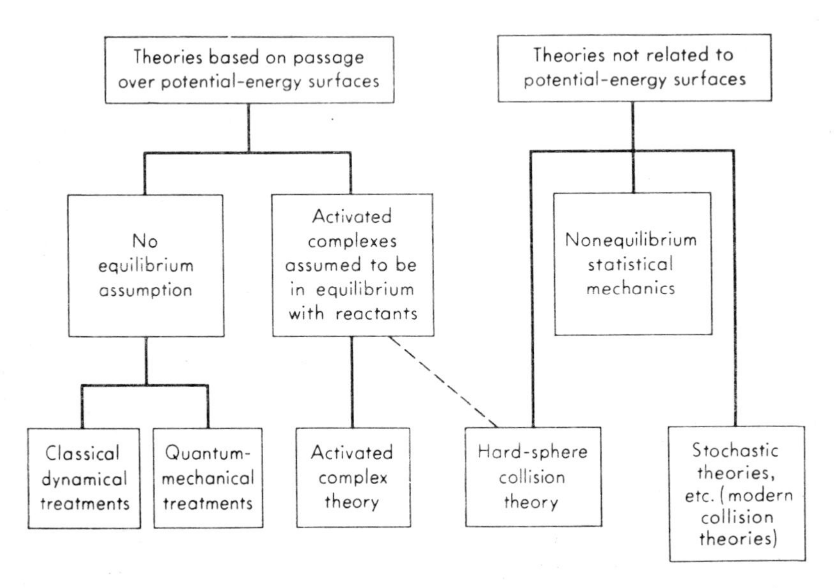

Figure 6-1. A survey of contemporary theories of chemical reaction rates (from Ref.[5])

The assumption of the decomposability of the non-stationary wave function $\Psi(\boldsymbol{r}, \boldsymbol{R}, t)$ into the product of the time independent function $\Psi(\boldsymbol{r}, \boldsymbol{R})$ and the time-dependent term leads[2-4] to the well-known form (6-2) of the product

$$\Psi(\boldsymbol{r}, \boldsymbol{R}, t) = \Psi(\boldsymbol{r}, \boldsymbol{R}) \exp\left(-\mathrm{i}\frac{E}{\hbar}t\right), \tag{6-2}$$

and to the problem of the solution of the corresponding time-independent Schrö-dinger equation for stationary states $\Psi(\boldsymbol{r}, \boldsymbol{R})$ (see Eq. (3-1)).

If, however, it is required to numerically describe the time behaviour of a given, chemically interesting system, it is necessary to employ a large number of strong approximations. The first is the Born-Oppenheimer approximation (see Chap. 3); further choice of approximations and methods for their solution leads to a wide variety of approaches. Fig. 6-1 depicts possible ways of classifying individual concepts and theories of chemical rate processes and relationships among them. However, from the point of view of contemporary quantum chemistry, only an approach based on the activated complex[6] (or transition state[7]) concept has a broad application, primarily for processes in the gaseous phase. For very simple systems, the concept of classical trajectories can also be used[3,5,8-12], gradually utilizing non-empirical potential hypersurfaces obtained from quantum-chemical calculations to a still greater degree.

Chap. 6.1 contains a brief summary of that part of the theory of chemical reactivity commonly used in contemporary numerical quantum chemistry, i.e. calculations of equilibrium and rate constants for chemical reactions in the gaseous phase. Here, considerations of rate processes are usually converted into calculations of equilibrium constants using equilibrium hypothesis (where this is plausible). This survey of conventional approaches (i.e. not considering isomerism of the reaction components or activated complexes) will form a basis in the remaining sections of chapter 6 for developing a suitably generalized concept which properly considers the phenomenon of chemical isomerism.

6.1.1 Calculation of Equilibrium Thermodynamic Characteristics

Quantum chemical calculations of equilibrium characteristics are based on direct combination of quantum chemistry and statistical thermodynamics, i.e. two fields that have been traditionally somewhat separated[13-17]. The use of quantum-chemical calculations as generators of the molecular characteristics required in statistical thermodynamic procedures instead of spectroscopic experiments[18-25] has enabled the calculation of the equilibrium characteristics independent of experimental information. The concept of the partition function forms a basis for all statistical thermodynamic calculations. Partition function Q of a set of N identical, indistinguishable species can be simply expressed in terms of partition function q' for a single species[26-30] assuming that the species are independent (this is ensured, for example, for an ideal gas because of its definition):

$$Q = (q')^N/N!\,, \tag{6-3}$$

where q' is given by

$$q' = \sum_j \exp(-e_j/kT)\,, \tag{6-4}$$

the summation is carried out over all the quantum states of the species and k and T have the usual significance. The ambiguity in the selection of the energy scale can be overcome by introducing a partition function q (in addition to q'), based on energies related to the ground-state energy e_0:

$$q = \sum_j \exp(-(e_j - e_0)/kT)\,. \tag{6-5}$$

Value e_0 has the significance of enthalpy, or energy, at absolute zero temperature, i.e. the zero-point molar energy H^0_O.

It then follows from statistical thermodynamics that the thermodynamic functions of the system can be expressed in the framework of the Gibbs canonical ensemble as a function of q and H^0_O; for example, it holds for the standard molar enthalpy H^0_T, entropy S^0_T, Gibbs function G^0_T, and heat capacity c^0_p at constant pressure p of one mole of particles of an ideal gas that[26-30]:

$$H^0_T = H^0_O + RT^2\left(\frac{\partial \ln q^0}{\partial T}\right)_p\,, \tag{6-6}$$

$$S^0_T = R\ln\left(\frac{q^0}{N_A}\right) + RT\left(\frac{\partial \ln q^0}{\partial T}\right)_p\,, \tag{6-7}$$

$$G^0_T = H^0_O - RT\ln\left(\frac{q^0}{N_A}\right), \tag{6-8}$$

$$c^0_p = 2RT\left(\frac{\partial \ln q^0}{\partial T}\right)_p + RT^2\left(\frac{\partial^2 \ln q^0}{\partial T^2}\right)_p\,, \tag{6-9}$$

where N_A denotes the Avogadro constant and q^0 is partition function q corresponding to the selected standard state (in Eqs. (6-6)–(6-9) q^0 is an explicit function of p; an alternative expression of these relationships in terms of partition function explicitly dependent on volume V is also often used). In this work, the standard state is generally that of an ideal gas at a pressure of 101 325 Pa (1 atm). Relationships (6-6)–(6-9) can also be used for the calculation of changes in the values of standard thermodynamic functions along a chemical reaction. Consider the reaction

$$\sum_{i=1}^{n} \nu_i A_i = 0\,, \tag{6-10}$$

leading to the establishment of equilibrium characterized by the equilibrium constant* K_p:

$$K_p = \prod_{i=1}^{n} p_i^{\nu_i} . \tag{6-11}$$

Considering the relationship between K_p and the value of ΔG_T^0 for a given process (6-10), Eq. (6-8) leads to the expression of equilibrium constant K_p in terms of the partition function q_i^0 of the components of reaction (6-10) and change ΔH_0^0 in the zero-point energy during this reaction:

$$K_p = \prod_{i=1}^{n} (q_i^0/N_A)^{\nu_i} \exp(-\Delta H_0^0/(RT)) . \tag{6-12}$$

It should be noted that a single approximation is necessary to obtain relationships (6-6)–(6-9) and (6-12) – the approximation of non-interaction between particles, e.g. the approximation of an ideal gas state.

The only general, proper method for the calculation of partition function (6-5) is direct summation over all the quantum states in the system. In practice, however, this approach is used only exceptionally[35]. The usual procedure involves the transfer to a simpler quantum-mechanical model whose energy spectrum can be obtained in closed form. Another possibility involves the method of quasiclassical partition functions[36,37] using the classical energy expression. This then leads to simple, frequently analytical formulae containing the molecular constants for the given system.

A key approximation used in the construction of partition function (6-5) is the assumption of the independence of the individual motions in the molecule – for which arguments were given by Born and Oppenheimer[38] – leading to separation of the overall partition function into partition functions q_t, q_r, q_v, q_e, and q_n for the corresponding translational, rotational, and vibrational motions, electronic excitations, and nuclear spin:

$$q = q_t q_r q_v q_e q_n . \tag{6-13}$$

In general, however, only the first three functions are employed. The electronic partition function is often approximated by the degeneracy factor for the corresponding ground state, or as a direct summation according to Eq. (6-5) for a limited

---

* In this work, the usual (although not quite correct) use of dimensioned equilibrium constants will be employed in place of rigorous[31,32], dimensionless constants introduced in terms of reduced quantities (e.g., ratios p_i/p^0, where p^0 designates the selected standard pressure); the term standard is also used for these dimensionless equilibrium constants[32]. This at present still common deviation from the rigorous procedure is not critical here, as the equilibrium constants of isomerizations (that are, in any case, dimensionless) will frequently be used here. The rigorous dimensionless character of the standard equilibrium constants is closely connected with the introduction of thermodynamic standard states. It should be noted that frequently insufficient attention is devoted to the specification of standard states (cf. e.g.[33,34]) in the current literature.

number of lowest states. Decomposition (6-13) assumes that the rotational and vibrational molecular constants are identical for all the electronic states. Correct consideration of the differences between the individual electronic states could require separate treatment for each state and execution of the remaining procedure according to the general approach described in Chap. 6.2 (i.e. a special case of the isomerism of reaction components).

Neglecting the contribution of nuclear spin leads[22,25], for quantities defined by the second law of thermodynamics, to the usual practical (virtual) values. Correct consideration of the relationship between rotational quantum states and nuclear spin states is, however, important for some molecules in connection with isomerism conditioned by nuclear spin, as discussed in Chap. 2.2. This is generally manifested in different partition functions for each type of nuclear spin isomer. From this point of view, the calculation of (overall) values of thermodynamic functions for molecules exhibiting isomerism conditioned by nuclear spin is again a special case of the problem of description of processes with isomerism of the reaction components, which will be discussed in Chap. 6.2. It should be noted that, when the rotational motion and the contribution of nuclear spin can be separated, it can be demonstrated[39] that the factor corresponding to nuclear spin exactly cancels out in the calculation of the equilibrium constants according to Eq. (6-12).

The commonly used translational partition function is based on the energy spectrum of a particle in a rectangular potential box; however, a number of mostly quite plausible simplifications must be carried out to obtain this function in closed form. It should, however, be noted that the discussion[40] has indicated that the results obtained for a rectangular prism cannot be automatically used for other types of reaction vessels and that deviations from the standard formula can be expected for certain extreme cases (e.g. cavities – see Chap. 6.5). In addition, a formal proof was recently given[41] of the fact that the form of the translational partition function depends on the shape of the vessel. Thus, it is known that such a dependence exists; however, at present it cannot be expressed in a closed (or at least utilizable) and, at the same time, sufficiently precise form. This is true of the conventional translational partition function used in the framework of Boltzmann statistics. Possible effects of the shape of the vessel could appear only at low temperatures, so that the rigorous approach to the whole problem is further complicated by the requirement that either Fermi-Dirac or Bose--Einstein statistics be used. However, for quantum-chemical applications, the partition function for the translation does not involve any problem, as its construction does not require any information other than the values of the universal constants and the molecular masses. It should be noted that, considered rigorously, this dependence on the molecular mass as input information (possibly combined[35] with the measured) values of the universal constants and the form of the Coulomb law following from observations) is the only point at which the concept of quantum chemical calculations of thermodynamic characteristics is, in principle, dependent on experiment.

The rotational partition function for a linear or nonlinear molecule is generally derived from the corresponding quantum mechanical solution for the rigid structure, possibly using the corresponding quasi-classical partition function. This leads[18-30] to a simple formula involving the principal moments of inertia of the corresponding rigid structure that can readily be derived[42,43] from its structural parameters. The conventionally used formulae, however, are only the high-temperature limits of corresponding infinite series for linear[44,45] or non-linear[45-49] rigid structures, i.e. the first terms. The symmetry requirements of the wave function for symmetric molecules[22] lead to realization of only part of the rotational quantum states compared to the analogous nonsymmetrical structure; this is reflected in the rotational partition function in the symmetry number. The symmetry number can readily be deduced[22] from the point group of symmetry of the corresponding structure or its rotational subgroup[22,50]. The symmetry numbers for transition states will be discussed in greater detail in Chap. 6.1.2.

The vibrational partition function is mostly constructed by utilizing the solution of the Schrödinger equation for a harmonic oscillator, leading to a simple closed result. This result can readily be transferred to an N-atom molecule with $3N$-6 vibrational degrees of freedom ($3N$-5 for a linear structure). The use of normal coordinates from the classical solution permits conversion of a ($3N$-6)-dimensional problem into $3N$-6 one-dimensional problems. The total partition function is then the simple product of the partition functions of the individual vibrational modes which can readily be constructed employing the corresponding vibrational frequencies. Aspects of vibrational analysis that are important in connection with quantum-chemical description of harmonic vibrational modes were discussed in Chap. 3.6.

Approximation of partition functions in the rigid rotator and harmonic oscillator (RRHO) approximation is[22,25,51] a common tool in the calculation of the thermodynamic functions of polyatomic molecules on the basis of spectral data. Simultaneously, it is a practical upper limit in the use of quantum chemical methods to generate molecular characteristics for the construction of partition functions for polyatomic molecules. Transition to a less rough approximation would require more detailed information on the behaviour of the potential energy hypersurface in place of the simple description in terms of the second derivatives of the energy at the minimum.

Corrections for vibrational anharmonicity, rotational-vibrational interaction and centrifugal distortion are the most important corrections to the RRHO approximation. Expressions for these corrections for diatomic molecules[22,25,51] are extensively employed in combination with molecular constants obtained from spectroscopic experiments, and combination with quantum chemical methods is also simple. The data given in Table 6-1 for the N_2 molecule[25] at a temperature of 2000 K yield information on the importance of the individual contributions for the thermodynamic functions of a diatomic molecule. It is apparent that, for most purposes, the data from the RRHO approximation can be considered quite satisfactory for diatomic molecules of this type. In fact, the agreement between[52] the HO approximation and

Table 6-1. Contributions[a] to the standard molar thermodynamic potential $-(G_T^\circ - H_0^\circ)/T$ and the molar heat capacity at constant pressure c_p° of the N_2 molecule at a temperature of $T = 2000$ K

Type of contribution	$-(G_T^\circ - H_0^\circ)/T$ (J K^{-1} mol^{-1})	c_p° (J K^{-1} mol^{-1})
Translational	169.088	20.786
Rigid rotational (RR)	48.647	8.314
Harmonic oscillation (HO)	1.720	6.611
RRHO approximation	219.455	35.711
Vibrational anharmonicity	0.008	0.096
Rotational-vibrational interaction	0.017	0.088
Rotational stretching	0.033	0.067
Total	219.513	35.962

[a] From Ref.[25]

Table 6-2. Comparison[a] of values of partition function q, standard molar thermodynamic potential $-(G_T^\circ - H_0^\circ)/T$, heat content function $H_T^\circ - H_0^\circ$, entropy S_T°, and heat capacity at constant volume c_v° obtained for the Xe_2 molecule from the exact summation[b] and from the HO approximation[b]

T (K)	q	$-(G_T^\circ - H_0^\circ)/T$ (J K^{-1} mol^{-1})	$H_T^\circ - H_0^\circ$ (J mol^{-1})	S_T° (J K^{-1} mol^{-1})	c_v° (J K^{-1} mol^{-1})
100[c]	4.87	13.14	803	21.17	6.02
	3.56	10.59	552	16.11	3.64
200	8.70	17.97	1130	23.64	1.71
	5.28	13.83	761	17.66	1.17
300	11.04	19.96	1201	23.97	1.82
	6.20	15.17	841	17.95	0.54

[a] From Refs.[52,53]
[b] The values from the exact summation and from the harmonic oscillator (HO) approximation are given in the upper and the lower line, resp.
[c] Normal boiling point of Xe is[52] 166 K.

the results of direct summation of the values obtained by solution of the Schrödinger equation for the corresponding anharmonic potential can be considered good even for the extremely anharmonic Xe_2 molecule. It is apparent from the data in Table 6-2 that the thermodynamic function values are of the same order in both cases. For precise calculations of diatomic molecules that are not in the $^1\Sigma$ electronic state, relationships are available[24,54,55] for the dependence of the rotational energy on the electronic state employed for precise calculation of the thermodynamic functions.

General relationships have also long been available for the correction of the thermodynamic functions of polyatomic molecules for vibrational anharmonicity[23,56]; however, broad application of these expressions is prevented by a lack of data from spectral measurements, or by the magnitude of the necessary quantum-chemical calculations. Thus they have so far been used only for calculation of the thermodynamic functions of some triatomic molecules[25,56,57] and, exceptionally, for polyatomic molecules[25]. The consequences of Fermi resonance[58], vibrational mode degeneracy for linear molecules[23], adiabatic corrections[59,60], problems in the calculation of the thermodynamic functions of vibronic systems[61], and other[25] fine corrections for the calculation of the values of thermodynamic functions have also been studied.

The treatments for description of the anharmonicity of vibrational motion in the calculation of thermodynamic functions are the subject of constant interest[62-66]. The two-parameter (D, α) Morse potential

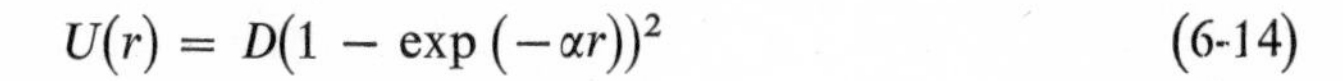

$$U(r) = D(1 - \exp(-\alpha r))^2 \tag{6-14}$$

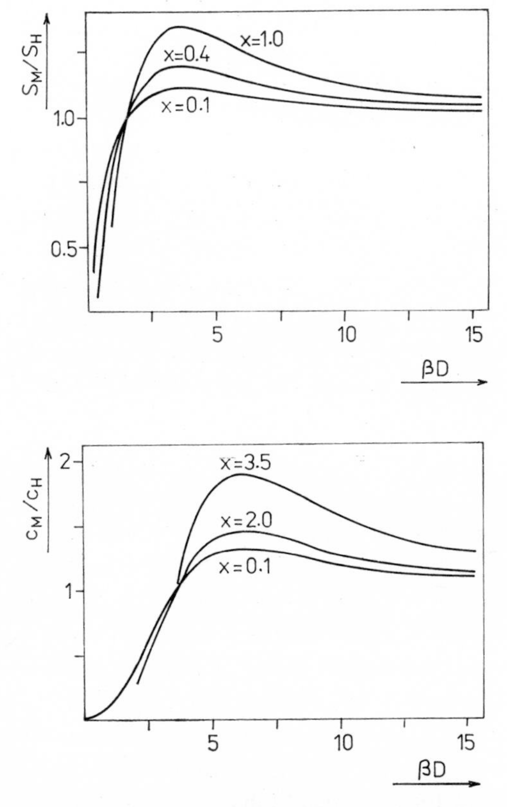

Figure 6-2. The entropy S and heat capacity c of a Morse (M) oscillator compared to those of a harmonic (H) oscillator; $x = h\nu/kT$, $\beta = 1/kT$ where ν and D denote the characteristics[64] of the Morse potential

is a useful and frequently employed[62,64] potential form, where r is the displacement from the equilibrium point; the corresponding Schrödinger equation with this potential has an analytical solution. A systematic study of the relationship between the thermodynamic functions obtained in the framework of the Morse and the corresponding harmonic potentials was carried out by Amorebieta and Colussi[64] for a broad region of potential parameters (6-14). The relationships are apparent from Fig. 6-2, giving the results for the entropy and heat capacity in terms of reduced parameters $h\nu/kT$ and D/kT, where ν is the corresponding[64] harmonic vibration frequency. These results indicate that marked differences between the two approaches appear only for very shallow potentials. Analytical results were also recently obtained by Witschel[65,66] for three-parameter potential (6-15) and for the potential with terms $k_{2n}x^{2n}$ ($n \geqq 2$):

$$U(r) = k_2 r^2 + k_3 r^3 + k_4 r^4 \tag{6-15}$$

in the framework of the quasi-classical partition function technique (results for the behaviour at high temperatures). An important characteristic of potential (6-15) is the fact that the single formula can describe both single- and double-well asymmetric potentials.

Internal free or restricted (hindered) rotations are examples of non-rigidity of the molecular skeleton that can make a considerable contribution to the values of the thermodynamic functions. Various partition functions, differing in their degree of sophistication, are available for the rotation of symmetric tops attached to a rigid frame[67,68], asymmetric tops attached to a rigid frame[69], and, in the most general case, rotating groups attached to a rotating group without a reference rigid frame[70] (compound rotation). The potential energy for restricted internal rotation is mostly approximated by the first few terms in the corresponding Fourier cosine series expansion (6-16), where Φ designates the angle of rotation and the choice of the value of k depends, among other things, on the symmetry of the problem

$$U(\Phi) = \sum_{k=1}^{\infty} \frac{V_k}{2} [1 - \cos(k\Phi)] \, . \tag{6-16}$$

Constants in this expansion (barriers to restricted rotation) V_k are quite commonly determined by contemporary quantum-chemical methods (e.g.[71,72]). This permits the inclusion of the contribution of internal rotation in the values of the thermodynamic functions evaluated on the basis of quantum-chemical information. Table 6-3 gives a numerical illustration of the importance of the contribution of internal rotation for the example of toluene[25,73] with the rotation of the methyl group approximated by free rotation (the anharmonicity correction given in Table 6-3 is empirical[73]). It is interesting in relation to the phenomenon of isomerism that a method has been developed[75] on the basis of Pitzer's suggestion[74] for the calculation of the thermodynamic functions of a molecule exhibiting hindered rotation with an asymmetric rotor, converting the problem to calculation of the thermodynamic functions of a mixture

Table 6-3. Contributions[a] to the standard molar entropy S°_T and the molar heat capacity at constant pressure c°_p of toluene at a temperature of $T = 1000$ K

Type of contribution	S°_T (J K^{-1} mol^{-1})	c°_p (J K^{-1} mol^{-1})
Translational	190.31	20.79
Overall rotational	122.18	12.47
Free internal rotational	19.83	4.16
Harmonic vibrations	209.04	223.50
Anharmonicity correction	1.89	4.04
Total	543.25	264.96

[c] From Refs.[25,73]

Table 6-4. Contributions[a] to the standard molar entropy S°_T of XeF_6 at temperature $T = 335$ K

Type of contribution	S°_T (J K^{-1} mol^{-1})
Translational and rotational	273.2
Vibrational	86.6
Pseudorotational	42.8
Total	402.6

[a] From Ref.[77]

of two conformational isomers. In addition to internal rotation, ways of including other types of molecular non-rigidity in the calculation of thermodynamic characteristics have been described. This is, primarily, the pseudo-rotational motion introduced in the classical work of Kilpatrick *et al.*[76] for the calculation of the thermodynamic functions of cyclopentane. Analysis of the cyclopentane conformation indicates that the angular strain forces formed in this five-membered ring as a result of deformation of the bonding angles are almost cancelled out by the torsional forces resulting from hydrogen-hydrogen repulsions. These interrelations between the individual components of the potential energy result in the fact that the puckering in the slightly puckered cyclopentane ring is of an indefinite type, as the potential energy is essentially the same for all possible small puckerings. The vibrational motion in this ring can be studied as a two-dimensional problem, revealing[76] that these two degrees of freedom are associated both with the ordinary vibrational motion and with one-dimensional rotation in which the phase of the puckering moves around the ring. At present, the thermodynamic properties of a number of substances (cf.[25]) have been obtained within the framework of the pseudo-rotation concept. For example, recently Pitzer and Bernstein[77] used this approach to explain the marked disagreement between the observed entropy value and data from the calculation for a rigid structure in the XeF_6 system. Suitable selection of the parameters of the pseudo-rotational motion led to complete agreement between the observed and calculated data (Table 6-4). This structural peculiarity of XeF_6 is connected with symmetry distortion ascribable to a pseudo-Jahn-Teller effect. Other types of statistical-mechanical treatment[78] of non--rigidity include four-membered rings ('ring puckering') developed with cyclobutane and the results[79] for the vibrational thermodynamic functions in a double-well symmetric potential, important (for example) for the thermodynamics of the inversion motion in molecules. It should, however, be noted that it was found that the simple harmonic oscillator approximation can readily be employed in the latter case[25].

The importance of internal rotation in the interconversion of rotamers justifies a detailed description here of a recently obtained result[80,81] for the partition function of free internal rotation, which also has a certain validity for restricted motion. It apparently holds, for an arbitrary partition function in Eq. (6-5) (for a nondegenerate ground state), that

$$\lim_{T \to 0} q = 1 \tag{6-17}$$

and, similarly, (for example) for the entropy from Eq. (6-7) calculated on the basis of this partition function:

$$\lim_{T \to 0} S_T^0 = 0 \,. \tag{6-18}$$

The conventional partition function q_{fir} for free internal rotation

$$q_{\text{fir}} = \frac{1}{n} \left(\frac{8\pi^3 I_r kT}{h^2} \right)^{1/2} = \left(\frac{\pi}{\sigma} \right)^{1/2} \tag{6-19}$$

(where I_r is the corresponding reduced moment of inertia, n is the internal rotation symmetry number and the other symbols have the usual significance) clearly does not satisfy general relationships (6-17) and (6-18) (similar to a number of other partition functions), as it holds in terms of dimensionless variable σ that

$$\lim_{\sigma \to \infty} q_{\text{fir}} = 0 \,, \tag{6-20}$$

$$\lim_{\sigma \to \infty} \delta_{\text{fir}} S_T^0 = -\infty \,, \tag{6-21}$$

where $\delta_{\text{fir}} S_T^0$ is the contribution of the free internal rotation to S_T^0. This erroneous behaviour of partition function (6-19) in the low temperature and/or reduced moment of inertia region(s) has attained importance in connection with the application to the study of small van der Waals molecules, where low values are typical for both parameters I_r and T.

The analysis[80,81] of the derivation of conventional formula (6-9) indicates that this erroneous behaviour is a result of the transition from rigorous summation to approximative integration. It holds rigorously that

$$q_{\mathrm{fir}} = 2\sum_{j=0}^{\infty} \mathrm{e}^{-\sigma j^2} - 1 = 2\lambda(\sigma) - 1 . \tag{6-22}$$

If integration is used in place of summation for calculation of function $\lambda(\sigma)$, an approximation is involved. Nonetheless, mathematical analysis yields Euler's summation formula (see for example[82]) for the determination of the difference resulting from replacing summation by integration; rigorously, it holds that

$$\lambda(\sigma) = \frac{1}{2}\left(\frac{\pi}{\sigma}\right)^{1/2} + \frac{1}{2} - 2\sigma \int_0^{\infty} P_1(x)\, \mathrm{e}^{-\sigma x^2} x \, \mathrm{d}x , \tag{6-23}$$

where $P_1(x)$ is a function closely related to the first Bernoulli polynomial and expressable for every nonintegral x, for example, as:

$$P_1(x) = - \sum_{n=1}^{\infty} \frac{\sin(2\pi n x)}{\pi n} . \tag{6-24}$$

The last term in Eq. (6-23) can be evaluated by use of the principle[83] of the Laplace method, so that in summary the following approximation q_1 can be obtained for the exact partition function of free internal rotation:

$$q_1 = \left(\frac{\pi}{\sigma}\right)^{1/2} + 1 + \mathrm{e}^{-\sigma} - \left(\frac{\pi}{\sigma}\right)^{1/2} \Phi(\sigma^{1/2}) , \tag{6-25}$$

where Φ designates the well-known error function

$$\Phi(z) = \frac{2}{\pi^{1/2}} \int_0^{z} \mathrm{e}^{-u^2} \, \mathrm{d}u . \tag{6-26}$$

Function (6-25) exhibits correct limiting behaviour and, although it is not an exact formula, it is the best possible approximation available.

Fig. 6-3 depicts the dependence of function q_1 and of the contributions $\delta_1(H_T^0 - H_0^0)/(RT)$ and $\delta_1 S_T^0/R$ of free internal rotation described in terms of q_1 to the standard molar heat content function and entropy and comparison with results obtained using conventional function q_{fir}. These results clearly indicate[80,81] that important changes in the values of the thermodynamic functions can be expected on transition from q_{fir} to approximation q_1, not only for van der Waals molecules at low temperatures, but also for (at least some) organic molecules at low and moderate temperatures.

The useful approximation proposed by Pitzer and Gwinn[68], whose reliability was recently broadly and successfully tested[84], represents a practical technique for transition from (6-19) to a partition function for hindered internal rotation. It is apparent that result (6-25) could be readily incorporated into this technique[81], permitting its applicability for extremely low values of I_r and T. It should also be noted, for the sake of completeness, that a concept of an effective (temperature-dependent) potential term was recently proposed[85] that replaces the conventional potential in the classical partition function for hindered internal rotation, leading to the same results as the semi-classical approach.

The above-mentioned fact that the RRHO approximation at present represents the upper limit to quantum chemical calculations of the thermodynamic characteristics of polyatomic molecules emphasizes the importance of the greatest possible knowledge of the quality of the RRHO approximation. Unfortunately, contemporary understanding of this problem is based on results obtained for relatively rigid systems consisting of only several atoms. These results themselves are very encouraging.

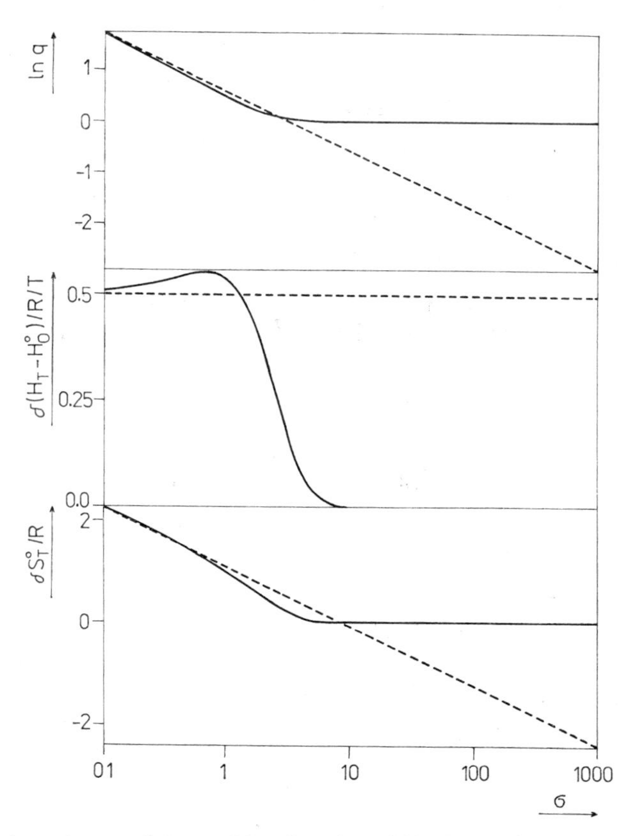

Figure 6-3. The dependences of the partition function of free internal rotation q and of its contributions δ to functions $(H_T^\circ - H_0^\circ)\, R^{-1}T^{-1}$ and S_T°/R calculated using the q_{fir} (Eq. (6-19), dashed lines) and q_1 (Eq. (6-25), solid lines) approaches on the dimensionless parameter σ (from Ref.[80])

Godnew[86] gives an estimation of 0.33 J K^{-1} mol^{-1} (0.1%) for the difference in the entropy term for H_2O at 1000 K, based on exact calculation and on the RRHO approximation; comparison[87] for NH_3 yields a value of 0.66 J K^{-1} mol^{-1} (0.3%) for the same data. Testing[88] of the RRHO approach for a group of triatomic molecules at temperatures of up to 6000 K yields results that are just as favourable. It was demonstrated in the works of Bron *et al.*[89–91] that the correction for deviations from the RRHO model is quite unimportant for isotope exchange equilibria. Another type of test of the quality of the RRHO approximation involves the study[92] of 26 equilibrium processes in the gaseous phase. The equilibrium constants for these processes were evaluated in terms of the RRHO approximation from experimental molecular constants (the values for term ΔH_0^0 were used from other experimental sources than the usual evaluation of the corresponding temperature dependence of the values of log K_p). The average difference between the log K_p value calculated in this way and that obtained experimentally was greater than 10% for only 5 reactions of the set studied and was always less than 30%. This test[92] is an important argument for the usefulness of the RRHO approximation for quantum-chemical study of the characteristics of chemical reactivity, considering the possible experimental inaccuracies in terms of log K_p and ΔH_0^0. The above-mentioned good results for the RRHO approximation in the calculation of the thermodynamic functions of polyatomic molecules are valid for relatively rigid molecules and do not automatically imply its usefulness for very non-rigid systems. The field[93] of quantum chemical studies of the thermodynamic characteristics of processes leading to formation of molecular clusters or van der Waals molecules[94] supplied up-to-date examples of such non--rigid species. When the results obtained for two-atom clusters (the Xe_2 molecule – Table 6-2) are applied to complexes of polyatomic molecules, it must be borne in mind that the number of markedly anharmonic vibrational degrees of freedom increases: van der Waals dimers generally exhibit six intermolecular vibrational modes. Simultaneously, however, the importance of the correction of the thermodynamic functions to the RRHO approximation depends on temperature. The fact that van der Waals molecules are most frequently encountered and have been studied experimentally at low and very low temperatures[94] leads to a reduction of errors in quantum chemical study of the thermodynamics of their formation using the RRHO approximation. However, the RRHO approximation can even be used for the calculation of the thermodynamic characteristics of non-rigid structures at high temperatures, e.g. see the results obtained by Strauss and Thiele[95] for the C_3 molecule. The potential controlling the distortion of the bond angle in this structure is markedly anharmonic and, moreover, implies marked non-rigidity of the whole structure with respect to this motion – cf. (for example) the *ab initio* SCF calculations of Liskow *et al.*[96] in Fig. 3-2. These results actually indicate that the non-rigidity of the C_3 molecule can readily be compared with that exhibited by van der Waals molecules. It was found at a temperature of 2400 K that[95] an error of 17.6 J K^{-1} mol^{-1} is involved in the use of the RRHO approximation compared with the best theoretical value, i.e. 5.2%.

The quality of the description of internal motion in molecules is not reflected only in the values of partition function q introduced by Eq. (6-5), but also in a change in the zero point energy ΔH_0^0. As discussed in Chap. 4.1, the existence of motion in the ground state leads to a quantum correction $\Delta_0^{(q)}$ to the classical value $\Delta E_0^{(cl)}$ derived from the relative positions of the minima on the corresponding energy hypersurface:

$$\Delta H_0^0 = \Delta E_0^{(cl)} + \Delta_0^{(q)} \, . \tag{6-27}$$

In terms of the harmonic normal vibrational frequencies ω_i, their degeneracy factors d_i and anharmonicity constants x_{ij}, the contribution of a single molecule, $\Delta_0'^{(q)}$, to the value of the overall quantum correction $\Delta_0^{(q)}$ is given by the expression[89,97,98]:

$$\Delta_0'^{(q)} = G_0 + \tfrac{1}{2}h \sum_i d_i \omega_i + \tfrac{1}{4}h \sum_i \sum_{j \geqq i} d_i d_j x_{ij} \, , \tag{6-28}$$

where G_0 is a constant term generally neglected in the calculation of thermodynamic characteristics; however, its possible importance in the precise calculation of equilibrium constants has been demonstrated[89,97,98]. In the quantum chemical description of reaction thermodynamics, the range of molecular characteristics that can be generated leads, however, to the necessity of limiting Eq. (6-28) to the second term. In the calculation of thermodynamic characteristics, quantum correction $\Delta_0^{(q)}$ is especially important in the determination of the equilibrium constant values. Numerical comparisons[89,90,97,98] have indicated good agreement between the precise values and results based on only the harmonic term in relationship (6-28).

In conclusion to this practical discussion of partition functions as the connecting factors between the macro- and microscopic characterization of the systems, closely related very general concepts will be mentioned. Recently, Löwdin[99,100] published a unification of quantum mechanics, quantum statistics, classical thermodynamics and (to a certain degree) classical mechanics using trace algebra. This is, at present, the most complete physical basis for the whole theory of chemical reactivity and a means of generalizing considerations[101]. The process of generalization appears most markedly in relationship to the entropy (see, e.g.[102,103]) where it makes special use of information theory[104].

6.1.2 Calculation of Rate Characteristics in the Framework of the Activated Complex Theory

The activated complex theory is quite closely connected with the ideas formulated in 1915 by Marcelin[105], i.e. that realization of a chemical reaction is based on crossing of a certain (critical) surface in the phase space of the reacting molecules. For application to quantitative calculation of the rate constants of chemical reactions, this concept was first utilized by Pelzer and Wigner[106]. This approach was simplified by the introduction of the equilibrium hypothesis, i.e. the assumption of a certain equilibrium between species, passing across this critical surface, and the reactants by Eyring[6] (the activated complex theory) and Evans and Polanyi[7] (the transition state theory). This approximation, also called the theory of absolute reaction rates, converts calculation of rate constant k of the chemical reaction to relationship (6-29):

$$k = \frac{kT}{h} K^{\neq} . \tag{6-29}$$

The problem is thus reduced to the calculation of equilibrium constant $K^{\neq}$ for the formation of the activated complex from the reactant(s), i.e. essentially the problem discussed in Chap. 6.1.1 (see Refs.[13-25]). However, compared with Eq. (6-12), there is a certain difference in the construction of the partition function of the activated complex in connection with exclusion of one vibrational degree of freedom. The thermodynamic consequences of this exclusion were discussed precisely by Guggenheim[107]. Arnot[108] later stated (and was immediately criticized[109]) that Eq. (6-29) should be corrected by a factor of 2; nevertheless, the form generally employed is that given above. Further development of the concept of an activated complex* is given, for example, in the reviews by Laidler and Polanyi[110], Hofacker[111], Christov[112], and Hirschfelder[113]. The critical surface separating the reactant region and the product region in configurational space can be selected[114-117] in a different manner than in the conventional activated-complex theory. In this connection, discussions[110,118-121] about the theory of absolute reaction rates based on the Gibbs function hypersurfaces are especially interesting.

A great deal of attention has been devoted[122-133] to the question of symmetry numbers for the construction of the rotational partition function of an activated complex. Even though this extensive discussion eventually demonstrated[132,133] that the conventional symmetry numbers are consistent with the formulation of the theory of absolute reaction rates, analysis of this classical concept was very fruitful. At the beginning of this debate, two reasons were given for not using symmetry numbers. It appeared[122,124] that they can lead to erroneous rate expressions for symmetric reactions and also that complications can occur when optically active species are present. It has been suggested that these difficulties can be overcome by replacing symmetry numbers by statistical factors[122-126]. For example, Schlag[122] recommended replacing ratios of conventional symmetry numbers by reaction path degeneracy factors derived from the group theory approach. This technique was then simplified by replacing this latter procedure by the direct count method[123]. A variation of this method involves the distinguishing of statistical factors for the forward and backward reactions[124]; other modifications have been suggested[125,126]. Special attention has been paid to the use of statistical factors for optically isomeric reaction pathways and

* It is interesting that, as recently pointed out by Hirschfelder[113], Eyring's classical work[6] was originally not accepted for publication by the Editor of the Journal of Chemical Physics.

optically isomeric activated complexes[127,128]. A further stage in the development of the problem is concerned with the discovery[129–131] that statistical factors need not always be an appropriate tool, as it has been found that the resultant ratio of forward and backward rate constants leads to erroneous equilibrium constant values. This apparent contradiction was finally removed[132,133] by demonstrating that the symmetry number method used in a certain, unambiguous, 'proper' way leads to adequate expressions for the rate formulae. Particular care is necessary in considerations of processes involving optically active species and for symmetric reactions, i.e. reactions where the reactants and products would be indistinguishable if atomic labelling were not used. A prototype bimolecular symmetric reaction is the process $H + H_2 \rightarrow$ $\rightarrow H_2 + H$; simultaneous consideration of the corresponding isotopically labelled process with formal utilization of symmetry numbers σ leads to the following scheme

$$\begin{array}{ccccc} H + D_2 & \rightarrow & H\text{—}D\text{—}D & \rightarrow & HD + D\,, \\ \sigma = 2 & & \sigma = 1 & & \sigma = 1 \end{array} \tag{6-30}$$

$$\begin{array}{ccccc} H + H_2 & \rightarrow & H\text{—}H\text{—}H & \rightarrow & H_2 + H\,, \\ \sigma = 2 & & \sigma = 2 & & \sigma = 2 \end{array} \tag{6-31}$$

indicating that (neglecting differences in the partition functions resulting from mass differences) reaction (6-30) is clearly favoured over reaction (6-31) by a factor of 2, although it would be expected that the rates should be identical under these conditions. This error results from routine application of the theory of absolute reaction rates for reaction (6-31). This theory determines the rate of passage through the transition

Table 6-5. Comparison[a] of quantum-mechanical (QM), transition-state-theory (TST), and TST corrected for QM tunnelling (TST/Q) values of the one-dimensional rate constant k for the collinear reaction $H + Cl_2 \rightarrow HCl + Cl$

T (K)	k (cm s^{-1})		
	QM	TST	TST/Q
300	1.8×10^3	1.23×10^3	1.89×10^3
400	4.8×10^3	3.85×10^3	4.72×10^3
500	9.0×10^3	7.85×10^3	8.75×10^3
600	1.4×10^4	1.28×10^4	1.37×10^4
700	1.9×10^4	1.85×10^4	1.92×10^4
800	2.4×10^4	2.45×10^4	2.52×10^4
900	2.8×10^4	3.07×10^4	3.13×10^4
1000	3.3×10^4	3.70×10^4	3.76×10^4

[a] From Ref.[136]

state in one direction; however, for reaction (6-31), passage in either direction leads to the (required) product. It thus follows from the identical nature of the reactants and products in reaction (6-31) that twice the rate constant for passage through the transition state in a single direction must be used for rate considerations (e.g. in determination of the lifetime of an H_2 molecule against H atom exchange). This rule then leads to the elimination of the reported paradox in reactions (6-30) and (6-31). The whole problem was based on the correct understanding of the quantities produced by the theory of absolute reaction rates.

The evaluation given by Laidler[5] in 1969 of the position of the activated complex theory in the calculation of rate constants is still valid: rates cannot be calculated more satisfactorily by any more detailed theory. Comparison[5] of the values of the rate constants calculated on the basis of the empirical characteristics of activated complexes with experimental kinetic data has demonstrated the usefulness of the theory of absolute reaction rates. Recent[134–141] comparison of quantum-mechanical calculations (or calculations in the framework of the classical trajectory method) with results obtained using the transition state theory are especially favourable. The agreement was found to be quite satisfactory from a practical point of view. The low temperature region is an exception; here the correction for the quantum--mechanical tunnelling effect can become important. Table 6-5 gives an example of a comparison of this kind for the collinear reaction[136] $H + Cl_2 \rightarrow HCl + Cl$; calculations were carried out for the one-dimensional rate constant (see for example[125]). Golden[145] also recently discussed the role and limitations[142] of the theory of absolute reaction rates.

The theory of absolute reaction rates was soon extended by the possibility[144] of correcting the rate constant values for the quantum-mechanical tunnelling effect (see e.g.[145,146]). Expressions are available in the literature for the probability of tunnelling through a one-dimensional barrier with various shapes[145–149]. The scheme for derivation of a correction for tunnelling combines the appropriate permeability (transmission probability) of barrier $G(W)$ for a particle with energy W with the thermal energy distribution given by the Boltzmann distribution. For example, for classical barrier height E in the semiclassical treatment of particle motion, the correction Γ to the rate constant for the tunnelling effect is given (see e.g.[146,149]) by the expression

$$\Gamma = \exp(E/kT) \int_0^\infty \frac{1}{kT} \exp(-W/kT)\, G(W)\, \mathrm{d}W. \tag{6-32}$$

The Wigner correction[148] is often employed, assuming a parabolic shape of the potential barrier:

$$\Gamma = 1 - \frac{1}{24}\left(\frac{h\nu^{\neq}}{kT}\right)^2, \tag{6-33}$$

where $\nu^{\neq}$ is the imaginary frequency of the normal mode of the activated complex decomposition. Provided the value of $\nu^{\ddagger}$ is not available from correctly carried out vibrational analysis of the corresponding activated complex, it can be approximated[148,149] using the curvature $A^{\ddagger}$ of the potential curve given by its second derivative (for details, see Ref.[129]):

$$(\nu^{\ddagger})^2 = \frac{A^{\ddagger}}{4\pi^2 m}, \tag{6-34}$$

where m should be[146] the corresponding reduced mass rather than the total mass. Correction factor Γ appears in relationship (6-29) as a multiplicative term. The Wigner correction for tunnelling is advantageous in that it needs an amount of information on the behaviour of the potential hypersurface identical to that required by the RRHO approximation. It has, however, a great disadvantage in that it does not differentiate between symmetric and asymmetric situations (cf.[150]). Eckart[147] utilized a more realistic shape for the potential barrier (also considering the asymmetric case). Shavitt[151] compared the corrections for tunnelling for the parabolic and the Eckart barrier; more detailed tables of values of the Eckart correction are given in Ref.[152]. Shin[153] gave a simplified expression for correction for tunnelling in the framework of the Eckart nonsymmetric potential, permitting simple numerical calculations. Recently, a further sophistication of calculations of the tunnelling correction has been considered, e.g. in Refs.[154–161].

The theory of absolute reaction rates can be included[162] in the statistical theories, avoiding solution of the dynamic problem by using statistical assumptions; while dynamic theories carry out this solution at either a quantum or a classical level, where the latter approach has undergone considerable development in recent years[8–12,162–164]. In the classical trajectories method, an overall picture of the dynamics is obtained through broad variation of the initial conditions (particle coordinates and momenta) for integration of the appropriate equations of motion. These are the classical equations of motion in the form (suitable for this purpose) of the Hamilton equations (see e.g.[165]):

$$\frac{\partial H}{\partial p_i} = \dot{q}_i\,; \quad \frac{\partial H}{\partial q_i} = -\dot{p}_i\,, \tag{6-35}$$

where $\{q_i\}$ is the selected set of (generalized) coordinates and $\{p_i\}$ is the corresponding set of generalized momenta; differentiation with respect to time is designated in the conventional manner [165]. This whole procedure coincides with quantum mechanics (chemistry) primarily only in the construction of the classical Hamiltonian H, for which the appropriate potential energy hypersurface is used; this is being taken ever more frequently from quantum-chemical calculations. In further stages, the classical approach is corrected for quantum effects, e.g. tunnelling or the probabilities of transition between hypersurfaces. Depending on varying initial conditions, a set of individual trajectories is obtained, i.e. the distribution of the results for the given reaction.

In spite of considerable development in experimental techniques towards state-to--state chemistry, nonetheless it is typical to pass from a set of individual trajectories to the condensation of results in terms of suitable averaging.

Classical reaction trajectories are a starting point for the development[117,141,166–174] of the concept of generalized transition states and the variational transition-state theory that finally permits transition from the classical to the quantum[166–179] transition-state theory. The conventional theory entails the localization of the transition state on a saddle point, while the transition state in the general concept can also be located outside the saddle point, e.g. stepwise at each point on the reaction coordinate between the reactants and products. This procedure permits full consideration of re-crossing effects (the conventional transition state theory completely agrees with classical dynamics only when each of the trajectories crosses the dividing surface between the reactants and products, passing through the saddle point only once). In addition, the corresponding rate constant for each generalized transition state is determined variationally by minimization using suitably chosen criteria. The classical variational theory leads to an upper limit for the exact classical (equilibrium) rate constant and thus permits testing of the conventional transition state theory. In addition, a more sophisticated description of the tunnelling effect has been developed for use in the variational transition-state theory and for motions in generalized transition states. The quantum transition-state theory formed in this way exhibits a very favourable approach[141] to the values of the experimentally observed rate constants.

6.1.3 Examples of Quantum-Chemical Calculations of Equilibrium and Rate Constants

Recently, a number[13,17,35,39,180–233] of calculations of equilibrium and rate constants have been carried out, including the determination of some other thermodynamic characteristics of chemical reactions on the basis of theoretically derived molecular parameters. These calculations were based on either quantum-chemical methods or empirical potentials. Examples of such calculations are listed in Table 6-6. Most of these processes took place in the gas phase; studies of heterogeneous equilibria[181,198] employed experimental thermodynamic information for description of a component in the solid phase. Various quantum-chemical methods were chosen depending on the character of the problem studied – from the HMO method to exact solution of the electronic problem. None of the studied equilibria or rate processes involving polyatomic components was studied by an approximation better than the RRHO approach, corresponding to the above-mentioned capabilities of numerical quantum chemistry in this respect. Conditions are, however, different for systems consisting of at most diatomic molecules – there is a large number of works[35,52,53,234–241] that properly consider anharmonicity and non-rigidity pheno-

Table 6-6. Selected examples[a] of theoretical calculations of equilibrium (K) and rate constants (k) (the activated-complex theory) of chemical reactions[a]

Characteristic	Reaction	Generating energy function	Ref.
K	$2\,LiCl(g) \rightleftarrows (LiCl)_2(g)$	Empirical potential	180
K	$nC(s) \rightleftarrows C_n(g)$; $n = 1-17$	Simple MO method	181
K	$H_2^+(g) + H_2(g) \rightleftarrows H_3^+(g) + H(g)$	*Ab initio* SCF	182
k	$H(g) + H_2(g) \rightarrow H_2(g) + H(g)$	*Ab initio* SCF CI	183
K	$K^+(g) + H_2O(g) \rightleftarrows KH_2O^+(g)$	Empirical potential	184
K	$2\,LiH(g) \rightleftarrows (LiH)_2(g)$	*Ab initio* SCF	185
K	Association of alkali$^+$ with O_2, N_2, CO_2	Empirical potential	186
K	$HCN(g) \rightleftarrows HNC(g)$	*Ab initio* SCF	187
k	Cyclohexane inversion	MINDO/2	188
K[b]	$p\text{-}H_2(g) \rightleftarrows o\text{-}H_2(g)$	Exact solution	35
K	$2\,H_2O(g) \rightleftarrows (H_2O)_2(g)$	*Ab initio* SCF	189
K	$2\,X(g) \rightleftarrows X_2(g)$; $X=CH_3$, NH_2, HCO	MINDO/2	190
K	$2\,X_2(g) \rightleftarrows (X_2)_2(g)$; X=N, O, Br, I	Empirical potential	191
K	$H_2(g) + H_3^+(g) \rightleftarrows H_5^+(g)$	CNDO/2, MINDO/2	13, 192
K	$n\,H_2O(g) \rightleftarrows (H_2O)_n(g)$; $n = 2-5$	Empirical potential	193
K	Association of nitrobenzene$^-$ with alkali$^+$	HMO	194
k	Ethyl-cation inversion	CNDO/2	195
K	$C_2H_2(g) + OH^-(g) \rightleftarrows C_2H^-(g) + + H_2O(g)$ $Li(g) + H_2O(g) \rightleftarrows LiOH(g) + H(g)$	*Ab initio* SCF	196
K	$2\,NF_2(g) \rightleftarrows N_2F_4(g)$	CNDO/2	197
K	$n\,C(s) \rightleftarrows C_n(g)$; $n = 4-7$	MINDO/2	198
K, k	*cis*-$N_2F_2(g) \rightleftarrows$ *trans*-$N_2F_2(g)$	CNDO/2	199, 200
K	$CH_3OH(g) + n\,H_2O(g) \rightleftarrows CH_3OH.$ $.n\,H_2O(g)$; $n = 1, 3, 6$	PCILO	201
K	$2\,NO(g) \rightleftarrows (NO)_2(g)$	*Ab initio* SCF	202
K	Equilibria containing AH_2, AH_3, and AH_4 and their cations; A = N, O	*Ab initio* SCF	203
K	$C_2H_5^+(g) + H_2(g) \rightleftarrows C_2H_7^+(g)$	MINDO/3; ab initio SCF and CEPA PNO	204
K	$2\,HF(g) \rightleftarrows (HF)_2(g)$ $2\,HCl(g) \rightleftarrows (HCl)_2(g)$	*Ab initio* SCF	205
k	$(HF)_2(g) \rightarrow (HF)_2(g)$ $(HCl)_2(g) \rightarrow (HCl)_2(g)$	*Ab initio* SCF	205
K	Lactam-lactim tautomeric equilibria	MINDO/3	206
K	$2\,H_2O(g) \rightleftarrows (H_2O)_2(g)$	*Ab initio* SCF CI	207, 208
k[c]	$H_2CCH_2F(g) \rightarrow H_2CCHF(g) + H(g)$	*Ab initio* SCF	209
k	$(H_2O)_2(g) \rightarrow (H_2O)_2(g)$	*Ab initio* SCF CI	211
k[c]	$H_2CCHF(g) \rightarrow HCCH(g) + HF(g)$ $H_2CCHF(g) \rightarrow H_2CC(g) + HF(g)$	*Ab initio* SCF	212

Table 6-6. (Cont.)

Characteristic	Reaction	Generating energy function	Ref.
K	$Na(g) + H_2O(g) \rightleftarrows NaOH(g) + H(g)$	*Ab initio* SCF	213
K	$MCl_4^{2-}(T_d)\,(g) \rightleftarrows MCl_4^{2-}(D_{4h})\,(g)$; M = Mn, Fe, Co, Ni	CNDO	214
K	*cis*-1-methylallyl$^-$(g) $\rightleftarrows$ *trans*--1-methylallyl$^-$(g)	MNDO	215
K	Equilibrium between two isomeric rectangular cyclobutadienes	*Ab initio* SCF	216
k	$H(g) + H_2CO(g) \rightarrow H_2(g) + HCO(g)$	*Ab initio* SCF CI	217
k	$H_2(g) + C_2H(g) \rightarrow H(g) + C_2H_2(g)$	*Ab initio* SCF CI	218

[a] According to Refs.[13,17]
[b] Higher quality partition functions than RRHO were used.
[c] The RRKM (see, e.g., Ref.[210]) theory of unimolecular reactions was used.

mena in calculation of thermodynamic characteristics on the basis of theoretical molecular characteristics. These frequently include studies of the thermodynamics of formation of diatomic clusters (most often rare gas dimers) on the basis of empirical[52,53,234–239,241] or quantum-chemical[240] potential curves.

From a methodical point of view, systematic studies of the applicability of the individual quantum chemical methods for generation of reliable information for the establishment of partition functions in the RRHO approximation are of very great importance (rather than studies of individual chemically interesting processes alone). The works[87,242] described such a study of the CNDO/2, INDO and MINDO/2 methods on a set of 12 four- to eleven-atom molecules. Dewar and Ford[220] carried out analogous testing of the MINDO/3 method on a set of 27 three- to twelve-atom molecules. In both cases, very good utilizability of the tested semiempirical methods for setting up of partition functions in the RRHO approximation was found. It was demonstrated that, even at higher temperatures, the term $T\,\Delta S_T^0$ is found with greater reliability than the term ΔH_T^0. In addition to equilibrium thermodynamic characteristics, work[220] also included a limited test of the calculation of activation parameters; this comparison of the MINDO/3 values for the activation entropies for six processes is given in Table 6-7. If the magnitude of the experimental errors given is considered, the agreement can be considered as very good. Although the very reasonable quality of the rotational partition functions could be expected considering the usual general reliability of semiempirical methods for calculations of molecular geometry[243], the results for the vibrational partition functions were not as apparent. The agreement here is the result of two factors[87,242]: the determination of the frequency values associated with bond angle deformation (important in the partition

Table 6-7. Comparison[a] of the MINDO/3 and observed activation entropies $\Delta S^{\ddagger}_{T}$

Rate process[b]	T (K)	$\Delta S^{\ddagger}_{T}$ (J K^{-1} mol^{-1})	
		MINDO/3	Observed[c]
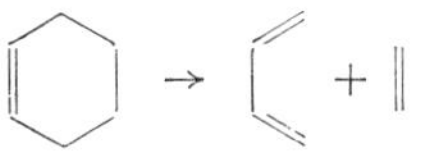 → + ‖	600	37.7	33.7
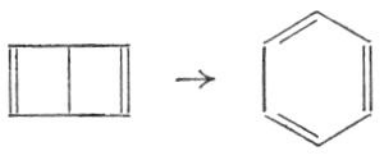 →	321	6.2	−20.9 ± 29.3[d]
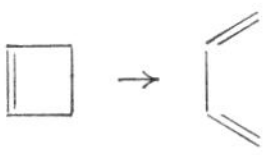 →	423	3.2	0.4 ± 3.8
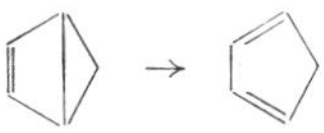 →	350	6.7	18.4 ± 3.8
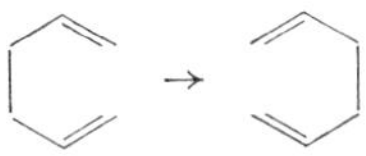 →	523	−71.1	−57.7 ± 4.2
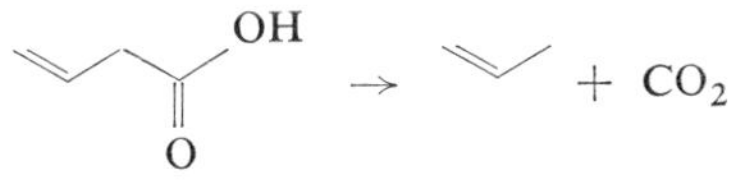 OH, O → + CO_2	650	−58.2	−42.7 ± 10.5

[a] From Ref.[220]
[b] Gas-phase processes.
[c] For sources of the observed values, see Ref.[220]
[d] Measurements in solution.

function because of their low values) is much more precise in the semi-empirical methods than the determination of frequencies connected with bond stretching. The latter values are found with a considerable error but, as they are overestimated (rather than underestimated) the vibrational partition function is not greatly affected by this error.

Although tests such as those carried out for semiempirical methods[87,220,242] have not yet been performed at the level of *ab initio* SCF calculations, conclusions can be drawn from knowledge of the relationship between the quality of the molecular parameters obtained from semiempirical and nonempirical calculations. The situation becomes particularly simple for the rotational partition function. Calculations of molecular geometry[244] at the *ab initio* SCF level are mostly comparable[245] with the quality of results obtained by the INDO method (see Table 3-6). Knowledge of the relationships between vibrational frequencies obtained on the basis of semi-empirical and nonempirical SCF calculations (Chap. 3.6) permits the assumption that, in general, the vibrational partition functions obtained from the *ab initio* SCF calculations need not be of worse quality then those found from semiempirical methods.

Summarizing, it appears that quantum chemical methods are quite useful for the calculation of partition functions in the RRHO approximation. However, in contrast to the entropy term, it should be noted that the precision of the calculation of ΔH_T^0 (or $\Delta E_0^{(cl)}$) can depend on the system studied, not only with semiempirical methods[246,247], but also for *ab initio* SCF calculations[248-253] (because of the effect of the limited length of the basis set and neglection of correlation effects – see Chap. 3.2). Nonetheless, the constant progress of quantum chemistry in the solution of this its central problem is promising. As the energy term has been treated in detail in a monograph[254], it is considered only superficially here.

6.2 Equilibrium Processes with Isomerism of the Components

The generally good level of applicability of quantum chemical methods for the calculation of the characteristics of equilibrium and rate processes, discussed in the above section, permitted transition from study of processes with extensive experimental information on the molecular properties of the components to equilibrium and rate processes for which little information was available on the reactants, products and/or activated complexes. Thus, gradually more studies appeared where investigation of the corresponding energy hypersurface demonstrated that at least one of the components of the process studied is not, rigorously, a chemical individual, but rather a mixture of at least two isomeric forms. This led to the decomposition of processes described experimentally by a single chemical equation into an appropriate number of partial processes. Each of these partial processes, whether equilibrium or rate, could be treated statistical-thermodynamically on the basis of its quantum chemical characteristics. However, the resultant partial equilibrium or rate characteristics could mostly not be obtained experimentally; only the summary (total, or effective) characteristics for the overall process, not distinguishing between individual isomeric forms, were available. (This, of course, does not imply the impossibility of obtaining these partial characteristics experimentally – it is only necessary to find a technique differing between the isomeric structures under the given conditions, for which rotational and NMR spectroscopy or chromatography seem especially promising.) A number of such systems treated in recent quantum-chemical studies (common chemical equilibria, equilibrium processes involving van der Waals molecules, or rate processes) have indicated that it is useful, and necessary, to generalize the conventional concept of chemical reactivity characteristics given in the previous chapter. This generalization would include possible isomerism of an arbitrary component of the equilibrium process or activated complex in a rate process. The conventional concept discussed in the previous chapter would then result from this general scheme for unit numbers of all isomers. Before systematically deriving the required relationships, several examples will be given of processes with component isomerism.

In a mass spectrometric study of carbon vapours, Drowart *et al.*[255] found the characteristics of the equilibrium process

$$4\,C\,(s) \rightleftarrows C_4\,(g)\,. \tag{6-36}$$

However, a later MINDO/2 theoretical analysis[198] of C_n (g) ($n = 4-7$) aggregates demonstrated that, in addition to the most stable form C_4 with point group of symmetry D_{2h}, there are an additional two isomers of higher energy with D_2 and T_d symmetry (Fig. 6–4). From a theoretical point of view, it would thus be useful to describe the thermodynamics of the three partial processes

$$4\,C\,(s) \rightleftarrows C_4\,(g;\ D_{2h})\,, \tag{6-37}$$

$$4\,C\,(s) \rightleftarrows C_4\,(g;\ D_2)\,, \tag{6-38}$$

$$4\,C\,(s) \rightleftarrows C_4\,(g;\ T_d)\,. \tag{6-39}$$

However, mass spectrometry does not permit the study of each of processes (6-37)

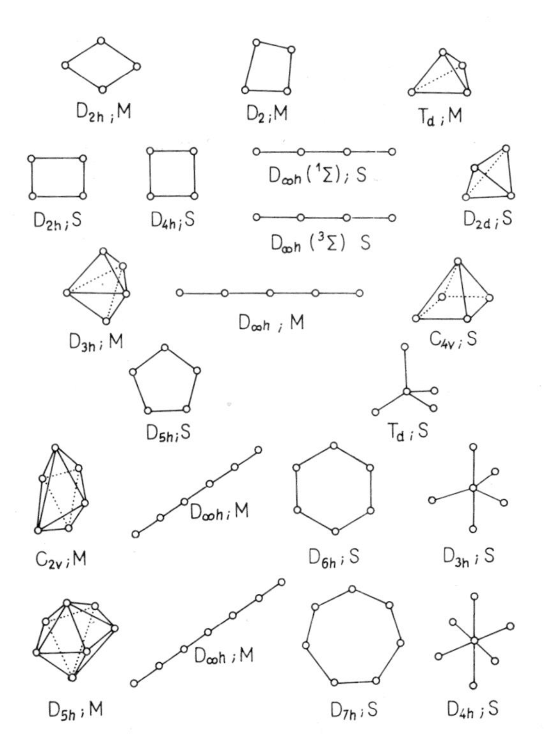

Figure 6-4. Schemes of stationary points found[198] on the MINDO/2 potential energy hypersurfaces of C_n ($n = 4-7$); M is a minimum, S a saddle point

to (6-39) separately, but only the description of the thermodynamics of overall process (6-36). It is therefore necessary also in theoretical description to pass from the characteristics of partial processes (6-37)–(6-39) to the summary values for the overall process (6-36) (assuming full equilibrium conditions). Isomerism was also found for all higher aggregates C_n studied in this work[198] (Fig. 6-4).

Nonequivalent structures with a summary formula of HCNO are not only the system for which the phenomenon of isomerism was first described (see Chap. 2.1), but also that for which the stability of the individual forms was apparently first found in theoretical calculations – in the very beginnings of quantum chemistıy Pauling and Hendricks[256] carried out a study of the relative stability of isomers on the basis of a simple electrostatic model. Half a century later, Poppinger *et al.*[257] carried out a detailed analysis of the singlet potential energy hypersurface for the HCNO system at an *ab initio* SCF level using three different basis sets, in an attempt to find all its minima, transition states, and reaction paths converting them. Depending on the choice of a basis set, the energy hypersurface was found to contain nine or seven minima and ten transition states allowing interconversion of structures corresponding to the energy minima. In spite of the large number of stationary points found, this number need not necessarily be exhaustive. The authors[257] limited the structure optimization to only some of the 38 possible, topologically different HCNO bonding isomers. Useful representation of the six-dimensional HCNO hypersurface permits transition to a two-dimensional surface that is a function of the bond angles $\Theta =$ $= \sphericalangle$ HCN and $\Phi = \sphericalangle$ CNO (Fig. 3-6). Apparently, a comprehensive study of an arbitrary equilibrium or rate process involving the HCNO molecule must consider this rather extensive isomerism.

The formation of van der Waals molecules is one type of process for which frequent isomerism of at least one reaction component can be expected. It has actually been

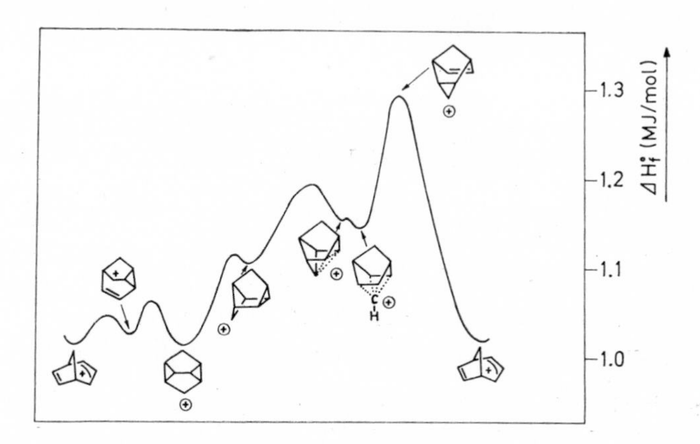

Figure 6-5. The connectivity and the magnitude of the MINDO/3 barriers[260] between selected C_8H_9 cations

found for the formation of the $(H_2)_2$ dimer:

$$2\,H_2\,(g) \;\rightleftarrows\; (H_2)_2\,(g) \tag{6-40}$$

that, on the basis of *ab initio* SCF calculations[258,259], this dimer is a mixture of four structures of comparable stability with symmetries D_{2h}, $D_{\infty h}$, C_{2v} and D_{2d}. Considering that even molecular hydrogen is a mixture of two forms (o-H_2 and p-H_2) (see, e.g.[35]), it became apparent that overall process (6-40) can be understood to be a result, for example, of 12 partial processes.

It is clear that the transition to a multiatomic system should generally be accompanied by a further increase in isomerism and a consequent increase in the complexity of description of the overall equilibrium and rate processes. Detailed study of this type is so far exceptional; an example is the analysis of the conditions in the $C_8H_9^+$ cation carried out by the MINDO/3 method[260], with the results summarized in Fig. 6-5. It is apparent that the three most favourable minima are energetically of comparable stability and are separated by such low barriers that not a single structure, but an equilibrium mixture of all three should be considered, even at room temperature.

6.2.1 A General Equilibrium Problem

Study[261] describes a solution of the problem for equilibrium process (6-10) in an ideal gas phase for isomerism of every reaction component. This equilibrium is described by the standard enthalpy change ΔH_T^0, standard entropy change ΔS_T^0, and standard change in the Gibbs function ΔG_T^0, for which it holds true, in terms of equilibrium constant K_p from Eq. (6-11), that

$$\Delta H_T^0 = RT^2\,\frac{\mathrm{d}\ln K_p}{\mathrm{d}T}\,, \tag{6-41}$$

$$\Delta S_T^0 = R\ln K_p + \Delta H_T^0/T\,, \tag{6-42}$$

$$\Delta G_T^0 = -RT\ln K_p\,. \tag{6-43}$$

Assume that each component A_k of reaction (6-10) is a mixture of j_k isomers $A_k^{(1)}$, $A_k^{(2)}, \ldots, A_k^{(j_k)}$, where $A_k^{(1)}$ designates the isomer with the lowest zero-point energy (if there are a greater number of such structures, then index 1 designates any one of them). This differentiation of components A_k into individual isomeric structures displays overall equilibrium (6-10) as the consequence of all possible partial equilibria, of which the set of $\prod_{i=1}^{n} j_k$ processes

$$\sum_{k=1}^{n} \nu_k A_k^{(i_k)} = 0\,, \quad (i_k = 1, 2, \ldots, j_k) \tag{6-44}$$

is selected. Relationships analogous to Eqs. (6-11) and (6-41)–(6-43) are valid for the partial thermodynamic characteristics of equilibrium (6-44):

$$K_{i_1,i_2,\dots,i_n} = \prod_{k=1}^{n} p_{k,i_k}^{\nu_k}, \tag{6-45}$$

$$\Delta H^0_{i_1,i_2,\dots,i_n} = RT^2 \frac{\mathrm{d}}{\mathrm{d}T} \ln K_{i_1,i_2,\dots,i_n}, \tag{6-46}$$

$$\Delta S^0_{i_1,i_2,\dots,i_n} = R \ln K_{i_1,i_2,\dots,i_n} + \Delta H^0_{i_1,i_2,\dots,i_n}/T, \tag{6-47}$$

$$\Delta G^0_{i_1,i_2,\dots,i_n} = -RT \ln K_{i_1,i_2,\dots,i_n}, \tag{6-48}$$

where p_{k,i_k} designates the partial pressure of isomer $A_k^{(i_k)}$ in the equilibrium mixture.

Provided that component A_k is not differentiated into individual isomers $A_k^{(i_k)}$, the number of chemical components of the system is n; considering the individual isomers, this increases to $\sum_{k=1}^{n} j_k$. Although $\prod_{k=1}^{n} j_k$ partial equilibria (6-44) are not all the equilibria conceivable for the system, nonetheless, all of the corresponding equilibrium constants (6-45) are not independent. For our purposes, it will be sufficient to consider the set of $1 - n + \sum_{k=1}^{n} j_k$ independent equilibrium constants:

$$\begin{aligned} K_{i_1} &= K_{i_1,1,1,\dots,1}, \quad (i_1 = 1, 2, \dots, j_1), \\ K_{i_2} &= K_{1,i_2,1,\dots,1}, \quad (i_2 = 1, 2, \dots, j_2), \\ &\vdots \\ K_{i_n} &= K_{1,1,1,\dots,i_n}, \quad (i_n = 1, 2, \dots, j_n). \end{aligned} \tag{6-49}$$

Processes corresponding to equilibrium constants (6-49) are connected with the following standard enthalpy and entropy changes:

$$\begin{aligned} \Delta H^0_{i_1} &= \Delta H^0_{i_1,1,1,\dots,1} = RT^2 \,\mathrm{d} \ln K_{i_1}/\mathrm{d}T, \\ \Delta H^0_{i_2} &= \Delta H^0_{1,i_2,1,\dots,1} = RT^2 \,\mathrm{d} \ln K_{i_2}/\mathrm{d}T, \\ &\vdots \\ \Delta H^0_{i_n} &= \Delta H^0_{1,1,1,\dots,i_n} = RT^2 \,\mathrm{d} \ln K_{i_n}/\mathrm{d}T, \end{aligned} \tag{6-50}$$

$$\begin{aligned} \Delta S^0_{i_1} &= \Delta S^0_{i_1,1,1,\dots,1} = R \ln K_{i_1} + \Delta H^0_{i_1}/T, \\ \Delta S^0_{i_2} &= \Delta S^0_{1,i_2,1,\dots,1} = R \ln K_{i_2} + \Delta H^0_{i_2}/T, \\ &\vdots \\ \Delta S^0_{i_n} &= \Delta S^0_{1,1,1,\dots,i_n} = R \ln K_{i_n} + \Delta H^0_{i_n}/T. \end{aligned} \tag{6-51}$$

Analogously, terms $\Delta G^0_{i_k}$ for the standard changes in the Gibbs function for the partial processes described by equilibrium constants (6-49) will be introduced. It is now necessary to find expressions for summary characteristics K_p, ΔH^0_T, ΔS^0_T, and ΔG^0_T in terms of partial characteristics K_{i_k}, $\Delta H^0_{i_k}$, $\Delta S^0_{i_k}$, and $\Delta G^0_{i_k}$.

Partial pressure p_k of component A_k is expressed in terms of the partial pressures p_{k,i_k} of the individual isomers $A_k^{(i_k)}$:

$$p_k = \sum_{i_k=1}^{j_k} p_{k,i_k} \,. \tag{6-52}$$

Suitable combination of Eqs. (6-11), (6-45), (6-49), and (6-52), and algebraic manipulation, leads to the relationship

$$K_p = K_{1,1,1,\dots,1}^{-(n-1)} \prod_{k=1}^{n} \left(\sum_{i_k=1}^{j_k} K_{i_k}^{1/\nu_k} \right)^{\nu_k} . \tag{6-53}$$

It follows from Eqs. (6-41), (6-50), and (6-53) that

$$\Delta H_T^0 = -(n-1)\, \Delta H_{1,1,1,\dots,1}^0 + \sum_{k=1}^{n} \sum_{i_k=1}^{j_k} w_{i_k} \, \Delta H_{i_k}^0 \,, \tag{6-54}$$

while relationships (6-42), (6-51), and (6-54) lead to

$$\Delta S_T^0 = -(n-1)\, \Delta S_{1,1,1,\dots,1}^0 + \sum_{k=1}^{n} \sum_{i_k=1}^{j_k} w_{i_k} (\Delta S_{i_k}^0 - R\nu_k \ln w_{i_k}) \,, \tag{6-55}$$

and, finally, Eqs. (6-54) and (6-55) yield

$$\Delta G_T^0 = -(n-1)\, \Delta G_{1,1,1,\dots,1}^0 + \sum_{k=1}^{n} \sum_{i_k=1}^{j_k} w_{i_k} (\Delta G_{i_k}^0 + RT\nu_k \ln w_{i_k}) \,, \tag{6-56}$$

where it holds for weight factors w_{i_k} that

$$w_{i_k} = K_{i_k}^{1/\nu_k} \Big/ \sum_{i_k=1}^{j_k} K_{i_k}^{1/\nu_k} \,. \tag{6-57}$$

It is apparent that weight factors (6-57) actually fulfil the normalization condition

$$\sum_{i_k=1}^{j_k} w_{i_k} = 1 \,. \tag{6-58}$$

In addition, it can be readily demonstrated that factors w_{i_k} are dependent only on the properties of isomers $A_k^{(i_k)}$. Consider the isomerization processes

$$A_k^{(1)} \rightleftarrows A_k^{(i_k)} \,, \quad (k = 1, 2, \dots, n;\ i_k = 1, 2, \dots, j_k) \,, \tag{6-59}$$

described by equilibrium constants $K_{1 \to i_k}$; in terms of the isomerization constants, Eq. (6-57) assumes the form

$$w_{i_k} = K_{1 \to i_k} \Big/ \sum_{i_k=1}^{j_k} K_{1 \to i_k} \,, \tag{6-60}$$

containing only the characteristics of the isomers of component A_k. If the changes in the standard thermodynamic functions $\Delta H^0_{1\to i_k}$, $\Delta S^0_{1\to i_k}$, $\Delta G^0_{1\to i_k}$ connected with processes (6-59) are introduced, Eqs. (6-53)–(6-56) assume the forms

$$K_p = K_{1,1,1,\ldots,1} \prod_{k=1}^{n} \left(\sum_{i_k=1}^{j_k} K_{1\to i_k} \right)^{\nu_k}, \tag{6-61}$$

$$\Delta H^0_T = \Delta H^0_{1,1,1,\ldots,1} + \sum_{k=1}^{n} \sum_{i_k=1}^{j_k} \nu_k w_{i_k} \Delta H^0_{1\to i_k}, \tag{6-62}$$

$$\Delta S^0_T = \Delta S^0_{1,1,1,\ldots,1} + \sum_{k=1}^{n} \sum_{i_k=1}^{j_k} \nu_k w_{i_k} (\Delta S^0_{1\to i_k} - R \ln w_{i_k}), \tag{6-63}$$

$$\Delta G^0_T = \Delta G^0_{1,1,1,\ldots,1} + \sum_{k=1}^{n} \sum_{i_k=1}^{j_k} \nu_k w_{i_k} (\Delta G^0_{1\to i_k} + RT \ln w_{i_k}). \tag{6-64}$$

The form of Eqs. (6-61)–(6-64) immediately indicates the value of the isomerism contribution to the summary thermodynamic characteristics compared with the values of the corresponding basic processes realized through the most stable structures in the individual sets of isomers (on the zero-point energy scale; alternatively, index 1 can be conventionally assigned to the most stable structures on the Gibbs function scale at a given temperature).

Relationships (6-53)–(6-57) or (6-60)–(6-64) represent a solution to the general equilibrium problem – they provide expressions for the summary thermodynamic quantities corresponding to the overall equilibrium process without distinguishing the individual isomers of the components (i.e. experimental values) in terms of the characteristics of the partial processes with individual isomeric structures (i.e. values obtained primarily from quantum-chemical studies).

We will now consider a common special case. It was apparent from the introductory examples that theoretical study mostly demonstrates isomerism of only one component for processes of the type

$$\sum_{k=1}^{r} \eta_k B_k \rightleftarrows C^{(i)}; \quad (i = 1, 2, \ldots, m), \tag{6-65}$$

which is clearly a special case of the general system discussed above, where it holds for the number of isomers of the individual components that $j_{B_1} = j_{B_2} = \ldots = j_{B_r} = 1$, $j_C = m$. For simplicity, let

$$\begin{aligned} K_i &= K_{1,1,\ldots,1,i}, \\ \Delta H^0_i &= \Delta H^0_{1,1,\ldots,1,i}, \\ \Delta S^0_i &= \Delta S^0_{1,1,\ldots,1,i}, \\ \Delta G^0_i &= \Delta G^0_{1,1,\ldots,1,i}, \quad (i = 1, 2, \ldots, m). \end{aligned} \tag{6-66}$$

This converts equations (6-53)–(6-57) or (6-60)–(6-64) to the form

$$K_p = \sum_{i=1}^{m} K_i , \tag{6-67}$$

$$\Delta H_T^0 = \sum_{i=1}^{m} w_i \, \Delta H_i^0 , \tag{6-68}$$

$$\Delta S_T^0 = \sum_{i=1}^{m} w_i (\Delta S_i^0 - R \ln w_i) , \tag{6-69}$$

$$\Delta G_T^0 = \sum_{i=1}^{m} w_i (\Delta G_i^0 + RT \ln w_i) , \tag{6-70}$$

$$w_i = K_i / K_p = K_{1 \to i} \Big/ \sum_{i=1}^{m} K_{1 \to i} . \tag{6-71}$$

In the derivation of the relationship determining the contributions of the partial processes to the corresponding summary values of the overall process, only some of the possible equilibria were considered (Eqs. (6-44) or (6-49)). If it were necessary to correctly describe the problem in the sense of simultaneous chemical equilibria[262,263], it would then be necessary to generate all the possible components permitted by the given set of atoms and then set up[262,263] an optimal set of linearly independent reactions. It should be noted that, in contrast to the general problem of simultaneous chemical equilibria (determination of the equilibrium composition of a given reaction system), it is necessary here to find the effective thermodynamic characteristics for a certain special class of simultaneous equilibria. It is apparent from the structure of the given derivation that, provided that transition from our reduced system of equilibria to a complete set in the sense[262,263] of the general problem of simultaneous chemical equilibria does not affect the values of the equilibrium constants (6-45) (ensured for a system in the ideal gas state), the selection has been carried out correctly. For completeness, it should also be noted that, for the theory of simultaneous chemical equilibria in situations where isomers are present, a technique has been introduced treating them as pseudospecies[263].

The main goal of the relationships found is to permit a rigorous comparison of theoretical and experimental thermodynamic characteristics for processes where the experimental arrangement, in contrast to the theoretical calculation, does not permit the distinguishing between individual isomeric structures. In this convolution connection, it is important whether the experiment was carried out under total equilibrium conditions (cf.[32,264]), as assumed in the derivation. In other words, it is important whether the rate processes establishing the equilibrium state were capable, under the experimental conditions, of actually attaining this equilibrium. This question can be answered in each individual case only after investigation of the barriers separating the individual isomers. It should be noted, however, that a whole important field of isomerism – rapid tautomeric equilibria (see Table 2-3 and also for example[265–267]) – reliably ensures equilibrium establishment. On the other hand, however, cases are well known (see for example[268–271]) where a specific generation of isomers is present, possibly even as preferential formation of less stable species[272,273]. Interesting possibilities for such selectivity are provided by the technique of infrared multiphoton excitations in an intense laser field. However, even if complete equilibrium is established under the experimental conditions (as assumed in the theoretical treatment), a further problem remains in relation to the interpretation of experimental results. Not all experimental techniques ensure indication (registration) of the individual isomers in a manner (always in the same way) proportional to their concentrations. While this is clearly true for mass spectrometry and pressure measurements, it is not ensured in some spectral techniques. It is necessary to consider the differences in the molar absorption coefficients of the individual isomeric structures for the given wavelength as it is typically not possible to estimate and consider these differences during experimental evaluations. In the limiting case (when the molar absorption coefficient equals zero) the contribution of the individual structure found theoretically would have to be suppressed. The only correct approach possible in such a case would involve the theoretical simulation of the particular spectrum and the evaluation of the relationships between the individual molar absorption coefficients. Formally, this fact can be included in Eq. (6-52) using additional weighting factors for the contributions of individual isomers to the partial pressure of the component that they form and to carry out the above derivation with this modification. In conclusion, however, it should be noted that experimental techniques are available that, under certain conditions, permit the reliable determination of the contents of the individual isomeric structures in the equilibrium mixture – an example is gas chromatography, in combination with (for example) NMR, or microwave spectroscopy.

Finally, the relationship of this general equilibrium problem to some earlier special results will be considered. The commonest case involving the consideration of the phenomenon of isomerism in the calculation of thermodynamic characteristics is an increase in the entropy of an equimolar mixture of optical isomers, derived in textbooks (e.g.[264]) from the entropy of mixing. This correction also follows[261] trivially from relationships (6-67)–(6-71). Similarly, the following examples are also included in the general scheme as special cases. In connection with the above-mentioned possibility of restricted rotation[74], Aston *et al.*[75] considered calculation of the thermodynamic functions of two conformers in equilibrium. This type of problem was further developed by Compton[274] in his calculation of the thermodynamic functions of conjugated compounds. The work by Heatley[275] considers an analogous problem for alkanes. Geometrical tautomerism was considered in the calculation of the thermodynamic characteristics of cyclic systems in works[276–278]. The importance of isomerism for the thermodynamics of formation of molecular (or atomic) complexes (clusters) is mentioned in the works[279,280] and is strongly emphasized in the work by Hoare[281]. The rotational isomeric state technique developed by Volkenshtein[282] and Flory[283,284], considering approximately isomerism caused by restricted rotation, is important for the thermodynamics of macromolecules. The literature contains a number of works evaluating weight factors for individual conformational chains with this type of motion (see for example Refs.[285-291]) which are important, *inter alia*, for biomolecules[292-294]. The contribution of isomerism (primarily rotational) to the values of thermodynamic functions is discussed in the works[295-306]; the systematic study of the thermodynamic properties of mixtures of rotamers carried out by Compton, Durig *et al.*[299-306] is essentially interesting. Although most studies of the consequences of the isomerism of the reaction components for the thermodynamics of chemical reactions are concerned with only the gaseous phase, attempts have been made to include the effect of the medium[201,298,307]. In conclusion, the useful concept of a stabilomer[308] will be mentioned; this species is defined as that isomer exhibiting the lowest Gibbs function value at room temperature.

6.2.2 Formulation at a Microscopic Level

The relationships derived in Chap. 6.2.1 using phenomenological thermodynamics are a useful tool. Nonetheless, it is interesting to interpret[309] them in terms of statistical mechanics and thus to consider the internal structure at a microscopic level. For every isomeric structure $A_s^{(i_s)}$ participating in equilibrium processes (6-44), consider the corresponding partition function q_{i_s}:

$$q_{i_s} = \sum_j \exp\left(-(e_j^{(i_s)} - e_0^{(i_s)})/kT\right), \quad (i_s = 1, 2, \ldots, j_s), \tag{6-72}$$

where $e_j^{(i_s)}$ designates the energy of the quantum state of isomer i_s, and $e_0^{(i_s)}$ is its ground state energy. The possibility of unambiguously introducing partition function q_{i_s} automatically assumes unambiguous assignment of the states in the whole isomeric system A_s to the individual isomeric structures $A_s^{(i_s)}$. In other words, it assumes independence of the motions of the individual isomers. In the framework of the Born-Oppenheimer approximation, this means decomposition of the hypersurface (or hypersurfaces if the isomers $A_s^{(i_s)}$ belong to different electronic states) for the potential energy of system A_s into isolated (i.e. separated by infinitely high barriers) domains corresponding to the individual isomers $A_s^{(i_s)}$. This property is exhibited, for example, by the hypersurface modified to satisfy the requirements (or assumptions) of the RRHO approximation. Relationships (6-44) and (6-52) are thus analogous, on a macroscopic level, to the microscopic assumption of separability.

Consider partition function q_s of component A_s, given by Eq. (6-5). If the problem were not separated into the consideration of individual isomeric structures, the use of partition functions q_s in general relationships (6-6)–(6-8) and (6-12) would lead to the complete thermodynamic description of process (6-10) and would permit direct comparison of theory and experiment. However, in some cases chemical utilizability and, in all cases, the state of numerical quantum chemistry leads us to distinguish the individual isomers $A_s^{(i_s)}$ of component A_s. Provided the assumption of complete independence of the individual isomers is valid, the relationship between partition functions q_s and functions q_{i_s} can be reduced to a simple sum:

$$q_s = \sum_{i_s=1}^{j_s} q_{i_s} \exp\left(-(e_0^{(i_s)} - e_0^{(s)})/kT\right); \quad (s = 1, 2, \ldots, n), \tag{6-73}$$

where $e_0^{(s)}$ designates the energy of the ground state of A_s, i.e. the minimum in set of values $e_0^{(i_s)}$.

If partition functions q_s are now used in form (6-73) to describe the thermodynamics of process (6-10) and the contributions of the individual isomers $A_s^{(i_s)}$ are separated in Eqs. (6-6)–(6-8) and (6-12), then, after suitable algebraic manipulation, relationships (6-53)–(6-56) or (6-61)–(6-64) can be reproduced, i.e. the relationships derived above on the basis of a phenomenological approach. Weight factors w_{i_s} can be expressed in terms of partition functions q_{i_s}:

$$w_{i_s} = \frac{q_{i_s} \exp\left(-(e_0^{(i_s)} - e_0^{(s)})/kT\right)}{\sum_{i_s=1}^{j_s} q_{i_s} \exp\left(-(e_0^{(i_s)} - e_0^{(s)})/kT\right)}. \tag{6-74}$$

As is clearly apparent from Eq. (6-74), the weight factor of isomer $A_s^{(i_s)}$ depends only on the parameters of the isomers of component A_s (cf. Eq. (6-60)); in fact, even the reference energy level $e_0^{(s)}$ exactly cancels out.

6.2.3 Limiting Properties of Summary Characteristics

In the evaluation of the effect of isomerism of the components of chemical processes on their summary thermodynamic characteristics, it is useful to know the limits within which the summary terms can vary. These limits can readily be found for the enthalpy term in Eq. (6-68). If the largest and smallest values, ΔH^0_{max} and ΔH^0_{min}, in the set of values ΔH_i^0 are introduced, then Eq. (6-58) can be used to demonstrate the validity of inequality

$$\Delta H^0_{min} \leqq \Delta H^0_T \leqq \Delta H^0_{max}. \tag{6-75}$$

Thus, the summary enthalpy term cannot assume values outside the interval $\langle \Delta H^0_{min}; \Delta H^0_{max} \rangle$. Considering that the more marked manifestations of isomerism of a given reaction component mostly assumes small energy differences between the most stable isomer and those with higher energy (i.e. similarity of values ΔH^0_{min} and ΔH^0_{max}), it can be expected that the enthalpy term will not be strongly affected by isomerism.

For the entropy term, however, it can be assumed (because of its physical significance) that the isomerism of the reaction components will have a stronger effect on the summary value. A rough estimate of the limits of the summary entropy term can be obtained from Eq. (6-69) using parameter[310]:

$$\lambda = K_{max}/K_{min}, \tag{6-76}$$

defined in terms of the largest and smallest values K_{max} and K_{min} in the set of equilibrium constants (6-66). Eq. (6-58) can readily be used to demonstrate the validity of the estimate

$$\Delta S^0_{min} + R(\ln m - \ln \lambda) \leqq \Delta S^0_T \leqq \Delta S^0_{max} + R(\ln m + \ln \lambda), \tag{6-77}$$

where ΔS^0_{max} and ΔS^0_{min} are the largest and smallest values in the set of ΔS^0_i values from Eq. (6-69). It can, however, be demonstrated that a stronger statement is valid for the limits of summary entropy term ΔS^0_T than that expressed by Eq. (6-77), viz. the relationship[93]:

$$\Delta S^0_{min} < \Delta S^0_T \leqq \Delta S^0_{max} + R \ln m, \quad (m > 1). \tag{6-78}$$

In order to demonstrate the validity of inequality (6-78), consider a function of m variables

$$f(w_1, w_2, \ldots, w_m) = \sum_{i=1}^{m} w_i \ln 1/w_i ; \quad 0 < w_i \leqq 1, \tag{6-79}$$

with binding condition

$$\sum_{i=1}^{m} w_i = 1. \tag{6-80}$$

The extreme behaviour of function (6-79) with binding condition (6-80) can be studied using the method of Lagrange (undetermined) multipliers; it follows that the extreme is attained only when

$$w_i = \frac{1}{m}, \quad (i = 1, \ldots, m). \tag{6-81}$$

After confirming that this extreme is a maximum, it can readily be demonstrated that Eq. (6-78) is valid. It should be noted that it would sometimes be more useful to use the lower limit in relationship (6-77). The estimates have demonstrated that the isomerism of a reaction component will have a more marked effect on the entropy term than on the enthalpy term. It should be noted that the upper limit found (Eq. (6-78)) is identical with the effect of mixing of m energetically and entropically identical isomers.

6.2.4 Examples of Quantum-Chemical Study of Chemical Equilibria with Isomerism of a Reaction Component

The literature contains a large number of theoretical works employing various quantum chemical methods that have demonstrated in a number of cases that the system studied is not a single chemical species but that it consists of two or more isomeric structures with relatively close energies. These studies mostly resulted from a need for more detailed theoretical characterization of a particular experimental situation; mostly, however, the discovery of isomerism was an unexpected side product. The theoretical study of less stable isomers is often inspired by observations in interstellar space[311,312]; a classical example is that of the HCN/HNC pair[187,313,314]. At a semi-empirical level, the works[198,199,269,315–324] are examples of the discovery or study of isomerism, while nonempirical results of this kind were obtained, for example, in the works[257,325–337]. These involve systems of only a few atoms; moreover, even such a thoroughly studied problem as the classical isomerism of ozone[326,338–342] cannot be considered as completely understood. The simple enols[243,344] are also small, and chemically very interesting, systems. Most information has been obtained on the structures of individual isomers and the depths of the corresponding minima on the potential energy hypersurface, but a complete basis for thermodynamic description is usually lacking. A number of results were obtained during the theoretical study of the relationships between rotamers[345–354]; these results are reviewed in Ref.[348]. Examples of quantum-chemical studies of the isomerism of inorganic compounds can be found in the review by Labarre[355]. Schleyer *et al.*[356–364] systematically studied structural problems encountered with organometallic compounds; a number of very interesting isomers were distinguished theoretically in these compounds. CNDO study of the equilibria between isomeric tetrachlorocomplexes[214] has been carried out up to the level of thermodynamic characteristics. Quantum chemical studies[260,365–373] have been carried out for the purpose of their utilization in the chemistry of gaseous ions and have indicated that considerable isomerism can exist even for relatively small cations. Radicals are a further type of species for which isomerism has been studied theoretically. For example, the work[551] employs a generalized BEBO method to study the 1, 2 hydrogen shift isomerizations of alkyl and alkenyl radicals.

Traditionally, attention has been paid to electronically excited states that formally represent isomers to the ground state structure[374–383] (provided that the molecular geometry is optimized). From the point of view of component isomerism in equilibrium processes it is essentially unimportant whether the isomerism occurs within one or more energy surfaces. It is apparent that the inclusion of the electronically excited states formally permits every chemical substance to be considered as a mixture of isomers. Generally, quantum-chemical calculations have demonstrated[374–383] marked differences in the geometries of the ground and electronically excited states. A number of cases are known[384] where the individual electronic states of triatomic molecules differ in their point group of symmetry. The consequences of isomerism for theoretically simulated electronic spectra (convolution of the summary spectrum as an appropriately weighted sum of the partial spectra) were discussed in work[269].

Study of the conformations of (primarily) organic molecules in terms of empirical potentials, especially using the consistent force field method (e.g.

Refs.[286,292,298,308,385–390]) has revealed a number of interesting examples of isomerism. There are a number of processes, many with practical importance, where isomerism of at least some of the components can be expected or is known from experiments, although possibilities for carrying out detailed theoretical studies are so far very limited. These include the cracking of petroleum, the preparation of high octane fuels, the effect of radiation on organic substances, processes in the field of polymer chemistry, etc.

In the mentioned cases, isomerism has been treated mostly only in terms of the position and depth of the individual hypersurface minima. The effect of temperature and the entropy term on the inter-isomeric stability and the height of the barrier separating the isomers have been analyzed only rarely. In addition to the determination of the barrier to intramolecular rearrangements it is sometimes also necessary to analyze conditions for an easy pathway for more complicated intermolecular or ionic rearrangements. Only this latter factor explains why the laboratory preparation of HNC or $CH_2=CH—OH$ is unsuccessful, although theoretical study indicated that the usual conditions for the existence of both hydrogen isocyanate[187,313,314] and vinyl alcohol[343] are fulfilled. The literature so far contains only two studies of chemical equilibria[197,198] for which the problem of isomerism of the reaction component was correctly treated on the basis of quantum chemical descriptions in the framework of Eqs. (6-67)–(6-71).

Work [197] is concerned with the study of the thermodynamics of dimerization (6-82) based on the statistical thermodynamic treatment of CNDO/2 molecular information. The results of quantum-chemical calculations confirmed that, as found earlier[391,392], N_2F_4 is a mixture of two isomers – *trans* and *gauche*. The energy difference between the two isomers and the height of the potential barrier separating them indicate[197] that they will be found in comparable amounts in an equilibrium mixture and that this equilibrium will be rapidly established at room temperature. These facts ensure

Table 6-8. Standard CNDO/2 enthalpy ΔH_T° and entropy ΔS_T° terms for the partial (to *trans*- or *gauche*-$N_2F_4(g)$) and overall (total) dimerizations of $NF_2(g)$[a]

T (K)[b]	ΔH_T°(kJ mol^{-1})			ΔS_T°(J mol^{-1} K^{-1})		
	trans	*gauche*	total	*trans*	*gauche*	total
298.15	−90.13	−94.58	−94.19	−183.7	−179.2	−177.1
373.15	−90.17	−94.69	−94.14	−183.9	−179.5	−177.0
423.15	−90.08	−94.65	−94.00	−183.6	−179.4	−176.6
523.1	−89.70	−94.34	−93.53	−182.8	−178.8	−175.6

[a] From Ref.[197]
[b] The limiting temperatures in the experimental studies quoted in Ref.[197]

that Eqs. (6-67)–(6-71) can be used in transition from the partial equilibrium characteristics to the summary values for process

$$2\,NF_2\,(g) \rightleftarrows N_2F_4\,(g)\,. \qquad (6\text{-}82)$$

Table 6-8 gives representative examples of weighting for this equilibrium. The numerical values given indicate that the isomerism of the reaction components is important, especially for the ΔS_T^0 term.

Work[198] deals with the isomerism of $C_n(g)$ aggregates; in addition to the above-mentioned example of three structures for the C_4 system (Eqs. (6-37)–(6-39)), two isomers were found for the C_5, C_6, and C_7 systems[198]. Here, the establishment of equilibrium was ensured by carrying out the experiment at a high temperature[255] (2400 K). Table 6-9 illustrates the transition from the partial thermodynamic characteristics, based[198] on the MINDO/2 quantum-chemical method, to summary values given by Eqs. (6-67)–(6-71). In this case, the isomerism of the reaction component is strongly reflected in both terms, ΔH_T^0 and ΔS_T^0.

Table 6-9. Standard MINDO/2 enthalpy ΔH_T^0 and entropy ΔS_T^0 terms for partial (D_{2h}, D_2, and T_d) and overall C_4-aggregate formations at a temperature of $T = 2400$ K[a]

Process	ΔH_T^0(kJ mol^{-1})	ΔS_T^0(J mol^{-1} K^{-1})	log K_p[b]
$4C(s) \rightleftarrows C_4(g;\, D_{2h})$	904.1	210.2	−8.700
$4C(s) \rightleftarrows C_4(g;\, D_2)$	952.2	213.6	−9.568
$4C(s) \rightleftarrows C_4(g;\, T_d)$	1301.5	205.5	−17.593
$4C(s) \rightleftarrows C_4(g)$[c]	909.9	213.6	−8.645

[a] From Ref.[198]
[b] K_p in atm; 1 atm = 101325 Pa.
[c] Overall process and its (total) characteristics.

6.2.5 *Inversion (Deconvolution) Equilibrium Problem*

So far, Eqs. (6-53)–(6-57) or (6-67)–(6-71) have been utilized for the transition from the partial characteristics to the summary values. This convolutional problem is primarily important for the theoretical generation of quantities that are experimentally attainable, with the aim of correct comparison of the theoretical results with experimental information. The formulae derived in Chap. 6.2.1 can, however, also be used to solve the opposite problem – the determination of the values of the partial characteristics on the basis of summary data for the corresponding overall process. The inversion (deconvolution) problem is especially important for the enthalpy or energy terms. Often the molecular parameters necessary for the construction of the partition function according to (6-5) are available, while data are lacking on the zero-point energies.

Consider processes of type (6-65) with isomerism of a single reaction component. Assume that the partition function yields values of changes in the standard thermodynamic potential, $\Delta\Phi_i^0$, as a function of temperature (see Eq. (6-8)) for all m possible realizations of process (6-65). Then m unknown values of the change in the zero-point energy $\Delta H_{i,0}^0$ can be obtained[261] on the basis of Eq. (6-67) by optimization of Eq. (6-83) as a least squares problem:

$$\Lambda = \sum_T \left[K_p - \sum_{i=1}^{m} \exp\left(\frac{\Delta\Phi_i^0}{R} - \frac{\Delta H_{i,0}^0}{RT} \right) \right]^2, \tag{6-83}$$

where the summation in Eq. (6-83) is understood to be over all the temperature values at which K_p was measured. It is apparent that at least m such temperatures are required (it should be recalled that terms $\Delta H_{i,0}^0$ are (by definition) independent of temperature). Optimization of expression (6-83) leads to the set of equations:

$$\frac{\partial \Lambda}{\partial\, \Delta H_{i,0}^0} = 0\,; \quad (i = 1, 2, \ldots, m)\,, \tag{6-84}$$

which can be solved numerically by, formally, the same approach as discussed in the chapter on the optimization of the molecular geometry. It is apparent that this approach is a generalization of the method commonly used for $m = 1$ and termed the 'third law analysis'.

The inversion equilibrium problem was utilized in work[393] to find the difference in the ground state energies between the *gauche*- and *trans*-N_2F_4 forms, i.e. a value which has met with considerable ambiguity in the literature[393]. The optimization calculation was carried out on the basis of experimental K_p values and experimental parameters required for construction of the thermodynamic potentials $\Delta\Phi_i^0$. The latter parameters, however, need not be derived from experiments but may result from quantum-chemical calculations. In particular, this approach can be conveniently used in connection with quantum-chemical methods that, in a given case, permit the reliable calculation of only the entropy (but not the enthalpy) term. The literature contains an example of such a connection with the inversion equilibrium problem[394] which will be discussed in Chap. 6.5.1.

6.3 Rate Processes with Parallel Isomerism of the Activated Complex

Consider a rate process in an ideal gas leading from reactants A_j to products B_j:

$$\sum_{j=1}^{n_A} n_j A_j \rightarrow \sum_{j=1}^{n_B} m_j B_j\,, \tag{6-85}$$

where n_j and m_j are the stoichiometric coefficients. Assume that there are n_x different (parallel) pathways for realizing rate process (6-85), each of which occurs through a single activated complex $X_i^{\ddagger}$ $(i = 1, 2, \ldots, n_x)$. The theory of absolute reaction rates reduces the calculation of rate constants k_i for these individual rate processes to determination of equilibrium constant $K_i^{\ddagger}$ in Eq. (6-29). The equilibrium hypothesis lying at the basis of the conventional concept of the absolute reaction rate theory directly implies that an equilibrium state is established, even when more than one type of activated complex is involved in the given rate process. A concept of the theory of absolute rates considering parallel isomerism in the activated complex can be formulated on the basis of relationships[107] for $K^{\ddagger}$ (there are certain differences[107] between the temperature dependences for $K^{\ddagger}$ and the conventional K), employing relationships valid for the general equilibrium problem. The following relationships hold[310] between the partial $(\Delta H_i^{\ddagger})$ and summary $(\Delta H_T^{\ddagger})$ activation enthalpy and between the partial $(\Delta S_i^{\ddagger})$ and summary $(\Delta S_T^{\ddagger})$ activation entropy:

$$\Delta H_T^{\ddagger} = \sum_{i=1}^{n_x} w_i^{\ddagger} \, \Delta H_i^{\ddagger} \,, \tag{6-86}$$

$$\Delta S_T^{\ddagger} = \sum_{i=1}^{n_x} w_i^{\ddagger} (\Delta S_i^{\ddagger} - R \ln w_i^{\ddagger}) \,, \tag{6-87}$$

where weight factors $w_i^{\ddagger}$ are given by

$$w_i^{\ddagger} = K_i^{\ddagger} / \sum_{i=1}^{n_x} K_i^{\ddagger} \,, \quad (i = 1, 2, \ldots, n_x) \,. \tag{6-88}$$

Work[310] also gives an expression of the classical Arrhenius parameters for overall rate process (6-85) in these terms.

When tunnelling corrections Γ_i for the corresponding partial rate processes realized through activated complexes $X_i^{\ddagger}$ are included, then the summary rate constant k for the overall rate process (6-85) is given by[200]

$$k = \sum_{i=1}^{n_x} k_i = \sum_{i=1}^{n_x} \Gamma_i \frac{kT}{h} K_i^{\ddagger} \,. \tag{6-89}$$

The summary tunnelling correction Γ can be obtained in terms of the partial corrections[200]:

$$\Gamma = \sum_{i=1}^{n_x} w_i^{\ddagger} \Gamma_i \,. \tag{6-90}$$

An analogous type of problem of the transition from microscopic (partial) to macroscopic overall kinetics must also be solved, e.g. in the framework of the classical trajectory concept (the weighted average problem[9]). Here the overall thermal rate constant $k(T)$ corresponding to the equilibrium reactants is connected with a differential rate per unit energy interval $k(E)$ (cf. e.g.[9]) by the relationship:

$$k(T) = \int_0^{\infty} [\exp(-E/kT)/Q] \, k(E) \, dE \,, \tag{6-91}$$

where the expression in square brackets can now be interpreted as a weight factor (Q designates the partition function). In order to determine $k(E)$, however, it is necessary to determine the reaction cross-section for all possible initial states of the reactants compatible with the given total energy E. It is apparent that suitable simplifications and approximations would lead from relationship (6-91) up to the level of the activated complex theory and our weighting relationship (6-89).

If parallel isomerism of the activated complex is considered, then the possibility of direct transition between the isomeric activated complexes along paths over the higher types of saddle points (i.e. points with at least two negative eigenvalues of the force constant matrix) must also be considered. Such transitions (including possible tunnelling) are conceivable, and the degree to which they can affect our weighting scheme should be analyzed. The greatest role, however, will clearly be played by the height of the energy barrier that must be overcome during these transitions. This is the first factor that can exclude such a transition. In addition, the equilibrium hypothesis itself can also be used; its formal application to this situation indicates that these interconversions between various parallel activated complexes should not affect the activated complex distribution itself[17].

So far, only isomerism in the activated complex has been considered; a further step in the generalization of rate processes (6-85) would involve extension of isomerism to the reactants A_j (or products B_j). This discussion could again be based on the formalism of the general equilibrium problem. It should, however, be noted (for the sake of completeness) that the concept of parallel activated complexes is different from the situation treated by the classical Curtin-Hammett principle[395,396]. This principle is applicable to reactions occurring between a common (isomeric) set of reactants and various sets of products. Thus, even though isomerism of the activated complex is essentially present, the products (in contrast to the reactants) are considered to be distinguishable. This is principally different from the situation considered in the previous paragraphs, assuming that the parallel reaction paths are indistinguishable, where only the overall reaction flux between the given reactants and products can be measured. In contrast to this situation, the Curtin-Hammett principle states that the differences in the rates of formation of the individual distinguishable products are controlled by differences in the corresponding transition states and not by differences in the primarily assignable (to these transition states) reactants. This result can readily be understood in the light of the proposed indistinguishability of the reactants connected with their rapid interconversion. Formally, the principle follows directly from application of the relevant summary partition function of the isomeric set of reactants in the framework of the conventional theory of absolute reaction rates.

An elementary example of the parallel isomerism of activated complexes involves a system containing two different activated complexes forming a pair of enantiomers. This situation (which drew considerable attention in connection with symmetry numbers – cf.[132]) is, of course, organically included in our weighting scheme. In the transition from the level of the partial rate constants to the overall value, the latter equals twice either of the two partial values. Similarly, transition from the partial to the overall entropy value is connected with an increase in the entropy by an increment of $R \ln 2$. It is interesting that when, in addition to the optical activity of the activated complex, this property is exhibited also by both reactants and products, under certain conditions[132] the theory of absolute reaction rates does not lead to the elementary rate constants but to a certain sum of them.

Quantum-chemical study of the mechanisms of a number of rate processes has demonstrated[397–406] that they can occur through more than a single type of activated complex. This often involves *cis-trans* isomerizations around —N=N— bond[397,399,401], occurring through inversion and rotation mechanisms with comparable probability but with different activated complex structures. The isomerism of the activated complex can also play a role in more complicated, multi-step mechanisms (e.g.[407–409]); correct description of these systems would, however, require modification or generalization of the weighting treatment (6-86)–(6-90). An important particular type of activated complex isomerism, i.e. sequential isomerism, will be discussed in the next chapter. Considerable isomerism can certainly be ex-

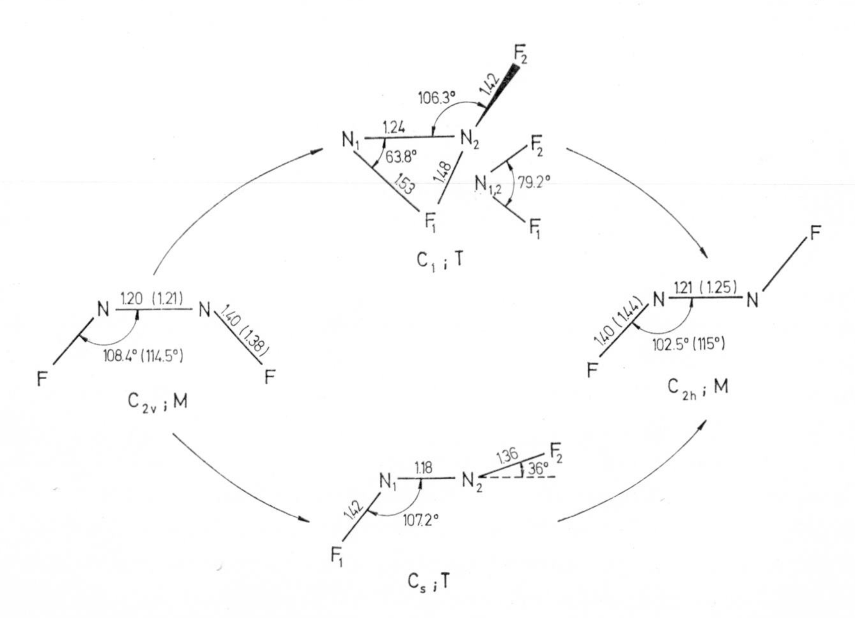

Figure 6-6. Structure of four stationary points found[199] on the CNDO/2 potential energy hypersurface of N_2F_2; M is a minimum, T a transition state (bond lengths in 10^{-10} m, experimental values in parantheses)

Table 6-10. Partial (through the activated complex of C_s or C_1 symmetry) and total CNDO/2 activation enthalpy $\Delta H_T^{\ddagger}$ and entropy $\Delta S_T^{\ddagger}$ of isomerization *cis*-N_2F_2(g) → *trans*-N_2F_2(g)[a]

T (K)	$\Delta H_T^{\ddagger}$ (kJ mol^{-1})			$\Delta S_T^{\ddagger}$ (J mol^{-1} K^{-1})		
	C_s	C_1	total	C_s	C_1	total
298.15	320.5	282.8	282.8	5.58	−2.69	−2.69
500	319.7	281.6	281.6	3.58	−5.74	−5.71
1000	316.7	278.9	280.1	−0.48	−9.39	−7.99
2000	309.0	272.5	280.7	−5.81	−13.82	−7.56
3000	300.7	265.0	278.4	−9.13	−16.84	−8.43

[a] From Refs.[199,200]

pected, foı example, in polymer kinetics. Theoretical study of such systems at present requires the use of simple models and the combination of quantum-chemical methods with an empirical approach. For example, work[398] is an example of a successful study of this kind, dealing with the kinetics of poly(vinylchloride) dechlorination.

Nonetheless, at present only two chemical processes[188,189] with activated complex isomerism are described theoretically in the literature to such a degree that it is possible[188,200,310] to carry out correct transition from the partial to the overall characteristics. The possibility that the isomerization

$$cis\text{-}N_2F_2\,(g) \quad \rightarrow \quad trans\text{-}N_2F_2\,(g) \tag{6-92}$$

proceeds through two mechanisms with different activated complexes and comparable potential barriers was demonstrated at the semiempirical[199,315,397] and nonempirical[399] levels. Study of the problem[199] by a modified CNDO/2 method demonstrated that the activated complex for the inversion mechanism has C_s symmetry, while rotation involves C_1 symmetry (Fig. 6-6). Table 6-10 and Fig. 6-7 give a review of partial and summary activation parameters over a wide temperature range. While the rate determining mechanism at lower temperatures is rotational, at temperatures above 1000 K the inversion mechanism also begins to be important. The effect on the $\Delta S_T^{\ddagger}$ term produced by the isomerism of the activated complex becomes quite pronounced at high temperatures. Process (6-92) is also the first example of a study[200] of the properties of the summary correction for tunnelling (Eq. (6-90)). Although the temperature interval of the experimental study and further factors[200] exclude any greater effect of the tunnelling correction on the rate constant values, nonetheless, this system can be considered[200] a suitable model for the study of qualitative relationships for the correction for tunnelling. The partial correction terms were found by the Wigner[148] method. While these partial terms exhibit the usual smooth decrease with increasing temperature, an anomalous shape with a local maximum (Fig. 6-8) was found for the summary term[200]. This type of behaviour is readily understandable on the basis of the form of Eq. (6-90) and can be generally expected under certain conditions. It should be noted that the summary $\Delta H_T^{\ddagger}$ and $\Delta S_T^{\ddagger}$ terms for process

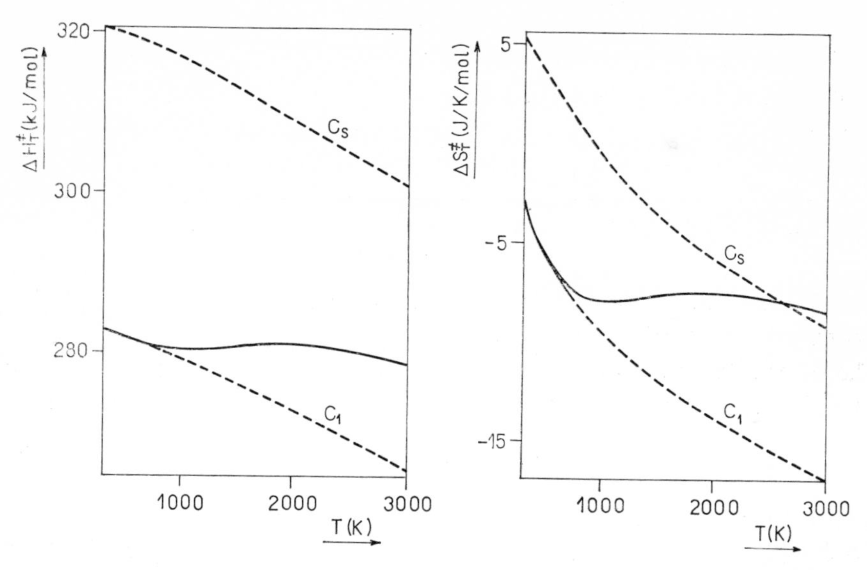

Figure 6-7. Temperature dependences[200] of the CNDO/2 activation enthalpy $\Delta H_T^{\ddagger}$ and entropy $\Delta S_T^{\ddagger}$ of the *cis-trans* isomerization of N_2F_2(g) for C_s and C_1 activated complexes (— — —) and for the overall activation process (———)

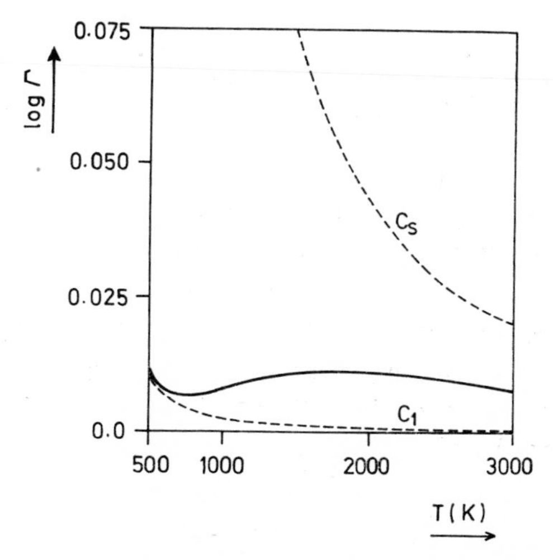

Figure 6-8. Temperature dependences of tunnel-effect corrections Γ for activated complexes C_s and C_1 (— — —) and for the overall activation process (———) in the *cis-trans* isomerization of N_2F_2(g) based on CNDO/2 calculations[200]

Table 6-11. Partial and total MINDO/2 activation enthalpy $\Delta H_T^{\ddagger}$ and entropy $\Delta S_T^{\ddagger}$ of cyclohexane isomerization $C_6H_{12}(g; D_{3d}) \rightarrow C_6H_{12}(g; D_2)$ at temperature $T = 298$ K

Rate process	$\Delta H_T^{\ddagger}$ (kJ mol^{-1})	$\Delta S_T^{\ddagger}$ (J mol^{-1} K^{-1})
Through an activated complex of C_s symmetry	26.36	25.02
Through an activated complex of C_2 symmetry	26.78	28.62
Overall process	26.60	32.75

[a] From Refs.[188,310]

(6-92) also have temperature dependences with local extremes (Fig. 6-7), although the corresponding partial quantities behave monotonously.

The other rate processes with the isomerism of the activated complex properly treated theoretically[310] is the isomerization of cyclohexane (see also Fig. 2-2):

$$C_6H_{12}\,(g; D_{3d}) \rightarrow C_6H_{12}\,(g; D_2)\,. \qquad (6\text{-}93)$$

It was found by the MINDO/2 method[188] that process (6-93) occurs through activated complexes with symmetry C_s and C_2. The weighting of the partial activation parameters at room temperature is given in Table 6-11. Isomerism has a substantial effect on the value of term $\Delta S_T^{\ddagger}$ under these conditions. The overall term is about one third greater than the partial value for the path with the lower potential barrier.

6.4 Rate Processes with Sequential Isomerism of the Activated Complex

If the isomeric activated complexes do not lie on several parallel pathways (as in the previous chapter) but are ordered on one common path connecting the reactants and products, this is called sequential activated complex isomerism. If n activated complexes are present in the sequence, they are separated by $n - 1$ local minima – intermediates. Recently, the simplest case of this phenomenon has been studied[410-414] – double sequential activated complex isomerism and its consequences for the convolution of theoretical and the deconvolution of experimental data. From a kinetic point of view, this situation (Fig. 6-9) is described by the scheme

$$A \underset{k_2}{\overset{k_1}{\rightleftarrows}} B \underset{k_3}{\overset{k_4}{\rightleftarrows}} C\,, \qquad (6\text{-}94)$$

where the k_i are the elementary rate constants for the partial rate processes. Intermediates (e.g.[415-417]) have long been important in organic chemistry and many of them have been experimentally very well characterized. Thus all the partial kinetic processes in scheme (6-94) could be distinguished and their characteristics determined. In contrast to this situation, here realizations of scheme (6-94) will primarily be discussed in which the intermediate did not appear or was not considered in the experimental study. It thus seemed appropriate to assume a simple concerted mechanism

$$A \rightarrow C \tag{6-95}$$

and to evaluate the primarily observed data in a conventional kinetic manner. Assume that the rate constant of process (6-95) is determined, for example, from the rate of formation of product C. There are then two possible ways of treating and condensing the observed data (see e.g.[418]). The first procedure employs the integrated form of the kinetic scheme for a single forward reaction, yielding the rate constant $k_{A \to C}^{(\rightarrow)}$:

$$k_{A \to C}^{(\rightarrow)} = \frac{1}{t} \ln \frac{c_A(0)}{c_A(0) - c_C(t)}, \tag{6-96}$$

where $c_A(0)$ designates the intial concentration of reactant A and $c_C(t)$ the concentration of product C at time t. A second possible procedure employs the integrated form of the kinetic scheme of the forward-reverse reactions, leading to rate constant $k_{A \to C}^{(\rightleftarrows)}$:

$$k_{A \to C}^{(\rightleftarrows)} = \frac{c_C(\infty)}{t c_A(0)} \ln \frac{c_C(\infty)}{c_C(\infty) - c_C(t)}, \tag{6-97}$$

where $c_C(\infty)$ designates the concentration of C at equilibrium. While treatment (6-97) can be used for all types of rate process (6-94), the use of procedure (6-96) assumes a negligible rate for the reverse reaction. On the other hand, (6-97) requires more information; in addition to $c_C(t)$, $c_C(\infty)$ must be known and this value need not always be available.

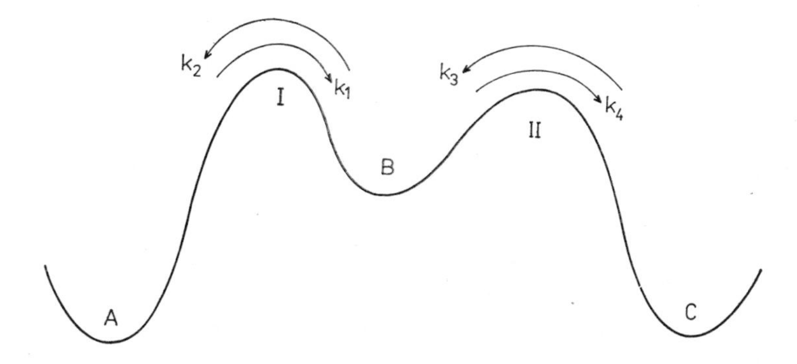

Figure 6-9. Reaction profile diagram for a rate process A → C realized through intermediate **B**, i.e. the process exhibiting double sequential isomerism of the activated complexes (I, II)

Recent quantum-chemical studies[419–428] have revealed a number of relatively simple chemical reactions for which a concerted mechanism could be assumed, i.e. (6-95); however, detailed theoretical study has demonstrated that these are non--concerted processes. The Cope rearrangement of 1,5-hexadiene is a good example:

$$\rightarrow \quad . \qquad (6\text{-}98)$$

This reaction was studied experimentally with suitable isotopic labelling[429–431] using both approach[430,431] (6-96) and approach[429] (6-97). No intermediate was demonstrated to be present, so that it seemed fully justified to assume a concerted mechanism (6-95). Nonetheless, isotopic labelling permitted the discovery and the distinguishing of another important characteristic of reaction (6-98), i.e. that it is realized along two parallel pathways[430,431] conventionally termed chair[429] and boat[430,431]. A later theoretical study[421,423] of process (6-98) by the MINDO/2 and MINDO/3 methods indicated that this process is realized through an intermediate, so that double sequential activated complex isomerism is involved (under a suitable isotopic labelling). Fig. 6-10 depicts the structural characteristics of the activated complexes and intermediates for both parallel reaction pathways found[423] by the MINDO/3 method.

The fact that intermediate B was found only theoretically in process (6-94) makes it necessary in the comparison of theoretical and experimental data to consider process

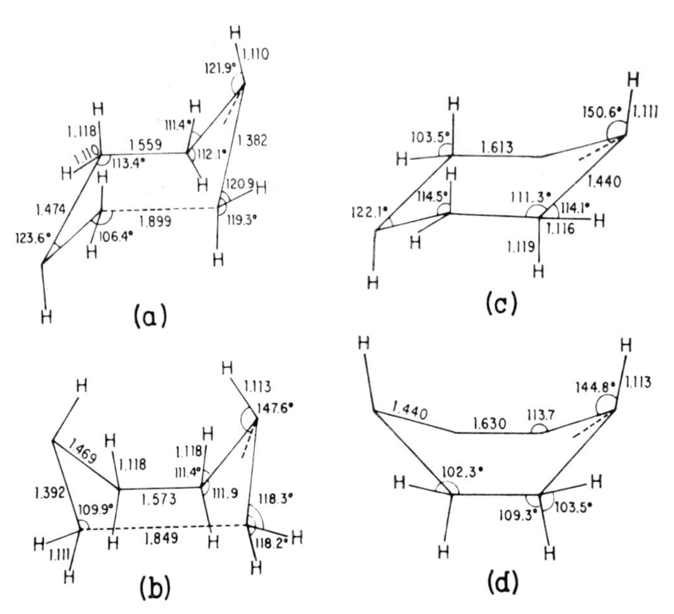

Figure 6 - 10. MINDO/3 geometries[423] for species involved in the Cope rearrangement of 1,5-hexadiene: (a), (b) chair and boat transition states; (c), (d) chair and boat intermediates

(6-95) as a convolution of the four partial processes included in (6-94). The elementary rate constants k_i appearing in (6-94) can be expressed in the usual manner in terms of the corresponding activation enthalpies $\Delta H_i^{\ddagger}$ and entropies $\Delta S_i^{\ddagger}$:

$$k_i = \frac{kT}{h} \exp\left(-\Delta H_i^{\ddagger}/(RT)\right) \exp\left(\Delta S_i^{\ddagger}/R\right). \tag{6-99}$$

Knowledge of rate constants k_i permits[432] the solution of kinetic scheme (6-94) in closed form, i.e. to give the time dependence for the concentrations of the individual species, e.g.:

$$c_{\mathrm{C}}(t) = c_{\mathrm{A}}(0) \frac{k_1 k_4}{u_1 u_2 (u_1 - u_2)} \left[u_2 \exp\left(-u_1 t\right) - u_1 \exp\left(-u_2 t\right) + u_1 - u_2\right], \tag{6-100}$$

where

$$u_{1,2} = \tfrac{1}{2}\{k_1 + k_2 + k_3 + k_4 \mp [(k_1 + k_2 + k_3 + k_4)^2 - \\ - 4(k_1 k_3 + k_1 k_4 + k_2 k_3)]^{1/2}\}. \tag{6-101}$$

When the time dependences of the individual concentrations are known, then it can be demonstrated[411] that the relationship of rate constants (6-96) and (6-97) (with the significance of the effective, overall values, considering the existence of an intermediate) to the elementary rate constants k_i and time t is given by

$$k_{\mathrm{A}\to\mathrm{C}}^{(\to)} = \frac{1}{t} \ln \frac{u_1 u_2 (u_2 - u_1)}{(u_2 - u_1)(u_1 u_2 - k_1 k_4) + k_1 k_4 \left[u_2 \exp\left(-u_1 t\right) - u_1 \exp\left(-u_2 t\right)\right]}, \tag{6-102}$$

$$k_{\mathrm{A}\to\mathrm{C}}^{(\rightleftarrows)} = \frac{k_1 k_4}{u_1 u_2 t} \ln \frac{u_2 - u_1}{u_2 \exp\left(-u_1 t\right) - u_1 \exp\left(-u_2 t\right)}. \tag{6-103}$$

Equations (6-102) and (6-103) represent exact solutions to problems at the rate constant level. It is, however, more common to compare theoretical and experimental kinetic data at the level of activation parameters. For example, the total, effective activation enthalpy, expressed in dependence upon the selection of procedure x (where $x \equiv \to$ or $\rightleftarrows$), is given by

$$\Delta H_{\mathrm{eff}}^{\ddagger(x)} = -RT + RT^2 \frac{\mathrm{d} \ln k_{\mathrm{A}\to\mathrm{C}}^{(x)}}{\mathrm{d}T}. \tag{6-104}$$

Thus, detailed calculation leads from formulae (6-101) and (6-102) to relationships for the effective activation enthalpy $\Delta H_{\mathrm{eff}}^{\ddagger(x)}$ and entropy $\Delta S_{\mathrm{eff}}^{\ddagger(x)}$. General formulae are available in Ref.[411] while a special case[410] for $k_1 = k_3$ and $k_2 = k_4$ (and a somewhat simplified form of Eq. (6-97) – replacement of the multiplication term connected with function ln by a simple factor of $\frac{1}{2}t$) is given in Chart 6-1.

Chart 6-1. Analytic formulae[410] for the time dependences of the effective values of activation enthalpies $\Delta H_{\text{eff}}^{\ddagger(\rightleftarrows)}$ and $\Delta H_{\text{eff}}^{\ddagger(\rightarrow)}$ and activation entropies $\Delta S_{\text{eff}}^{\ddagger(\rightleftarrows)}$ and $\Delta S_{\text{eff}}^{\ddagger(\rightarrow)}$ for $k_1 = k_3$ and $k_2 = k_4$.

$$\Delta H_{\text{eff}}^{\ddagger(\rightleftarrows)} = -RT + \frac{k_1}{k_{\text{eff}}^{(\rightleftarrows)}}\left(\frac{1}{2} - \frac{1 - e^{-2k_2t}}{P_1}\right)(RT + \Delta H_1^{\ddagger}) -$$

$$- \frac{k_1}{k_{\text{eff}}^{(\rightleftarrows)} P_1}((2k_2t + 1)\, e^{-2k_2t} - 1)(RT + \Delta H_2^{\ddagger}),$$

where

$$P_1 = 4k_2t + 2k_1t(1 - e^{-2k_2t}),$$

$$\Delta H_{\text{eff}}^{\ddagger(\rightarrow)} = -RT + \frac{k_1}{k_{\text{eff}}^{(\rightarrow)}}\left[1 - \frac{1}{tP_2}((k_2 + (k_1^2 + 3k_1k_2 + 2k_2^2)\, t) \,.\right.$$

$$\left. .\, e^{k_1t} - k_2\, e^{-2k_2t})\right](RT + \Delta H_1^{\ddagger}) +$$

$$+ \frac{k_2}{k_{\text{eff}}^{(\rightarrow)} tP_2}[k_1\, e^{k_1t} - (k_1 + (k_1 + 2k_2)\, k_1t)\, e^{-2k_2t}]\,(RT + \Delta H_2^{\ddagger}),$$

where

$$P_2 = (k_1 + 2k_2)^2/2 + (k_1 + 2k_2)(k_1 + k_2)\, e^{k_1t} - \frac{k_1}{2}(k_1 + 2k_2)\, e^{-2k_2t},$$

$$\Delta S_{\text{eff}}^{\ddagger(\rightleftarrows)} = \frac{1}{T}\Delta H_{\text{eff}}^{\ddagger(\rightleftarrows)} + R \ln\left(\frac{h}{kT} k_{\text{eff}}^{(\rightleftarrows)}\right),$$

$$\Delta S_{\text{eff}}^{\ddagger(\rightarrow)} = \frac{1}{T}\Delta H_{\text{eff}}^{\ddagger(\rightarrow)} + R \ln\left(\frac{h}{kT} k_{\text{eff}}^{(\rightarrow)}\right),$$

$$k_1 = \frac{kT}{h} e^{\Delta S_1^{\ddagger}/R}\, e^{-\Delta H_1^{\ddagger}/(RT)},$$

$$k_2 = \frac{kT}{h} e^{\Delta S_2^{\ddagger}/R}\, e^{-\Delta H_2^{\ddagger}/(RT)},$$

where T denotes the temperature and h, k, and R are the fundamental constants.

Now, the behaviour of the effective terms for the Cope rearrangement (6-96) will be illustrated, using interconversion through the boat intermediate. Parameters $\Delta H_1^{\ddagger}$ and $\Delta S_1^{\ddagger}$ were taken from the MINDO/3 calculation[423] while parameters $\Delta H_2^{\ddagger}$ and $\Delta S_2^{\ddagger}$ (which could not be determined in the MINDO/3 calculation[423] with sufficient precision because of the presence of orbital degeneracy) were estimated[412].

This parameter selection ensured that the ratio of the concentration of the intermediate to that of the reactant or product at equilibrium was less than 1 : 400. Using this selection of partial terms, the effective rate characteristics $k^{(x)}_{A\to A}$, $\Delta H^{\ddagger(x)}_{eff}$ and $\Delta S^{\ddagger(x)}_{eff}$ were studied over a broad time interval, including the experimental interval[430]. Fig. 6-11 depicts conditions for rate constants $k^{(\to)}_{A\to A}$ and $k^{(\rightleftarrows)}_{A\to A}$. In the time interval studied[430] the values of both these effective rate constants are very similar, but with increasing time term the difference systematically increases. The difference from the partial value k_1 is distinct – even within the time interval[430] both effective constants are more than twice smaller. Outside this interval a principal qualitative difference appears between the two curves. While function $k^{(\to)}_{A\to A}$ exhibits a maximum, rate constant $k^{(\rightleftarrows)}_{A\to A}$ monotonously increases to the limiting value $k_1k_2/(k_1 + 2k_2)$. It is principally important that the $k^{(x)}_{A\to A}$ values are relatively insensitive to time changes within the experimental time interval. Provided that the usual errors in kinetic measurements are considered, it is apparent that, in the particular case[430] of process (6-98), the existence of an intermediate could actually not be inferred from the time dependence of the measured rate constant, as no deviations from a concerted behaviour were apparent. This indicates that simple analysis of the time dependence of the measured rate constants need not indicate the existence of an intermediate, although the values of these rate constants can be significantly affected by the existence of an intermediate.

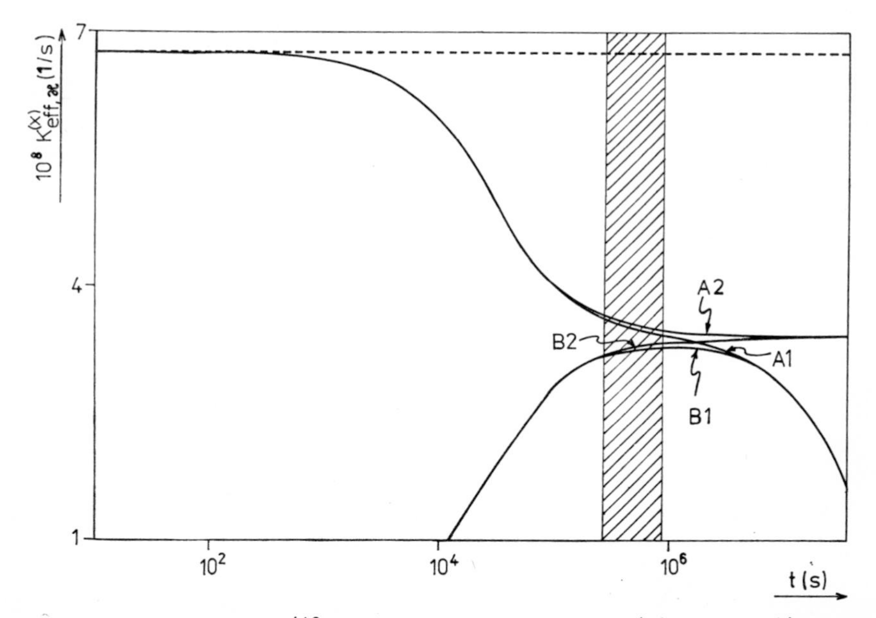

Figure 6-11. Time dependences[413] of the effective rate constants $k^{(\to)}_{eff,\varkappa}$ (1) and $k^{(\rightleftarrows)}_{eff,\varkappa}$ (2) for the boat pathway of the Cope rearrangement of 1,5-hexadiene at 532.15 K; A – $\varkappa = 1$, B – $\varkappa = 0$, where $\varkappa$ denotes the interference factor; the position of the MINDO/3 value[423] of k_1 employed and the experimental time region[430] are indicated by the dashed line and by the hatching, respectively

This fact emphasizes the necessity for close cooperation between experimental and theoretical approaches in the demonstration of the existence of an intermediate and in the interpretation of the measured effective rate constants.

Consider the behaviour of the effective activation enthalpy and entropy $\Delta H^{\ddagger(x)}_{\text{eff}}$ and $\Delta S^{\ddagger}_{\text{eff}}$ (Fig. 6-12). Here there is a marked difference in the theoretical and experimental capabilities (compared with conditions for the effective rate constants $k^{(x)}_{\text{A}\rightarrow\text{A}}$). Whilst the time dependence of the $k^{(x)}_{\text{A}\rightarrow\text{A}}$ value can be found (at least in principle) from experiment, this is not possible for the activation parameters (at least in the framework of the conventional approach). Thus, the analysis of the observed data yields only a single time independent value for each term, i.e. the activation enthalpy and entropy, which represent the time average parameters. Thus, the possibility of studying the time dependences of $\Delta H^{\ddagger(x)}_{\text{eff}}$ and $\Delta S^{\ddagger(x)}_{\text{eff}}$ on the basis of theoretical data

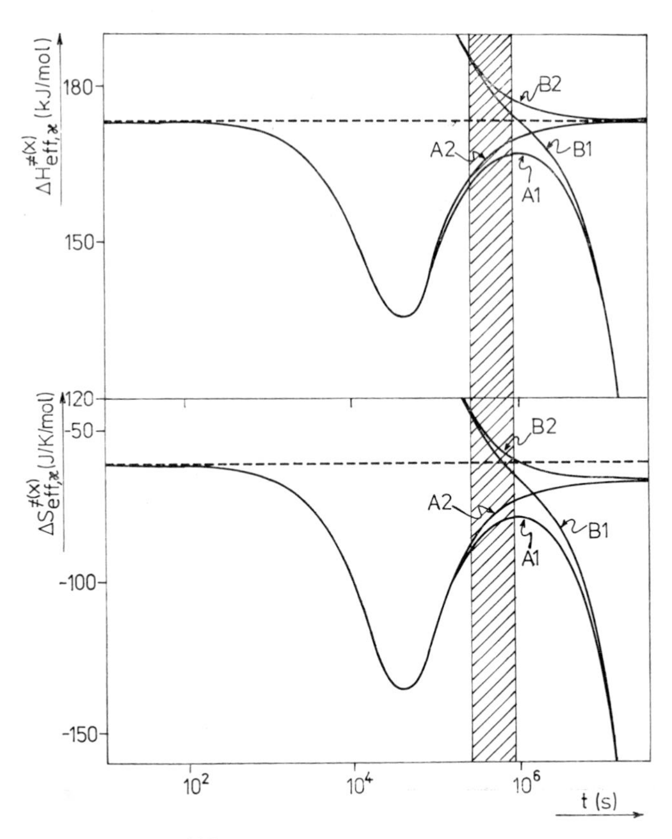

Figure 6-12. Time dependences[413] of the effective activation enthalpies and entropies $\Delta Y^{\ddagger(\rightarrow)}_{\text{eff},\varkappa}$ (1) and $\Delta Y^{\ddagger(\rightleftarrows)}_{\text{eff},\varkappa}$ (2) ($Y = H$ or S) for the boat pathway of the Cope rearrangement of 1,5-hexadiene at 532.15 K; A — $\varkappa = 1$, B — $\varkappa = 0$, where $\varkappa$ denotes the interference factor; the position of the MINDO/3 values[423] of $\Delta Y^{\ddagger}_{1}$ terms ($Y = H$ or S) used and the experimental time region [430] are indicated by the dashed lines and by the hatching, respectively

for the elementary rate processes is a considerable improvement. In our particular case (6-98), terms $\Delta H_{\text{eff}}^{\ddagger(\rightarrow)}$ and $\Delta H_{\text{eff}}^{\ddagger(\rightleftarrows)}$ are quite similar within the experimental interval[430] and a larger difference appears with longer times. Deviations from values $\Delta H_1^{\ddagger}$ are large at low time values (tens to hundreds of kJ/mol), while they vary from 1 to 13 kJ/mol within the interval[430]. The greatest changes, i.e. the most pronounced consequences of the existence of an intermediate, are encountered for quantity $\Delta S_{\text{eff}}^{\ddagger(x)}$. While term $\Delta S_{\text{eff}}^{\ddagger(x)}$ is higher than term $\Delta S_1^{\ddagger}$ at the beginning of the reaction by more than 300 J K^{-1} mol^{-1}, at large time values both effective entropy terms have a value lower than $\Delta S_1^{\ddagger}$. For quantity $\Delta S_{\text{eff}}^{\ddagger(\rightleftarrows)}$ this decrease attains a limiting value of about 5 J K^{-1} mol^{-1}, while a systematic decrease occurs for quantity $\Delta S_{\text{eff}}^{\ddagger(\rightarrow)}$. In addition, the time dependence of quantity $\Delta S_{\text{eff}}^{\ddagger(x)}$ is also more marked within the interval[430]: compared with about 5% for quantities $k_{A\rightarrow A}^{(x)}$ and $\Delta H_{\text{eff}}^{\ddagger(x)}$, the relative change in the effective entropy is about 40%. This indicates that the consequences of the existence of an intermediate that is distinguished only theoretically will typically appear most strongly for the activation entropy.

In conclusion, attempts to compare theory directly with experiment for processes with sequential activated complex isomerism by a straightforward comparison of experimentally derived analogues of quantities $k_{A\rightarrow A}^{(x)}$, $\Delta H_{\text{eff}}^{\ddagger(x)}$ and $\Delta S_{\text{eff}}^{\ddagger(x)}$ with quantum-chemical values k_1, $\Delta H_1^{\ddagger}$ and $\Delta S_1^{\ddagger}$ can be completely misleading (depending on the time interval employed, on the technique of treating the experimental kinetic data, and on the stability of the intermediate). Instead, it is usually necessary to pass from partial theoretical characteristics to the corresponding overall (observed) effective values. It should be noted, for completeness, that a certain simplification of the whole procedure would result from the use of the usual steady-state approximation[418], which, however, can be applied only for longer times[411].

The concept of activated complex sequential isomerism was recently extended[413] to include interfering intermediates. It is assumed that the concentration of product C is observed by a technique for which activated complex B is also active. Consequently, in actual fact concentration $c_C(t)$ would not be measured, but a certain effective concentration $c_{\text{eff}}(t)$:

$$c_{\text{eff}}(t) = c_C(t) + \varkappa c_B(t)\,, \tag{6-105}$$

where coefficient $\varkappa$ is termed the interference factor. It is apparent that, for the situation discussed in the previous paragraphs, $\varkappa = 0$; i.e. a non-interfering intermediate was present. Provided that, for example, a spectroscopic technique is used to determine concentration $c_C(t)$ or $c_{\text{eff}}(t)$, then $\varkappa$ is given by the ratio of the corresponding molar absorption coefficients and can, in principle, vary in the interval $\langle 0; \infty)$. Interference is, however, most important when the intermediate and product are very similar, i.e. when $\varkappa \approx 1$. Even though process (6-98) is apparently a process with a non-interfering intermediate ($\varkappa \approx 0$), it has been utilized for illustrative purposes as a model for situations with an interfering intermediate setting, formally, $\varkappa = 1$ (Figs. 6-11 and 6-12). It was found that the difference between cases $\varkappa = 0$ and $\varkappa = 1$

appears at the level of the effective rate constant and activation parameters at shorter times, while these differences gradually disappear at longer times.

Processes of the A → A type are generally symmetric reactions[132] (irregardless of the presence of an intermediate), and cannot be observed directly; in experimental studies the reactant and product must be distinguished by isotopic labelling. It is apparent that the theory of absolute reaction rates essentially yields different values for the elementary rate constants for processes with and without isotopic labelling. While differences in the moments of inertia and vibrational frequencies need not be critical, the formal use of the concept of a symmetry number[132] can cause substantial differences. It is clear that, in our comparisons, the theoretical elementary rate constant used should, strictly speaking, be calculated for the particular type of isotopic labelling used in the corresponding experiment. It should be noted in general that, for certain types of rate considerations for processes of the A → A type without isotopic labelling, it can be necessary[132] to consider not the rate constant following directly from the theory of absolute reaction rates but twice this value.

Although the theoretically determined presence of an intermediate in process (6-98) can lead to considerable complications in kinetic analysis and interpretation, it follows that even a more complex situation than that considered here can in fact take place. Komornicki and McIver[421] and Dewar *et al.*[423] pointed out that a distinct chair and boat character of the reaction pathway may be an artefact of the closed-shell approach (not permitting a reliable description of the highly biradical character of the intermediate). To describe these effects, Dewar *et al.*[423] extended their MINDO/3 calcu-

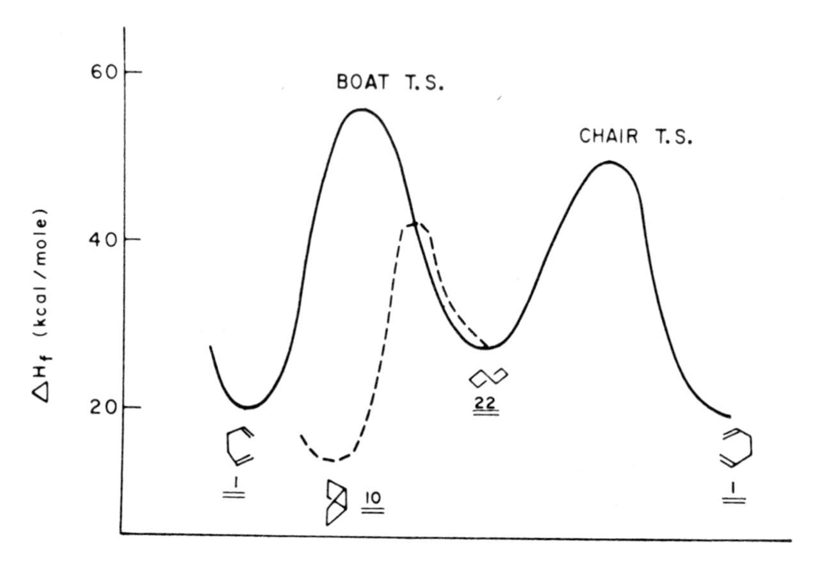

Figure 6-13. A possible shape[423] of the rigorous potential surface for the 1,5-hexadiene (1) rearrangement, with designation of the twist boat conformer (22) of the 1,4-cyclohexylene biradical and bicyclo[2.2.0]hexane (10)

lations to include a contribution from configuration interaction with the two lowest double excited configurations. These calculations then indicated the possibility that these two intermediates can collapse into a common species. This would then lead to the hypersurface schematically depicted in Fig. 6-13. The hypersurface would exhibit a large crater containing the twist-boat structure of the 1,4-cyclohexylene biradical (Fig. 6-14). The transition states are then represented by cols on the edge of this crater. The structure and energetics of the chair and boat type transition states should not be affected by configuration interaction[423]; however, it was discovered in this approximation that the biradical structure can also be convertible to bicyclo[2.2.0]-hexane through a third type of transition state (Fig. 6-14). If this complex picture were accurate, it is clear that comparison of experimental and theoretical data for process (6-98) would be even more complicated.

It is apparent in general that, above a certain size of the (organic) reaction system, the simultaneous presence of parallel and sequential isomerism of the activated complex must be assumed, even though the observation technique need not (in general) permit the distinguishing of all these isomers (for illustration of a relatively simple system for which the theory [433] nonetheless indicates the possibility of a number of

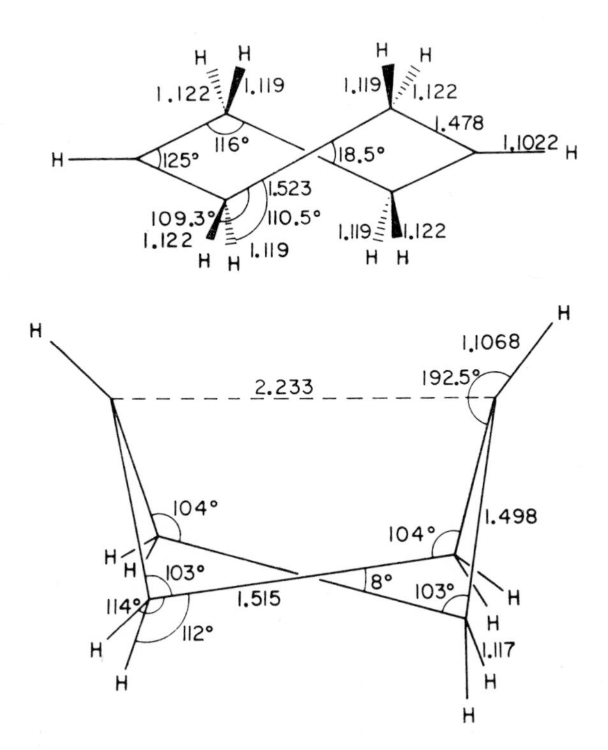

Figure 6-14. MINDO/3 (with inclusion of CI with the two lowest double excited configurations) geometries[423] of the twist boat conformer of the 1,4-cyclohexylene biradical (above) and of the transition state for conversion of 10 to 22 (see Figure 6-13)

simultaneous reaction paths, see Fig. 6-15). This demonstrates the justification and usefulness of further development of the general concept of the theory of chemical reactivity, including all the multifaceted possibilities of activated complex isomerism, and possibly including the isomerism of the reactants and products.

The concept of activated complex isomerism is an approach to rate processes including more than one saddle point on the potential energy hypersurface that retains the original features of the absolute reaction rate theory. However, these problems could now be discussed at a more modern level, i.e. using the master equation concept[552], the theory of stochastic processes[553] or the interpretation of isomerization dynamics using the quantum chaos approach[554]. Nonetheless, the unified statistical theory invented by Miller[546] and at present being developed[547–550] further is especially important in relation to the concept of activated complex sequential isomerism. This theory includes in a single scheme both rate processes with an intermediate and the direct reaction mechanism. It is apparent[411,414] from the unsatisfactory limiting behaviour of the formula in Chart 6-1 for $k_2 \to \infty$ that con-

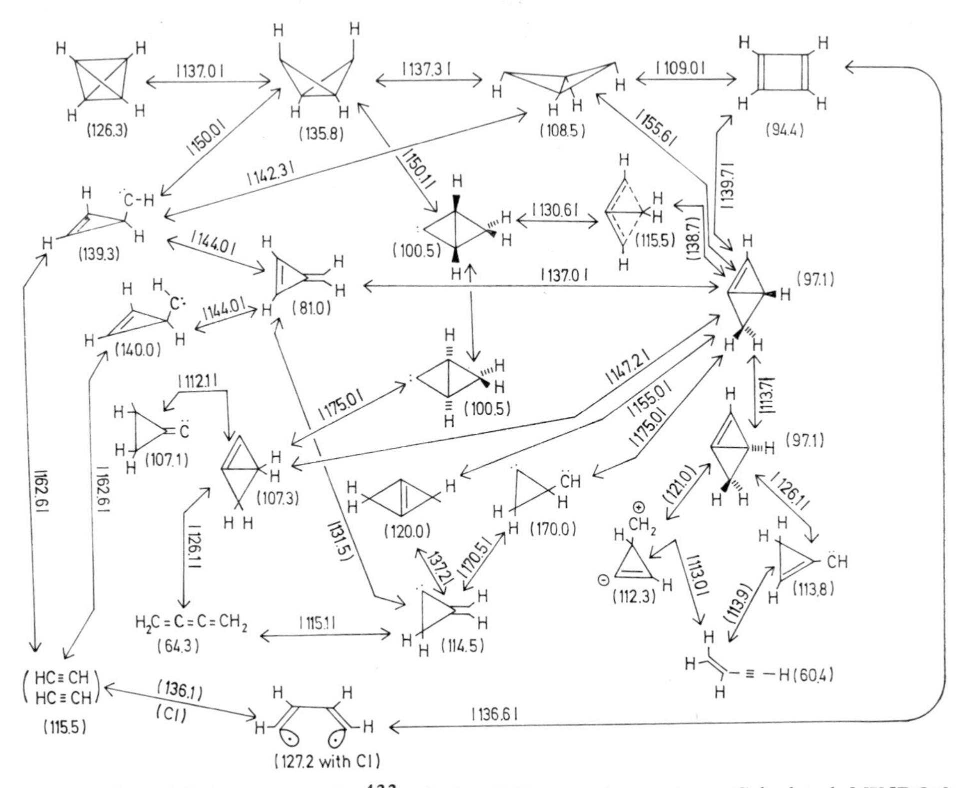

Figure 6-15. Global representation[433] of the C_4H_4 reaction system. Calculated MINDO/3 heats of formation at 25 °C (kcal mol^{-1}; 1 kcal = 4.184 kJ) of stable species are shown in parentheses under the formulae and of transition states in square brackets above the double-headed arrow linking the species involved

tinuous reduction to the single barrier situation has to be ensured for very low stability intermediates by modifying the treatment of activated complex sequential isomerism in the sense of the unified statistical theory.

6.5 Isomerism of Multiparticle Clusters

6.5.1 Examples of Theoretical Study of Gas-Phase Complexes

Isomerism of a reaction component is a rule rather than an exception in the field of weak intermolecular interactions (see for example[434-437]), especially in the formation of multiparticle (and especially multimolecular) clusters[93,94,281,437]. In addition, this isomerism is often rich, not limited to only two structures, and the energy differences between the individual forms are small. Thus, isomerism in multimolecular clusters must[93] be considered as an essential and general characteristic of these forms. In addition to the above-mentioned possibility of isomerism[258,259] of the $(H_2)_2$ dimer (Eq. (6-40)), theoretical study has also demonstrated that isomerism exists, for example, for the following aggregates (mostly dimers) of: Cl_2 (refs.[438-440]), NO (Refs.[202,394,441,442]), CO_2 (Refs.[443-445]), H_2CO (Ref.[446]), CH_3OH (Refs.[447,448]), CH_3COCH_3 (Ref.[446]), CF_3CH_2OH (Refs.[449,450]), NH_3 (Ref.[451]), as well as for heteromolecular clusters, e.g.[201,452-457] (CH_3OH.n H_2O, HF—HCl, HF—ClF, etc.). Most frequently, quantum-chemical studies have dealt with the oligomers of water, $(H_2O)_n$ in connection, *inter alia*, with the problem of anomalous water[458]. *Ab initio* SCF calculations[459] have indicated that these aggregates exhibit isomerism for $n = 2-8$. Results in the literature concerning possible dimer isomerism were not unambiguous; while a larger number of possible structures were found at a certain methodological level[459-461], a detailed work[193] found that isomerism is possible only for $n \geqq 4$. Finally, a recent systematic study of the presence of isomerism for $(H_2O)_2$ using the *ab initio* SCF CI 14-dimensional potential hypersurface[460] indicated that isomerism does not exist in this substance[207,208].

Table 6-12. Standard enthalpy ΔH°_T and entropy ΔS°_T terms[a] for the partial (to the cage or C_2 structure) and overall (total) water pentamerizations at temperature T

T (K)	ΔH°_T(kJ mol^{-1})			ΔS°_T(J mol^{-1} K^{-1})		
	cage	C_2	total	cage	C_2	total
100	−111.5	−90.0	−111.5	−457.3	−436.8	−457.3
400	−109.8	−86.1	−109.0	−463.6	−432.2	−461.3
700	−94.7	−70.3	−84.0	−436.0	−402.9	−415.8

[a] From Refs.[193,309]

The formation of isomeric multimolecular clusters is a special case of equilibria with component isomerism. Therefore, all the results of Chap. 6.2.1 are valid here, especially Eqs. (6-67)–(6-71). The heights of the barriers separating the individual isomeric structures permits the ready attainment of equilibirum, but draw into question the plausibility of assumptions concerning the separability of the motions of

Table 6-13. Temperature dependences[a] of weights w_P, w_{S_4}, and w_A of pyramid (P), S_4 cyclic (S_4), and asymmetric cyclic (A) water tetramers

T (K)	w_P	w_{S_4}	w_A
100	0.997	0.003	1×10^{-8}
400	0.624	0.297	0.078
700	0.334	0.267	0.399

[a] From Refs.[93,193]

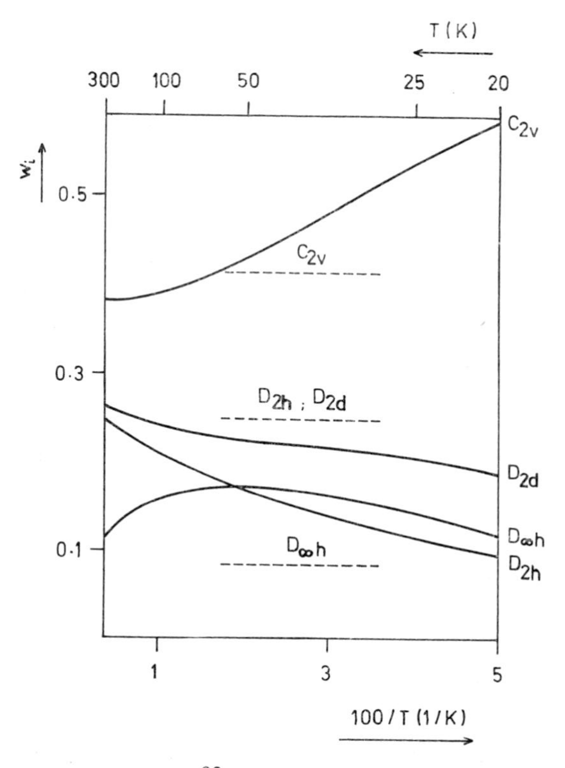

Figure 6-16. Temperature dependence[93] of weights w_i of $(H_2)_2$ with D_{2h}, $D_{\infty h}$, D_{2d} and C_{2v} symmetries; approximate temperature-independent weights[463] are indicated by the dashed lines

the isomers. At present, the literature[93,309,445,455,457,462] contains a thermodynamically correct treatment of the problem of isomerism leading from partial to summary terms ΔH_T^0 and ΔS_T^0 for the $(H_2)_2$, $(NO)_2$, $(H_2O)_4$, $(CO_2)_2$, HF.HCl, HF.ClF, $(H_2O)_5$, $CH_3OH.3\,H_2O$, and $CH_3OH.6\,H_2O$ clusters. Table 6-12 gives an illustration of the weighting treatment[193,309] using the $(H_2O)_5$ cluster. While at a temperature of 100 K the summary quantities are practically determined by the values corresponding to cage-type isomer, at a temperature of 700 K the stabilities of both these structures are very similar. This is manifested in a marked effect on both the summary ΔH_T^0 and ΔS_T^0 terms.

In multimolecular clusters, even the stability order of the individual structures may change on a change in the temperature. This is reflected in the temperature dependence[93] of the weight factors of the three isomers of $(H_2O)_4$ found in work[193] (Table 6-13). While the importance of the asymmetric cyclic tetramer is quite negligible at the beginning of the temperature interval studied, at the end of this interval this is the most stable structure. This example points out the difficulties encountered by attempts to employ only a single 'most stable' structure during the theoretical studies of clusters.

Calculation of the weight factors for the individual isomers according to (6-74) requires a range of information that is rarely available as the output of any quantum-chemical study. The literature contains proposals[463,464] for the calculation of weight factors w_i from purely geometric considerations. However, they do not agree well with the properly determined values – Fig. 6-16 gives such an example[93] for the four isomeric structures of $(H_2)_2$. In addition, w_i values found in this simple way are temperature independent. From the point of view of the extent of quantum-

Table 6-14. Temperature dependence[a] of the weight w_{trans} of *trans*-$(NO)_2$ in an equilibrium mixture with *cis*-$(NO)_2$ for various approximations[b] to vibrational motion

T (K)	w_{trans}		
	'Exact'	$q_v = 1$	$q_v = 1$; $\Delta_0^{(q)} = 0$
50	0.688	0.683	0.040
100	0.608	0.577	0.159
200	0.597	0.520	0.287
298.15	0.609	0.502	0.341
500	0.629	0.486	0.389

[a] From to Ref.[309]
[b] 'Exact' refers to values obtained using vibrational partition function q_v and quantum correction $\Delta_0^{(q)}$ for the zero-point motion, both in the harmonic-oscillator approximation.

-chemical calculations, the vibrational part of the partition function becomes most critical. Table 6-14 gives an example of the importance of the individual contributions to the weight factor[309] of *trans*-$(NO)_2$; neglection of vibrational partition function q_v leads to reasonable values. However, if quantum correction $\Delta_0^{(q)}$ is neglected in ΔH_0^0 (Eq. (6-27)), the quality of the w_{trans} value becomes much worse. It is thus apparent[309] that it is advisable to base the calculation of w_i on the, at present, usual extent of the output of quantum-chemical studies (only the depth and geometry of the minimum) only when the contribution of term $\Delta_0^{(q)}$ can be expected to be negligible.

While the necessity to work using the RRHO approximation to the partition function is a constant limitation in theoretical studies of clusters and their isomerism, it should be noted that this approximation affects the values of the standard thermodynamic terms most, rather than the weight factor w_i. In the latter case, it can be expected that there will be considerable cancellation between the numerator and the

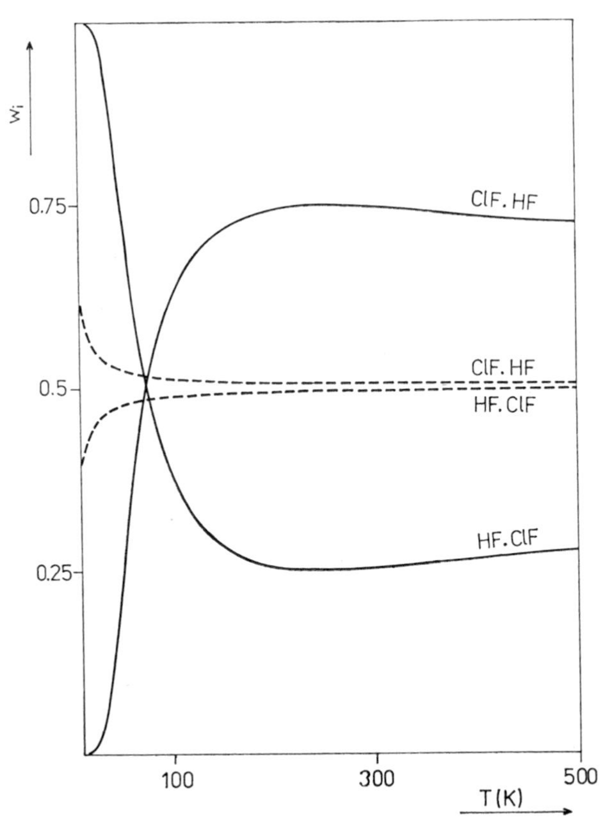

Figure 6-17. Temperature dependences[80] of weights w_i of the minimum-energy structures HF.ClF and ClF.HF; simple (Boltzmann) configurational factors (see text) are indicated by the dashed lines

denominator in Eq. (6-74). If the RRHO approximation is employed, it is preferable to consider possible cluster symmetry in terms of the symmetry number of the coresponding rigid structure. Although the appropriate permutation-inversion group should generally be used also for construction of the symmetry numbers for non-rigid, fluxional molecules[457,465,466], this is not true with the utilization of RRHO for description of clusters, as *de facto* only one part of the potential well is employed in setting up the vibrational partition function. This is effectively reflected in a reduction in the number of vibrational states in comparison with the real non-rigid situation.

A surprisingly low temperature for stability exchange was found[80,81,457,467] for the two isomeric structures in the HF—ClF system. The newest results[80] have shown that equimolarity is attained in an equilibrium mixture at a temperature of 75 K. (Fig. 6-17). The character of the temperature dependence of w_i explains why only one of the two structures could be observed in the experiments[468] (with an effective temperature estimated[469] at 10 K). Fig. 6-17 gives the w_i values together with the course of the simple Boltzmannn factors (also called simple configurational factors[281]), based only on the depths of the potential minima. It is apparent that the latter factors cannot, in general, be used as approximations to w_i. The relationships between the partial and overall standard thermodynamic terms are apparent from Fig. 6-18; the oscillating behaviour of the overall term is especially interesting. Considering the generally nonequilibrium conditions[470,471] in the type of experimental technique employed[468] (jet expansion), unfortunately the thermodynamic characteristics could

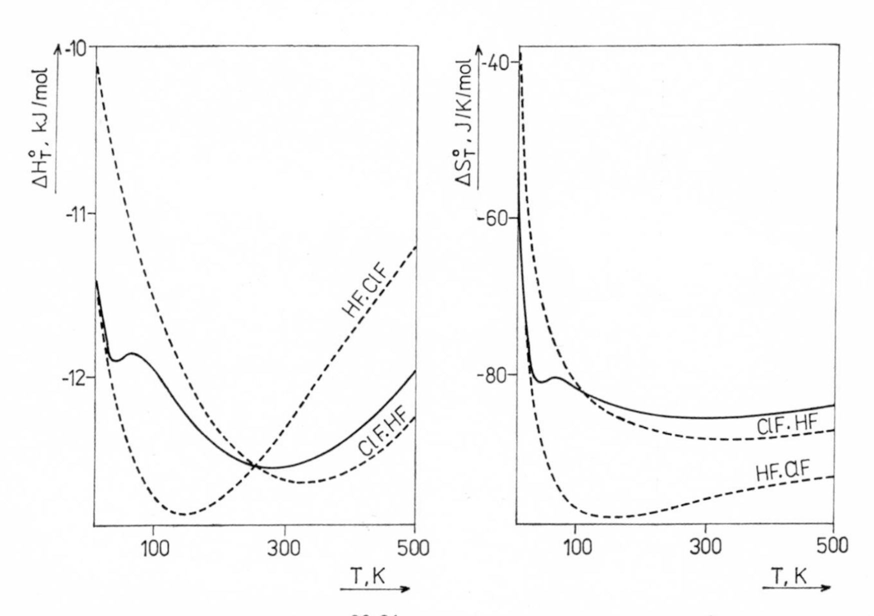

Figure 6-18. Temperature dependences[80,81] of the standard enthalpy ΔH_T^0 and entropy ΔS_T^0 for the partial (to HF.ClF or ClF.HF) and overall (———) associations of HF and ClF

not be found from these observations. Spectroscopic study of the HF—ClF system under equilibrium conditions seems impracticable because of the corrosiveness of this medium.

The $(CO_2)_2$ system can be considered as the first whose theoretically predicted isomerism could be experimentally verified[443-445]. For this system *ab initio* calculations[443,444] demonstrated the existence of two isomeric structures – parallel (P) and T forms. Although the P form has a potential energy of about 4 kJ mol^{-1} lower, it was found[445] (Fig. 6-19) that the interplay of all the parameters considered led to the attainment of equimolarity at a temperature of about 230 K, i.e. relatively close to the normal sublimation point of solid CO_2. Although data are at present lacking on the potential barrier separating the two isomers, this system seems to be the most suitable for (the first ever) experimental verification of theoretically predicted isomerism in gas-phase complexes.

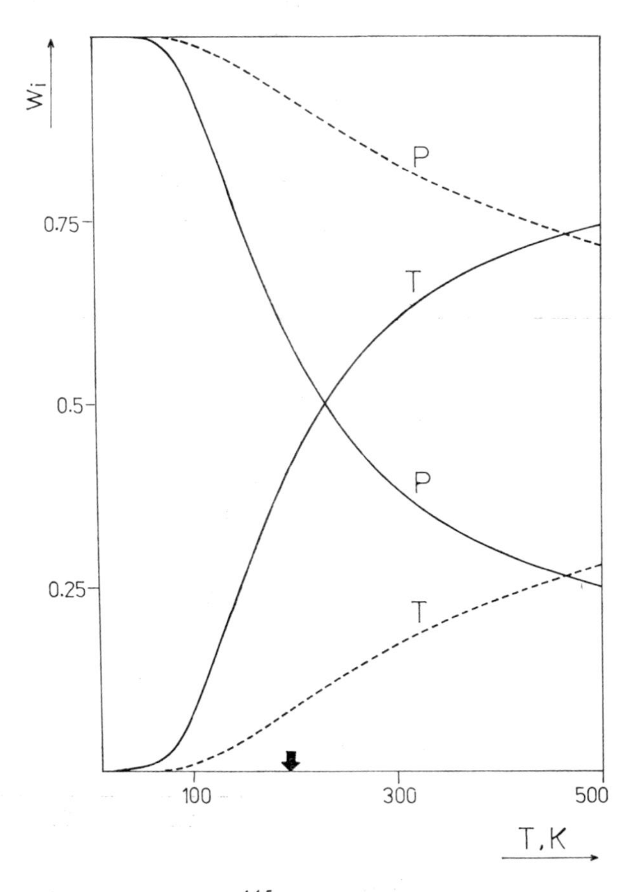

Figure 6-19. Temperature dependences[445] of weights w_i of the parallel (P) and T forms of $(CO_2)_2$; simple (Boltzmann) configurational factors (see text) are indicated by the dashed lines; the arrow indicates the normal sublimation point of CO_2

At present, cluster studies are typically concentrated on localization of energy minima. Potential barriers between them are studied less frequently, even though they determine the possibility of interisomer equilibration and (primarily) permit estimation of whether the isomeric structure will, in principle, be distinguished under particular observation conditions or whether fluxional behaviour must be expected.

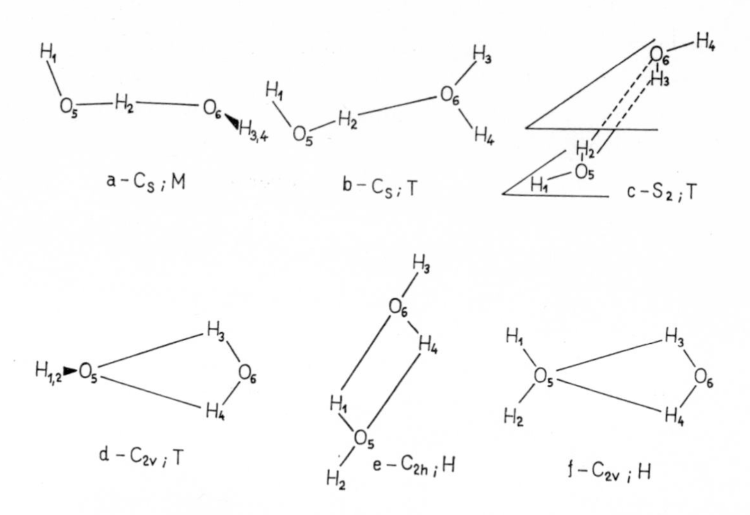

Figure 6-20. Schemes of the stationary points found[207,208] on the *ab initio* SCF CI hypersurface[460] of $(H_2O)_2$; M denotes a minimum, T a transition state, H a higher type of stationary point

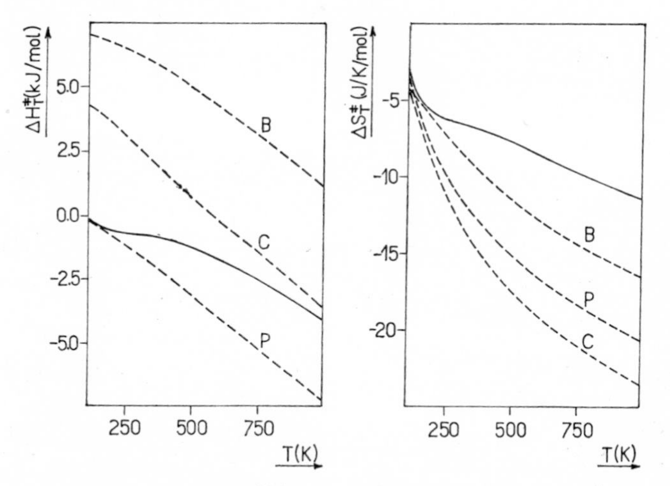

Figure 6-21. Temperature dependences[211] of activation enthalpy $\Delta H_T^{\ddagger}$ and entropy $\Delta S_T^{\ddagger}$ of the water dimer interconversion for C_s planar-linear (P — Figure 6-20b), S_2 closed (C — Figure 6-20c), and C_{2v} bifurcated (B — Figure 6-20d) transition states and for the overall activation process (———)

Work[211] on the symmetric rate process

$$(H_2O)_2\ (g) \rightarrow (H_2O)_2\ (g) \tag{6-106}$$

is an example of study of the potential barrier in autoisomerization. Although the possibility of isomerism of $(H_2O)_2$ was rejected at the energy minimum level in the framework of the *ab initio* SCF CI hypersurface, triple parallel isomerism of the activated complex was simultaneously demonstrated for process (6-106) (cf. Fig. 6-20). The interrelationships between the partial activation terms of process (6-106) corresponding to realization through the individual transition states found (Fig. 6-20b, c, d) and the overall terms are apparent from Fig. 6-21 and reflect considerable differences between the two types of quantities for higher temperatures.

A study of microclusters[281,472–483] in the framework of phenomenological potentials lies at the boundary between the gaseous and condensed state. Because of

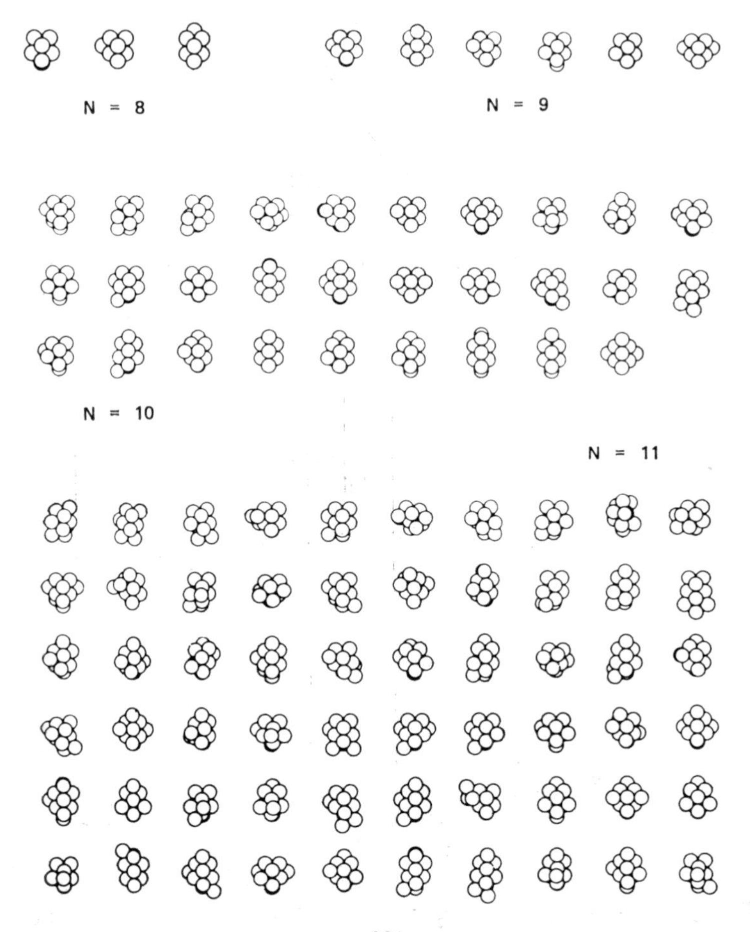

Figure 6-22. Octahedral Lennard-Jones isomers[281] for $N = 8$ to 11

the simple form of the interaction potential, multiparticle clusters of relatively large dimensions can be studied where the isomerism of these clusters is essentially greater (cf. Fig. 6-22). For example, a total of 988 isomers was found for the 13-particle cluster controlled by the Lennard-Jones potential[281]; however, transition to the Morse potential led to a decrease to 36 isomers, which is nonetheless a very substantial structural abundance with massive thermodynamic consequences. For example, Burton[477] demonstrated that the proper consideration of configurational effects for a 49-atom cluster at 60 K leads to a 500–3000-fold increase in the equilibrium concentration. Although this type of study is at present carried out using the RRHO approximation, in the future orientation to the anharmonic and non-rigid schemes can be expected (cf.[281,484,485]).

The inversion equilibrium problem is of special importance in cluster studies (Chap. 6.2.5). The usual manner of studying clusters is at present limited[436] to the SCF approximation. Neglection of the correlation energy connected with this approach need not have a great effect[441] on the geometry and force fields, but leads to a marked deterioration[394] of energy differences between individual isomers. These differences can, however, be found if structural information from quantum-chemical calculations is combined with the inversion equilibrium problem, permitting[202,394,441], for example, detailed description of the energetics and thermodynamics of $(NO)_2$ clusters.

6.5.2 *Clusters and Real Gases*

Non-ideality in the gas phase is mostly expressed (see e.g.[486]) in terms of the second and higher virial coefficients. Theoretically, it is in principle possible to separate[234] the virial coefficients (based on sufficiently realistic potentials) into two components: the first is connected with the collision between free molecules and the second with the formation of the corresponding multimolecular clusters. Model calculations[234,235] have demonstrated that, at low temperatures, the value of the second virial coefficient B_2 is determined primarily by dimer formation. This fact is important in relation to the utilization of quantum-chemical methods. While calculation of the contribution of molecular translation in the potential field formed by intermolecular interactions requires knowledge of the whole hypersurface of interaction energy, the behaviour in the minima is sufficient for the determination of the contribution of cluster formation. Thus, it is possible to use the virial coefficient values at low temperatures to generate information on clusters and to test quantum-chemical methods.

Work[487] describes a concept of non-ideality in the gaseous phase as a result of formation of multimolecular clusters, thoroughly considering any possible isomerism. This leads to a recursion formula[487] permitting the expression of the arbitrarily high virial coefficient B_n as a function of the equilibrium constants for the formation of the individual isomers of two- to n-molecular clusters. The results[202] obtained in

Table 6-15. Contributions[a] of the formation of $(NO)_2$ clusters to the second virial coefficient B_2 of nitric oxide at low temperatures

T (K)	B_2(cm^3 mol^{-1})			
	cis-$(NO)_2$	*trans*-$(NO)_2$	total-$(NO)_2$	observed[486]
121.72[b]	−87.8	−131.1	−218.9	−224.4
135.08	−28.4	−41.9	−70.3	−150.9
141.30	−18.0	−26.4	−44.4	−141.7

[a] From Ref.[202]
[b] The normal boiling point of nitric oxide is[a] 121.36 K.

the theoretical calculation of B_2 for nitric oxide (Table 6-15) are an example of this approach to the problem of non-ideality of the gaseous phase. It is apparent that the B_2 values close to the normal boiling point can be readily described in terms of this concept.

For completeness, it should be noted that this quasi-chemical approach should be formally differentiated from the approach to virial coefficients based[488] on the grand canonical ensemble partition function. The concept of a cluster introduced in this connection by Mayer (see for example[30] – cluster or group integrals) is not identical with the concept of a multimolecular cluster employed here. The concept introduced by Mayer[30] is far more general and led to the formulation of the cluster diagram technique, widely used in contemporary chemical physics (see, e.g.[489]).

6.5.3 *The Cluster Concept of the Liquid State*

Recently, the cluster concept formulated by Scheraga *et al.*[494–497] on the basis of an original idea by Frank and Wen[498] has been used for description of the liquid state (see e.g.[490]) or solvation effects (see, e.g.[434,436,491–493]) in connection with quantum-chemical methods. This concept considers a liquid to be a mixture of molecular clusters, e.g. water[497] as a mixture of $(H_2O)_2$ to $(H_2O)_9$ clusters. The discovery of a very low monomer content[497] for an arbitrary relevant temperature, varying from about 0.7% at 0 °C to about 1.6% at 100 °C, is interesting here. Results[459] concerning the dependence of the stability of $(H_2O)_n$ forms on number n suggest that considerations should be limited to small clusters. Interactions between clusters are approximated in this concept in the framework of the hard spheres model. The cluster model permitted[496,497] the successful description of the thermodynamic properties of the liquid state of water.

The cavity concept, introduced by Eley[499] and developed by Pierotti[500,501] and used to describe gas solubility in liquids[201,502] can be considered a special case of the cluster model. The use of quantum-chemical methods to determine the term connected with the interaction of the dissolved molecule and the cavity formed by solvent molecules permitted the elimination of the strong empirical element from the cavity concept. In general, there are many ways in which a cavity can be modelled, affected by the choice of the number of solvent molecules and by their mutual arrangement. In this connection the concept of cavity isomerism was introduced[201] and a technique for weighting the contributions of the individual cavity forms to the overall values of the thermodynamic characteristics of the solvation process was described[201]. Table 6-16 gives an illustration for the hydration of methanol on the basis of PCILO results[201]. The corresponding weighting procedure is again a special case of the general equilibrium problem discussed in Chap. 6.2.1.

Table 6-16. Partial and total PCILO[a] standard[b] hydration enthalpy ΔH°_h, entropy ΔS°_h, and the Henry's Law constant K_H at temperature $T = 298.15$ K

Process[c]	ΔH°_h(kJ mol^{-1})	ΔS°_h(J mol^{-1}K^{-1})	$1/K_H$[d]
$n = 1$	2.93	−118.0	2.09×10^{-7}
$n = 3$	−29.97	−118.0	1.22×10^{-1}
$n = 6$	−41.87	−118.0	1.48×10^{1}
Total	−41.78	−117.6	1.49×10^{1}

[a] From Ref.[201]
[b] Solution standard state: solution at unit mole fraction.
[c] The cavity was modelled by the water shell in the $CH_3OH.nH_2O$ cluster.
[d] K_H in atm; 1 atm = 101325 Pa.

The, at present, very rare possibility of transition from the cluster concept to the more rigorous statistical thermodynamic theory of liquids assumes the possibility of quantum-chemical calculations of the hypersurface for at least the pair interaction potential. If this type of information is available, it is frequently used in connection with the Monte Carlo technique; studies[503–515] mostly based on the *ab initio* calculations of the pair interaction potential of water carried out by Clementi *et al.* (see e.g.[506]) are examples of this approach. In the framework of the Monte Carlo method[516] the averaged value $\langle F \rangle$ of arbitrary quantity F is found by calculation of integral

$$\langle F \rangle \equiv \int F(\xi)\, P(\xi)\, \mathrm{d}\xi \tag{6-107}$$

by random generation of configurations, mostly using the technique described by Metropolis *et al.*[517]. In Eq. (6-107) $P(\xi)$ designates the configurational probability density in the ensemble considered and ξ is the vector representing the configuration (its form and the significance of the symbolic integral depend[516] on the choice of ensemble). It is apparent that the description of the interaction potential in terms of isolated minima separated by infinitely high barriers reduces the Monte Carlo method to a level corresponding to the approach used in this work for weighting of the contributions of the individual isomeric structures. It should, however, be emphasized that these two techniques do not become identical even for the given simplified form of the interaction potential. The conventional Monte Carlo approach involves only the Boltzmann term, with the potential energy, while the rotational-vibrational energy is assumed to be identical at all points in the configurational space. As can be seen in Figs. 6-17 and 6-19, this assumption can be misleading. Consideration of differences in rotational-vibrational motions would be possible within the framework of the Monte Carlo method, e.g. by transfer from the potential energy hypersurface to the Gibbs function hypersurface. The study of microclusters[281,518–523] by the Monte Carlo technique has also recently become more frequent.

The molecular dynamics method[516] represents an alternative to the Monte Carlo technique. This method is based on solution of the Newton equations of motion for a given set of particles and thus also permits the study of its time evolution. Thus, this method can yield both static properties and information on non-equilibrium properties; an example is the recent molecular dynamics calculations carried out on water[555] using the *ab initio* SCF CI potential[460]. This transition to nonequilibrium thermodynamics (see, e.g.[556]) is important, among other reasons, because some experimental techniques generate clusters under conditions far removed from equilibrium (see e.g. the jet expansion technique[470]).

6.5.4 *Isomerism of Adsorption Complexes*

In the study of heterogeneous catalysis and sorption, the quantum-chemical approach typically involves the approximation of the whole solid-phase surface by a relatively small set of particles: a cluster. This cluster approach has been established as an accepted alternative technique (see, for example[524–527]) to the methods of solid state physics. The potential energy hypersurface describing the interactions of the cluster with the substrate (mostly represented in terms of hypersurface stationary points) is then used as a starting point for the study of real processes corresponding to the model used. This cluster approach has provided a good deal of evidence[528–541] that catalytic reaction or sorption processes can be realized not through a single but through several different isomeric complexes of the adsorbate – solid adsorbent type. In a similar manner as in the analogous situations discussed above, it can be expected that these individual isomeric structures will only rarely be distinguishable experimentally; thus, the observed quantities will again most probably have a convolutional nature. Contemporary theoretical studies have pointed to two special types of isomerism of adsorption complexes. One of these, termed[542] site-caused isomerism, results when a number of different (rigid) clusters are used[524–538] to model different adsorption sites on the solid surface, where only a single adsorption complex is formed on each cluster. The second important type, termed adsorbate-caused isomerism[542], appears when there are several different arrangements of the adsorbed molecule on one (fixed) adsorption site modelled by the given (single) cluster[539-541]. In general,

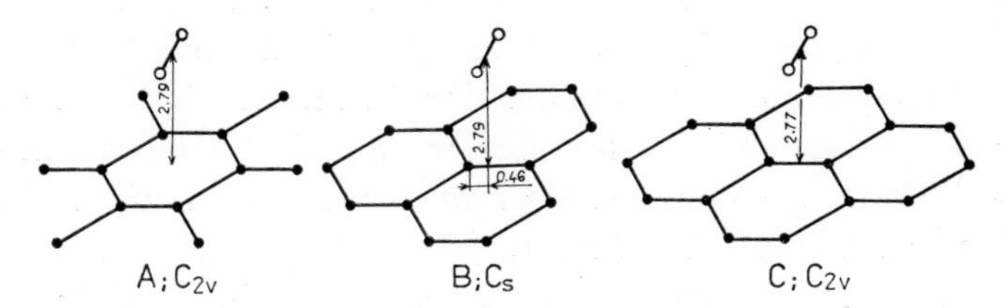

Figure 6-23. Optimum structures[543] of model adsorption complexes for the physical adsorption of O_2 on graphite (○ — O atoms, ● — C atoms); structural characteristics in 10^{-10} m

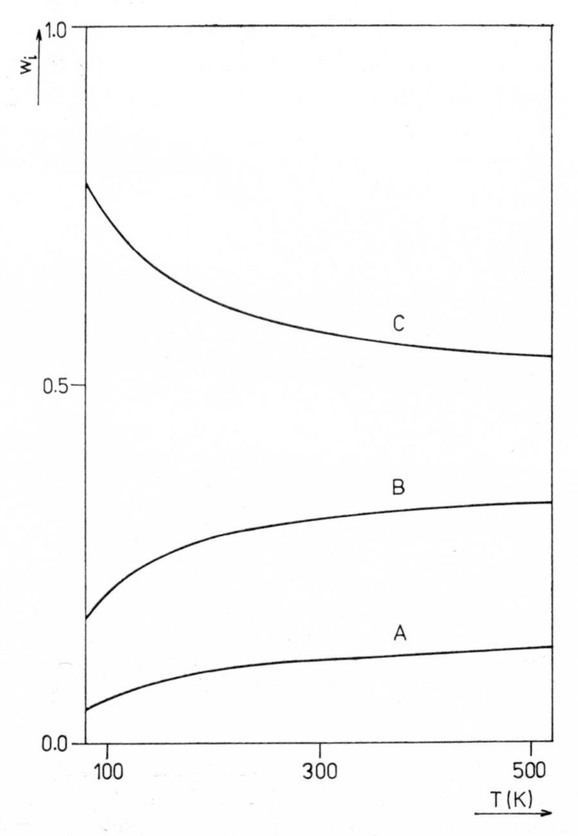

Figure 6-24. Temperature dependences[543] of the weight factors w_i of the adsorption complexes A, B and C (see Figure 6-23)

however, it is also necessary to consider the simultaneous presence of various combinations of both of these particular types of isomerism.

Approximations included in the conventional cluster approach also clearly limit the type of description during transition from the potential energy terms to the Gibbs function. It is thus necessary to employ the localized adsorption concept, as the otherwise more sophisticated concept of mobile adsorption would require knowledge of the whole potential energy hypersurface. It was demonstrated[34,542–545] that, with suitable selection of starting assumptions, the isomerism of adsorption complexes of these two types with transition from the partial to the total characteristics can be treated in a manner essentially identical with that given in Eqs. (6-67)–(6-71). There are, however, important differences[35,542–545] in the manner of construction of the partition function for the adsorbed molecule (suppression of translation and rotation, configurational contribution considering the distinguishability of adsorption sites, and selection of a standard state).

Fig. 6-23 depicts the results of a study[543] of the (physical) sorption of O_2 on graphite assuming site-caused isomerism. The considered three types of clusters model adsorption above the centre of the hexagon, above the C atom, and above the middle of the C—C bond. The weight factors for the resulting adsorption complexes A, B, and C indicate (Fig. 6-24) that structure C is actually predominant only at the

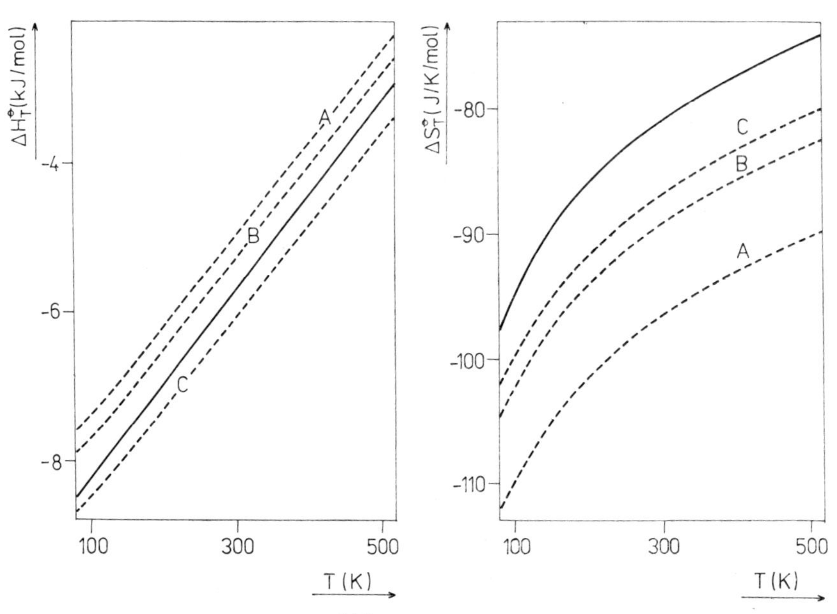

Figure 6-25. Temperature dependences[543] of the partial standard enthalpy and entropy terms for adsorption processes leading to complexes A, B and C (see Figure 6-23) in the physical adsorption of O_2 on graphite and the corresponding overall (———) values ΔH_T^θ and ΔS_T^θ; standard states — ideal gas at 101 325 Pa and the state of a surface concentration of 10^{20} molecules/m^2

lowest temperatures. However, above a temperature of 210 K the weight factors for all three structures are of the same order, where only about half of the adsorption complexes present at equilibrium are of type C. Although no interchange occurs here in the stabilities of the individual adsorption complexes, nonetheless the consequences of isomerism of the adsorption complex for the values of the thermodynamic characteristics are considerable (Fig. 6-25). The deviation of the total enthalpy term from the partial value for structure C is quite small at the lowest temperatures, but increases with increasing temperature. This difference is, however, significant for the entropy term; for example, above a temperature of 110 K it is greater than 5 J K^{-1} mol^{-1}. This illustrative example indicates that in gas-solid interactions the isomerism of the adsorption complex can greatly affect comparison of the theoretical and observed data as well as the interpretation of the latter terms.

6.6 Summary

Above a certain complexity of chemical processes or a certain size of the components, isomerism of the reaction components (both equilibrium and rate processes) can be considered to be a general characteristic of chemical reactions. Such isomerism has been distinguished for quite a large number of systems, but has been properly treated only for a very few theoretically studied processes. Considering, however, that contemporary quantum-chemical studies are concerned mainly with relatively small systems and also that isomerism need not be necessarily assumed *a priori* (so that the search is limited to finding the first suitable structure), it can be expected that in the future examples will appear of much more extensive isomerism of the reaction components that is distinguishable only at a theoretical level. In any case, existing examples of complete studies of the consequences of such isomerism of the reaction components convincingly demonstrate that the simulation of multi-isomer characteristics in terms of those for (any) single structure can no longer be considered satisfactory.

The concept of the isomerism of reaction components is developing in close connection with the development of the theory of chemical reactivity. Thus, it can be expected that, similar to the theory of chemical reactions developed on several levels differing in the degree of sophistication of the theoretical description and in their ability to yield numerical data for real systems, the problem of isomerism of the components of chemical processes will also be treated at various methodical levels including, for example, the Gibbs function hypersurface or the density matrix formalism. A successful example of such a parallel approach is the problem of sequential isomerism of activated complexes where, in addition to the practical approach discussed here, a methodologically far more advanced unified statistical theory concept is being developed[546–550].

The existence of the isomerism of reaction components can be understood as a new, independent argument for more intense cooperation between theory and experiment.

This is a result of the limited capabilities (in some situations) of experimental techniques in the distinguishing and chracterization of systems with complex isomeric compositions. On the other hand, because of the approximative nature of any theoretical approach it is necessary ot carry out comparison with existing obesrved data in the best way possible. Such cooperation can be a further stimulant for the development of a completely general concept of chemical reactivity considering all types of isomerism of the reaction components.

REFERENCES

(1) DIRAC, P. A. M.: 1929, Proc. Roy. Soc. *A 123*, pp. 714—733.

(2) DAVYDOV, A. S.: 1965, Quantum Mechanics, Pergamon Press, Oxford.

(3) NIKITIN, E. E.: 1970, Theory of Elementary Atomic — Molecular Processes in Gases, Khimija, Moscow, (in Russian).

(4) MCWEENY, R.: 1972, Quantum Mechanics: Principles and Formalism, Pergamon Press, Oxford.

(5) LAIDLER, K. J.: 1969, Theories of Chemical Reaction Rates, McGraw-Gill, New York.

(6) EYRING, H.: 1935, J. Chem. Phys. *3*, pp. 107—115.

(7) EVANS, M. G., and POLANYI, M.: 1935, Trans. Faraday Soc. *31*, pp. 875—894.

(8) POLANYI, J. C., and SCHREIBER, J. L.: 1974, in Physical Chemistry. An Advanced Treatise, Vol. VIA, Ed. W. Jost, Academic Press, New York, pp. 383—487.

(9) LEVINE, R. D., and BERNSTEIN, R. B.: 1974, Molecular Reaction Dynamics, Oxford University Press, New York.

(10) NIKITIN, E. E., and ZÜLICKE, L.: 1978, Selected Topics of the Theory of Chemical Elementary Processes, Springer-Verlag, Berlin.

(11) CHRISTOV, S. G.: 1980, Collision Theory and Statistical Theory of Chemical Reactions, Springer-Verlag, Berlin.

(12) 1981, Potential Energy Surfaces and Dynamics Calculations, Ed. D. G. Truhlar, Plenum Press, New York.

(13) SLANINA, Z.: 1975, Radiochem. Radioanal. Lett. *22*, pp. 291—298.

(14) ČÁRSKY, P., SLANINA, Z., and ZAHRADNÍK, R.: 1975, Chem. Listy *69*, pp. 529—543.

(15) CLEMENTI, E.: 1976, Bull. Soc. Chim. Belg. *85*, pp. 969—975.

(16) BERAN, S., ČÁRSKY, P., HOBZA, P., PANCÍŘ, J., POLÁK, R., SLANINA, Z., and ZAHRADNÍK, R.: 1978, Usp. Khim. *47*, pp. 1905—1932.

(17) SLANINA, Z.: 1981, Advan. Quantum Chem. *13*, pp. 89—153.

(18) GIAUQUE, W. F.: 1930, J. Am. Chem. Soc. *52*, pp. 4808—4815.

(19) RODEBUSH, W. H.: 1931, Chem. Rev. *9*, pp. 319—346.

(20) KASSEL, L. S.: 1936, Chem. Rev. *18*, pp. 277—313.

(21) WILSON JR., E. B.: 1940, Chem. Rev. *27*, pp. 17—38.

(22) GODNEW, I. N.: 1963, Berechnung thermodynamischer Funktionen aus Moleküldaten, VEB Deutscher Verlag der Wissenschaften, Berlin.

(23) WOOLLEY, H. W.: 1956, J. Res. Nat. Bur. Stand. *56*, pp. 105—110.

(24) ČERNÝ, Č.: 1961, Chem. Listy *55*, pp. 1115—1136.

(25) FRANKISS, S. G., and GREEN, J. H. S.: 1973, in Chemical Thermodynamics. Specialist Periodical Reports, Vol. 1, Ed. M. L. McGlashan, The Chemical Society, London, pp. 268—316.

(26) ASTON, J. G., and FRITZ, J. J.: 1959, Thermodynamics and Statistical Thermodynamics, Wiley, New York.

(27) Hill, T. L.: 1960, An Introduction to Statistical Thermodynamics, Addison-Wesley, Reading, Mass.
(28) Andrews, F. C.: 1963, Equilibrium Statistical Mechanics, Wiley, New York.
(29) Hála, E., and Boublík, T.: 1970, Einführung in die statistische Thermodynamik, Vieweg, Braunschweig.
(30) Mayer, J. E., and Mayer, M. G.: 1977, Statistical Mechanics, 2nd Ed., Wiley — Interscience, New York.
(31) Schütz, A.: 1976, Chem. Listy *70*, pp. 1233—1243.
(32) McGlashan, M. L.: 1979, Chemical Thermodynamics, Academic Press, London.
(33) Slanina, Z.: 1983, Chem. Phys. Lett. *95*, pp. 553—554.
(34) Slanina, Z.: 1984, React. Kinet. Catal. Lett. *26*, pp. 179—182.
(35) Kosloff, R., Levine, R. D., and Bernstein, R. B.: 1974, Mol. Phys. *27*, pp. 981—992.
(36) Glasstone, S.: 1946, Theoretical Chemistry, van Nostrand, New York.
(37) Landau, L. D., and Lifshits, E. M.: 1976, Statistical Physics, Vol. 1, Nauka, Moscow, (in Russian).
(38) Born, M., and Oppenheimer, R.: 1927, Ann. Phys., 4. Folge *84* (der ganzen Reihe *389*), pp. 457—484.
(39) Slanina, Z.: 1971, Thesis, Charles University, Prague, Appendix D-4.
(40) Ref.[39], Appendix D-2.
(41) Slanina, Z.: 1985, Int. J. Quantum Chem., *27*, pp. 691—697.
(42) Hirschfelder, J. O.: 1940, J. Chem. Phys. *8*, p. 431.
(43) Ref.[39], Appendix D-3.
(44) Mulholland, H. P.: 1928, Proc. Cambridge Phil. Soc. *24*, pp. 280—289.
(45) Kassel, L. S.: 1933, J. Chem. Phys. *1*, pp. 576—585.
(46) Viney, I. E.: 1933, Proc. Cambridge Phil. Soc. *29*, pp. 142—148.
(47) Viney, I. E.: 1933, Proc. Cambridge Phil. Soc. *29*, p. 407.
(48) Gordon, A. R.: 1934, J. Chem. Phys. *2*, pp. 65—72.
(49) Stripp, K. F., and Kirkwood, J. G.: 1951, J. Chem. Phys. *19*, pp. 1131—1133.
(50) Slanina, Z., and Zahradník, R.: 1976, Chemical Reactivity and Spectroscopies, SPN Publishing House, Prague, p. 44, (in Czech).
(51) 1971, JANAF Thermochemical Tables, NSRDS — NBS 37, Ed. D. R. Stull, and H. Prophet, U.S. Department of Commerce, Washington.
(52) Shin, H. K.: 1977, Chem. Phys. Lett. *47*, pp. 225—230.
(53) Shin, H. K.: 1977, Private Communication.
(54) 1962, Thermodynamic Properties of Individual Compounds, Vol. I, Ed. Glushko V.P., Acad. Sci. USSR Publ. House, Moscow, p. 48, (in Russian).
(55) Baehr, H. D., Hartmann, H., Pohl, H.-C., and Schomäcker, H.: 1968, Thermodynamische Funktionen idealer Gase für Temperaturen bis 6000 °K, Springer-Verlag, Berlin.
(56) Pennington, R. E., and Kobe, K. A.: 1954, J. Chem. Phys. *22*, pp. 1442—1447.
(57) Isaacson, A. D., Truhlar, G. D., Scanlon, K., and Overend, J.: 1981, J. Chem. Phys. *75*, pp. 3017—3024.
(58) Woolley, H. W.: 1955, J. Res. Nat. Bur. Stand. *54*, pp. 299—308.
(59) Kleinman, L. I., and Wolfsberg, M.: 1973, J. Chem. Phys. *59*, pp. 2043—2053.
(60) Bardo, R. D., Kleinman, L. I., Raczkowski, A. W., and Wolfsberg, M.: 1978, J. Chem. Phys. *69*, pp. 1106—1111.
(61) Crotov, S. S., Ischenko, A. A., and Ivashkevich, L. S.: 1979, Int. J. Quantum Chem. *16*, pp. 973—983.
(62) Ishida, T.: 1974, J. Chem. Phys. *61*, pp. 3009—3016.
(63) Wolfsberg, M.: 1979, J. Chem. Phys. *70*, pp. 5322—5323.
(64) Amorebieta, V. T., and Colussi, A. J.: 1981, Chem. Phys. Lett. *82*, pp. 530—533.

(65) Witschel, W.: 1981, Z. Naturforsch. *36a*, pp. 481—488.
(66) Witschel, W.: 1982, Chem. Phys. Lett. *86*, pp. 558—562.
(67) Pitzer, K. S.: 1937, J. Chem. Phys. *5*, pp. 469—472.
(68) Pitzer, K. S., and Gwinn, W. D.: 1942, J. Chem. Phys. *10*, pp. 428—440.
(69) Pitzer, K. S.: 1946, J. Chem. Phys. *14*, pp. 239—243.
(70) Kilpatrick, J. E., and Pitzer, K. S.: 1949, J. Chem. Phys. *17*, pp. 1064—1075.
(71) Pople, J. A.: 1974, Tetrahedron *30*, pp. 1605—1615.
(72) Combs, L. L., and Holloman, M.: 1975, J. Phys. Chem. *79*, pp. 512—521.
(73) Scott, D. W., Guthrie, G. B., Messerly, J. F., Todd, S. S., Berg, W. T., Hossenlopp, I. A., and McCullough, J. P.: 1962, J. Phys. Chem. *66*, pp. 911—914.
(74) Pitzer, K. S.: 1937, J. Chem. Phys. *5*, pp. 473—479.
(75) Aston, J. G., Szasz, G., Woolley, H. W., and Brickwedde, F. G.: 1946, J. Chem. Phys. *14*, pp. 67—79.
(76) Kilpatrick, J. E., Pitzer, K. S., and Spitzer, R.: 1974, J. Am. Chem. Soc. *69*, pp. 2483 to 2488.
(77) Pitzer, K. S., and Bernstein, L. S.: 1975, J. Chem. Phys. *63*, pp. 3849—3856.
(78) Rathjens Jr., G. W., Freeman, N. K., Gwinn, W. D., and Pitzer, K. S.: 1953, J. Am. Chem. Soc. *75*, pp. 5634—5642.
(79) Pitzer, K. S.: 1939, J. Chem. Phys. *7*, pp. 251—255.
(80) Slanina, Z.: 1982, J. Phys. Chem., *86*, pp. 4782—4786.
(81) Slanina, Z.: 1983, Ber. Bunsenges. Phys. Chem. *87*, pp. 28—32.
(82) Knopp, K.: 1957, Theory and Application of Infinite Series, Blackie, London.
(83) Erdélyi, A.: 1956, Asymptotic Expansions, Dover Publ., New York.
(84) Isaacson, A. D., and Truhlar, D. G.: 1981, J. Chem. Phys. *75*, pp. 4090—4094.
(85) Stratt, R. M., and Desjardins, S. G.: 1982, J. Chem. Phys. *76*, pp. 5134—5144.
(86) Ref.[22], p. 203.
(87) Slanina, Z., Berák, P., and Zahradník, R.: 1977, Collect. Czech. Chem. Commun. *42*, pp. 1—15.
(88) McBride, B. J., and Gordon, S.: 1961, J. Chem. Phys. *35*, pp. 2198—2206.
(89) Bron, J., and Wolfsberg, M.: 1972, J. Chem. Phys. *57*, pp. 2862—2869.
(90) Bron, J., and Hacker, S. O.: 1973, Can. J. Chem. *51*, pp. 2765—2768.
(91) Chen, C. L., Bopp, P., and Wolfsberg, M.: 1982, J. Chem. Phys. *77*, pp. 579—580.
(92) Slanina, Z., and Zahradník, R.: 1974, Collect. Czech. Chem. Commun. *39*, pp. 729—735.
(93) Slanina, Z.: 1979, Advan. Mol. Relax, Interact. Proces. *14*, pp. 133—148.
(94) Blaney, B. L., and Ewing, G. E.: 1976, Annu. Rev. Phys. Chem. *27*, pp. 553—586.
(95) Strauss, H. L., and Thiele, E.: 1967, J. Chem. Phys. *46*, pp. 2473—2480.
(96) Liskow, H. H., Bender, C. F., and Schaefer III, H. F.: 1972, J. Chem. Phys. *56*, pp. 5075 to 5080.
(97) Wolfsberg, M.: 1969, Advan. Chem. Ser. *89*, pp. 185—191.
(98) Wolfsberg, M., Massa, A. A., and Pyper, J. W.: 1970, J. Chem. Phys. *53*, pp. 3138—3146.
(99) Löwdin, P.-O.: 1977, Int. J. Quantum Chem. *12*, Suppl. 1, pp. 197—266.
(100) Löwdin, P.-O.: 1981, Int. J. Quantum. Chem. *20*, pp. 775—776.
(101) Linderberg, J.: 1977, Int. J. Quantum Chem. *12*, Suppl. 1, pp. 267—276.
(102) Dutta, M.: 1968, Physics Today, pp. 75—79.
(103) Mead, C. A.: 1982, Advan. Chem. Phys. *51*, pp. 113—175.
(104) Skagerstam, B.-S. K.: 1975, J. Stat. Phys. *12*, pp. 449—462.
(105) Marcelin, M. R.: 1915, Ann. Phys. Paris *3*, pp. 120—231.
(106) Pelzer, H., and Wigner, E.: 1932, Z. Phys. Chem. *B 15*, pp. 445—471.
(107) Guggenheim, E. A.: 1937, Trans. Faraday Soc. *33*, pp. 607—614.
(108) Arnot, C. L.: 1972, J. Chem. Educ. *49*, pp. 480—482.

(109) MAHAN, B. H.: 1974, J. Chem. Educ. *51*, pp. 709—711.
(110) LAIDLER, K. J., and POLANYI, J. C.: 1965, Progr. React. Kinet. *3*, pp. 1—61.
(111) HOFACKER, G. L.: 1969, Int. J. Quantum Chem., Quantum Chem. Symp. *3*, pp. 33—37.
(112) CHRISTOV, S. G.: 1974, Ber. Bunsenges. Phys. Chem. *78*, pp. 537—548.
(113) HIRSCHFELDER, J. O.: 1982, Ber. Bunsenges. Phys. Chem. *86*, pp. 349—355; cf. also URRY, D. W.: 1982, Int. J. Quantum Chem., Quantum Biol. Symp. *9*, pp. 1—3.
(114) HORIUTI, J.: 1938, Bull. Chem. Soc. Japan *13*, pp. 210—216.
(115) KECK, J. C.: 1967, Advan. Chem. Phys. *13*, pp. 85—121.
(116) KOEPPL, G. W.: 1974, J. Am. Chem. Soc. *96*, pp. 6539—6548.
(117) TRUHLAR, D. G., and GARRETT, B. C.: 1980, Accounts Chem. Res. *13*, pp. 440—448.
(118) ELIASON, M. A., and HIRSCHFELDER, J. O.: 1959, J. Chem. Phys. *30*, pp. 1426—1436.
(119) SZWARC, M.: 1962, in The Transition State. Spec. Publ. No. 16, The Chemical Society, London, p. 25.
(120) TWEEDALE, A., and LAIDLER, K. J.: 1970, J. Chem. Phys. *53*, pp. 2045—2052.
(121) AGMON, N., and LEVINE, R. D.: 1977, Chem. Phys. Lett. *52*, pp. 197—201.
(122) SCHLAG, E. W.: 1963, J. Chem. Phys. *38*, pp. 2480—2482.
(123) SCHLAG, E. W., and HALLER, G. L.: 1965, J. Chem. Phys. *42*, pp. 584—587.
(124) BISHOP, D. M., and LAIDLER, K. J.: 1965, J. Chem. Phys. *42*, pp. 1688—1691.
(125) JOHNSTON, H. S.: 1966, Gas Phase Reaction Rate Theory, Ronald Press, New York.
(126) ELLIOTT, C. S., and FREY, H. M.: 1968, Trans. Faraday Soc. *64*, pp. 2352—2368.
(127) GOLD, V.: 1964, Trans. Faraday Soc. *60*, pp. 738—739.
(128) MARCUS, R. A.: 1965, J. Chem. Phys. *43*, pp. 2658—2661.
(129) MURRELL, J. N., and LAIDLER, K. J.: 1968, Trans. Faraday Soc. *64*, pp. 371—377.
(130) MURRELL, J. N., and PRATT, G. L.: 1969, Trans. Faraday Soc. *66*, pp. 1680—1684.
(131) BISHOP, D. M., and LAIDLER, K. J.: 1969, Trans. Faraday Soc. *66*, pp. 1685—1687.
(132) POLLAK, E., and PECHUKAS, P.: 1978, J. Am. Chem. Soc. *100*, pp. 2984—2991.
(133) COULSON, D. R.: 1978, J. Am. Chem. Soc. *100*, pp. 2992—2996.
(134) TRUHLAR, D. G., and KUPPERMANN, A.: 1971, Chem. Phys. Lett. *9*, pp. 269—272.
(135) BAER, M., HALAVEE, U., and PERSKY, A.: 1974, J. Chem. Phys. *61*, pp. 5122—5131.
(136) TRUHLAR, D. G., MERRICK, J. A., and DUFF, J. W.: 1976, J. Am. Chem. Soc. *98*, pp. 6771 to 6783.
(137) ESSÉN, H., BILLING, G. D., and BAER, M.: 1976, Chem. Phys. *17*, pp. 443—449.
(138) MILLER, W. H.: 1976, Accounts Chem. Res. *9*, pp. 306—312.
(139) CONNOR, J. N. L., JAKUBETZ, W., and LAGANÀ, A.: 1979, J. Phys. Chem. *83*, pp. 73—78.
(140) TRUHLAR, D. G.: 1979, J. Phys. Chem. *83*, pp. 188—199.
(141) PECHUKAS, P.: 1982, Ber. Bunsenges. Phys. Chem. *86*, pp. 372—378.
(142) LIN, S. H., and EYRING, H.: 1972, Proc. Nat. Acad. Sci. U. S. A. *69*, pp. 3192—3194.
(143) GOLDEN, D. M.: 1979, J. Phys. Chem. *83*, pp. 108—113.
(144) GLASSTONE, S., LAIDLER, K. J., and EYRING, H.: 1941, The Theory of Rate Processes, McGraw-Hill, New York.
(145) LÖWDIN, P.-O.: 1965, Advan. Quantum Chem. *2*, pp. 213—360.
(146) BELL, R. P.: 1973, The Proton in Chemistry, Chapman and Hall, London; 1980, The Tunnel Effect in Chemistry, Chapman and Hall, London.
(147) ECKART, C.: 1930, Phys. Rev. *35*, pp. 1303—1309.
(148) WIGNER, E.: 1932, Z. Phys. Chem. *B19*, pp. 203—216.
(149) BELL, R. P.: 1959, Trans. Faraday Soc. *55*, pp. 1—4.
(150) DE LA VEGA, J. R.: 1982, Accounts Chem. Res. *15*, pp. 185—191.
(151) SHAVITT, I.: 1959, J. Chem. Phys. *31*, pp. 1359—1367.
(152) JOHNSTON, H. S., and RAPP, D.: 1961, J. Am. Chem. Soc. *83*, pp. 1—9.
(153) SHIN, H.: 1963, J. Chem. Phys. *39*, pp. 2934—2936.

(154) TRUHLAR, D. G., and KUPPERMANN, A.: 1971, J. Am. Chem. Soc. *93*, pp. 1840—1851.
(155) JAKUBETZ, W.: 1979, J. Am. Chem. Soc. *101*, pp. 298—307.
(156) MILLER, W. H.: 1979, J. Am. Chem. Soc. *101*, pp. 6810—6814.
(157) GARRETT, B. C., and TRUHLAR, D. G.: 1979, J. Phys. Chem. *83*, pp. 2921—2926.
(158) GRAY, S. K., MILLER, W. H., YAMAGUCHI, Y., and SCHAEFER III, H. F.: 1981, J. Am. Chem. Soc. *103*, pp. 1900—1904.
(159) AGRESTI, A., BACCI, M., and RANFAGNI, A.: 1981, Chem. Phys. Lett. *79*, pp. 100—104.
(160) CERJAN, C. C., SHI, S., and MILLER, W. H.: 1982, J. Phys. Chem. *86*, pp. 2244—2251.
(161) CRIBB, P. H., NORDHOLM, S., and HUSH, N. S.: 1982, Chem. Phys. *69*, pp. 259—266.
(162) CONNOR, J. N. L.: 1976, Chem. Soc. Rev. *5*, pp. 125—148.
(163) POLANYI, J. C.: 1972, Accounts Chem. Res. *5*, pp. 161—168.
(164) GOLDENBERG, M. JA., and KUZNECOV, N. M.: 1982, Usp. Khim. *51*, pp. 1075—1090.
(165) LEECH, J. W.: 1958, Classical Mechanics, Methuen, London,
(166) GARRETT, B. C., and TRUHLAR, D. G.: 1979, J. Phys. Chem. *83*, pp. 200—203.
(167) GARRETT, B. C., and TRUHLAR, D. G.: 1979, J. Phys. Chem. *83*, pp. 1079—1112.
(168) GARRETT, B. C., and TRUHLAR, D. G.: 1979, J. Am. Chem. Soc. *101*, pp. 4534—4548.
(169) GARRETT, B. C., and TRUHLAR, D. G.: 1979, J. Am. Chem. Soc. *101*, pp. 5207—5217.
(170) GARRETT, B. C., and TRUHLAR, D. G.: 1980, J. Am. Chem. Soc. *102*, pp. 2559—2570.
(171) GARRETT, B. C., and TRUHLAR, D. G.: 1980, J. Chem. Phys. *72*, pp. 3460—3471.
(172) GARRETT, B. C., TRUHLAR, D. G., and MAGNUSON, A. W.: 1981, J. Chem. Phys. *74*, pp. 1029—1043.
(173) SKODJE, R. T., TRUHLAR, D. G., and GARRETT, B. C.: 1981, J. Phys. Chem. *85*, pp. 3019 to 3023.
(174) GARRETT, B. C., TRUHLAR, D. G., and MAGNUSON, A. W.: 1982, J. Chem. Phys. *76*, pp. 2321—2331.
(175) MILLER, W. H.: 1971, J. Chem. Phys. *55*, pp. 3146—3149.
(176) MILLER, W. H.: 1974, J. Chem. Phys. *61*, pp. 1823—1834.
(177) McLAFFERTY, F. J., and PECHUKAS, P.: 1974, Chem. Phys. Lett. *27*, pp. 511—514.
(178) MILLER, W. H.: 1975, J. Chem. Phys. *62*, pp. 1899—1906.
(179) KUPPERMANN, A.: 1979, J. Phys. Chem. *83*, pp. 171—187.
(180) BERKOWITZ, J.: 1958, J. Chem. Phys. *29*, pp. 1386—1394.
(181) PITZER, K. S., and CLEMENTI, E.: 1959, J. Am. Chem. Soc. *81*, pp. 4477—4485.
(182) SCHWARTZ, M. E., and SCHAAD, L. J.: 1967, J. Chem. Phys. *47*, pp. 5325—5334.
(183) SHAVITT, I.: 1968, J. Chem. Phys. *49*, pp. 4048—4056.
(184) SEARLES, S. K., and KEBARLE, P.: 1969, Canad. J. Chem. *47*, pp. 2619—2627.
(185) KOLLMAN. P., BENDER, C. F., and ROTHENBERG, S.: 1972, J. Am. Chem. Soc. *94*, pp. 8016—8020.
(186) SPEARS, K. G.: 1972, J. Chem. Phys. *57*, pp. 1850—1858.
(187) BOOTH, D., and MURRELL, J. N.: 1972, Mol. Phys. *24*, pp. 1117—1122.
(188) KOMORNICKI, A., and MCIVER, JR., J. W.: 1973, J. Am. Chem. Soc. *95*, pp. 4512—4517.
(189) BRAUN, C., and LEIDECKER, H.: 1974, J. Chem. Phys. *61*, pp. 3104—3113.
(190) ZAHRADNÍK, R., SLANINA, Z., and ČÁRSKY, P.: 1974, Collect. Czech. Chem. Commun. *39*, pp. 63—70.
(191) SLANINA, Z., HOBZA, P., and ZAHRADNÍK, R.: 1974, Collect. Czech. Chem. Commun. *39*, pp. 228—235.
(192) ZAHRADNÍK, R., HOBZA, P., SLANINA, Z., and ČÁRSKY, P.: 1975, Chem. Listy *69*, pp. 897 to 914.
(193) OWICKI, J. C., SHIPMAN, L. L., and SCHERAGA, H. A.: 1975, J. Phys. Chem. *79*, pp. 1794 to 1811.

(194) KRYGOWSKI, T. M., LIPSZTAJN, M., and RADZIKOWSKI, P.: 1975, J. Mol. Struct. *28*, pp. 163—169.
(195) HEIDRICH, D., and STROMEYER, M.: 1976, Z. Chem. *16*, pp. 152—153.
(196) ČÁRSKY, P., ZAHRADNÍK. R., and KOZÁK, I.: 1976, Chem. Phys. Lett. *41*, pp. 165—167.
(197) SLANINA, Z., SCHLANGER, J., and ZAHRADNÍK, R.: 1976, Collect. Czech. Chem. Commun. *41*, pp. 1864—1874.
(198) SLANINA, Z., and ZAHRADNÍK, R.: 1977, J. Phys. Chem. *81*, pp. 2252—2257.
(199) SLANINA, Z.: 1977, Chem. Phys. Lett. *50*, pp. 418—422.
(200) SLANINA, Z.: 1978, Collect. Czech. Chem. Commun. *43*, pp. 2358—2367.
(201) HOBZA, P., BOČEK, K., HOFFMANN, H.-J., SLANINA, Z., and ZAHRADNÍK, R.: 1978, Collect. Czech. Chem. Commun. *43*, pp. 665—675.
(202) SLANINA, Z.: 1978, Collect. Czech. Chem. Commun. *43*, pp. 1974—1989.
(203) KELLÖ, V., URBAN, M., ČÁRSKY, P., and SLANINA, Z.: 1978, Chem. Phys. Lett. *53*, pp. 555—559.
(204) KÖHLER, H.-J., and LISCHKA, H.: 1978, Chem. Phys. Lett. *58*, pp. 175—179.
(205) HOBZA, P., ČÁRSKY, P., and ZAHRADNÍK, R.: 1979, Collect. Czech. Chem. Commun. *44*, pp. 3458—3463.
(206) KREBS, C., HOFFMANN, H.-J., KÖHLER, H.-J., and WEISS, C.: 1980, Chem. Phys. Lett. *69*, pp. 537—539.
(207) SLANINA, Z.: 1980, J. Chem. Phys. *73*, pp. 2519—2521.
(208) SLANINA, Z.: 1980, Collect. Czech. Chem. Commun. *45*, pp. 3417—3435.
(209) KATO, S., and MOROKUMA, K.: 1980, J. Chem. Phys. *72*, pp. 206—217.
(210) ROBINSON, P. J., and HOLBROOK, K. A.: 1972, Unimolecular Reactions, Wiley-Interscience, New York.
(211) SLANINA, Z.: 1981, Advan. Mol. Relax. Interact. Proces. *19*, pp. 117—128.
(212) KATO, S., and MOROKUMA, K.: 1981, J. Chem. Phys. *74*, pp. 6285—6297.
(213) KOZÁK, I., and ČÁRSKY, P.: 1981, Collect. Czech. Chem. Commun. *46*, pp. 2146—2148.
(214) PELIKÁN, P., and LIŠKA, M.: 1982, Collect. Czech. Chem. Commun. *47*, pp. 1556—1569.
(215) DEWAR, M. J. S., and NELSON, D. J.: 1982, J. Org. Chem. *47*, pp. 2614—2618.
(216) SCHAAD, L. J., HESS, JR., B. A., and EWIG, C. S.: 1982, J. Org. Chem. *47*, pp. 2904—2906.
(217) HARDING, L. B., and SCHATZ, G. C.: 1982, J. Chem. Phys. *76*, pp. 4296—4297.
(218) HARDING, L. B., SCHATZ, G. C., and CHILES, R. A.: 1982, J. Chem. Phys. *76*, pp. 5172—5173.
(219) O'HARE, P. A. G., BATANA, A., and WAHL, A. C.: 1973, J. Chem. Phys. *59*, pp. 6495—6501.
(220) DEWAR, M. J. S., and FORD, G. P.: 1977, J. Am. Chem. Soc. *99*, pp. 7822—7829.
(221) KÖHLER, H.-J.: 1979, Z. Chem. *19*, pp. 235—236.
(222) ENGEL, G., KÖHLER, H.-J., and WEISS, C.: 1979, Tetrahedron Lett. *32*, pp. 2975—2978.
(223) ČÁRSKY, P., ZAHRADNÍK, R., URBAN, M., and KELLÖ, V.: 1979, Chem. Phys. Lett. *61*, pp. 85—87.
(224) ČÁRSKY, P., and ZAHRADNÍK, R.: 1979, Int. J. Quantum Chem. *16*, pp. 243—256.
(225) RAFFENETTI, R. C., and PHILLIPS, D. H.: 1979, J. Chem. Phys. *71*, pp. 4534—4540.
(226) BOECK, G., KÖHLER, H.-J., and WEISS, C.: 1980, Z. Chem. *20*, p. 426.
(227) BOECK, G., KÖHLER, H.-J., and WEISS, C.: 1980, Z. Chem. *20*, pp. 455—456.
(228) SCHATZ, G. C., and WALCH, S. P.: 1980, J. Chem. Phys. *72*, pp. 776—778.
(229) WALCH, S. P., WAGNER, A. F., DUNNING JR., T. H., and SCHATZ, G. C.: 1980, J. Chem. Phys. *72*, pp. 2894—2896.
(230) SCHATZ, G. C., WALCH, S. P., and WAGNER, A. F.: 1980, J. Chem. Phys. *73*, pp. 4536—4547.
(231) SHIBATA, M., ZIELINSKI, T. J., and REIN, R.: 1980, Int. J. Quantum Chem. *18*, pp. 323—329.
(232) BASILEVSKY, M. V., WEINBERG, N. N., and ZHULIN, V. M.: 1981, Theor. Chim. Acta *59*, pp. 373—385.
(233) SANA, M.: 1981, Int. J. Quantum Chem. *19*, pp. 139—161.

(234) STOGRYN, D. E., and HIRSCHFELDER, J. O.: 1959, J. Chem. Phys. *31*, pp. 1531—1545.
(235) STOGRYN, D. E., and HIRSCHFELDER, J. O.: 1960, J. Chem. Phys. *33*, pp. 942—943.
(236) LECKENBY, R. E., and ROBBINS, E. J.: 1966, Proc. Roy. Soc. *A 291*, pp. 389—412.
(237) MAHAN, G. D., and LAPP, M.: 1969, Phys. Rev. *179*, pp. 19—27.
(238) BOUCHIAT, C. C., BOUCHIAT, M. A., and POTTIER, L. C. L.: 1969, Phys. Rev. *181*, pp. 144—165.
(239) MAHAN, G. D.: 1970, J. Chem. Phys. *52*, pp. 258—265.
(240) POULAT, C., LARSEN, S. Y., and NOVARO, O.: 1975, Mol. Phys. *30*, pp. 645—648.
(241) SHIN, H. K.: 1977, Chem. Phys. Lett. *49*, pp. 193—196.
(242) SLANINA, Z.: 1974, Thesis, Czechoslovak Academy of Sciences, Prague.
(243) DASHEVSKII, V. G.: 1973, Usp. Khim. *42*, pp. 2097—2129.
(244) DASHEVSKII, V. G.: 1974, Usp. Khim. *43*, pp. 491—518.
(245) POPLE, J. A.: 1973, in Energy, Structure and Reactivity, Ed. D. W. Smith and W. B. McRae, Wiley, New York, pp. 51—68.
(246) WIBERG, K. B.: 1968, J. Am. Chem. Soc. *90*, pp. 59—63.
(247) ISAACS, N. S.: 1969, Tetrahedron *25*, pp. 3555—3566.
(248) SNYDER, L. C., and BASCH, H.: 1969, J. Am. Chem. Soc. *91*, pp. 2189—2198.
(249) HOPKINSON, A. C., YATES, K., and CSIZMADIA, I. G.: 1972, Theor. Chim. Acta *23*, pp. 369 to 377.
(250) BOTSCHWINA, P., and MEYER, W.: 1977, Chem. Phys. *20*, pp. 43—52.
(251) GEORGE, P., TRACHTMAN, M., BRETT, A. M., and BOCK, C. W.: 1977, Int. J. Quantum Chem. *12*, pp. 61—81.
(252) KARLSTRÖM, G., JÖNSSON, B., ROOS, B. O., and SIEGBAHN, P. E. M.: 1978, Theor. Chim. Acta *48*, pp. 59—74.
(253) ČÁRSKY, P., ZAHRADNÍK, R., HUBAČ, I., URBAN, M., and KELLÖ, V.: 1980, Theor. Chim. Acta *56*, pp. 315—328.
(254) HURLEY, A. C.: 1973, Advan. Quantum Chem. *7*, pp. 315—334.
(255) DROWART, J., BURNS, R. P., DEMARIA, G., and INGHRAM, M. G.: 1959, J. Chem. Phys. *31*, pp. 1131—1132.
(256) PAULING, L., and HENDRICKS, S. B.: 1926, J. Am. Chem. Soc. *48*, pp. 641—651.
(257) POPPINGER, D., RADOM, L., and POPLE, J. A.: 1977, J. Am. Chem. Soc. *99*, pp. 7806—7816.
(258) KOCHANSKI, E., ROOS, B., SIEGBAHN, P., and WOOD, M. H.: 1973, Theor. Chim. Acta *32*, pp. 151—159.
(259) JASZUŃSKI, M., KOCHANSKI, E., and SIEGBAHN, P.: 1977, Mol. Phys. *33*, pp. 139—146.
(260) JEFFORD, C. W., MAREDA, J., PERLBERGER, J.-C., and BURGER, U.: 1979, J. Am. Chem. Soc. *101*, pp. 1370—1378.
(261) SLANINA, Z.: 1975, Collect. Czech. Chem. Commun. *40*, pp. 1997—2004.
(262) HOLUB, R., and VOŇKA, P.: 1975, The Chemical Equilibrium of Gaseous Systems, Academia, Prague.
(263) SMITH, W. R.: 1980, in Theoretical Chemistry: Advances and Perspectives, Vol. 5, Ed. H. Eyring, and D. Henderson, Academic Press, New York, pp. 185—259.
(264) JANZ, G. J.: 1967, Thermodynamic Properties of Organic Compounds, Academic Press, New York.
(265) PULLMAN, B., and PULLMAN, A.: 1971, Advan. Heterocycl. Chem. *13*, pp. 77—159.
(266) KWIATKOWSKI, J. S., and PULLMAN, B.: 1975, Advan. Heterocycl. Chem. *18*, pp. 199—335.
(267) MINKIN, V. I., OLEKHNOVICH, L. P., and ZHDANOV, YU. A.: 1981, Accounts Chem. Res. *14*, pp. 210—217.
(268) DEPUY, C. H., BIERBAUM, V. M., FLIPPIN, L. A., GRABOWSKI, J. J., KING, G. K., and SCHMITT, R. J.: 1979, J. Am. Chem. Soc. 101, p. 6443.

(269) Slanina, Z., and Grabowski, Z. R.: 1979, Collect. Czech. Chem. Commun. 44, pp. 3441 to 3451.
(270) Buechele, J. L., Weitz, E., and Lewis, F. D.: 1981, J. Am. Chem. Soc. *103*, pp. 3588—3589.
(271) Ausloos, P. J., and Lias, S. G.: 1981, J. Am. Chem. Soc. *103*, pp. 6505—6507.
(272) Buechele, J. L., Weitz, E., and Lewis, F. D.: 1979, J. Am. Chem. Soc. 101, pp. 3700—3701.
(273) Chang, J. S., Baldwin, A. C., and Golden, D. M.: 1979, J. Chem. Phys. 71, pp. 2021 to 2024.
(274) Compton, D. A. C.: 1977, J. Chem. Soc., Perkin Trans II, pp. 1307—1311.
(275) Heatley, F.: 1972, J. Chem. Soc., Faraday Trans. *II 68*, pp. 2097—2109.
(276) Beckett, C. W., Pitzer, K. S., and Spitzer, R.: 1947, J. Am. Chem. Soc. *69*, pp. 2488 to 2495.
(277) McCullough, J. P., Finke, H. L., Hubbard, W. N., Good, W. D., Pennington, R.E., Messerly, J. F., and Waddington, G.: 1954, J. Am. Chem. Soc. 76, pp. 2661—2669.
(278) Scott, D. W., and Crowder, G. A.: 1967, J. Chem. Phys. 46, pp. 1054—1062.
(279) Roos, S. D., Labes, M. M., and Schwarz, M.: 1956, J. Am. Chem. Soc. 78, pp. 343—345.
(280) Orgel, L. E., and Mulliken, R. S.: 1957, J. Am. Chem. Soc. *79*, pp. 4839—4846.
(281) Hoare, M. R.: 1979, Advan. Chem. Phys. *40*, pp. 49—135.
(282) Volkenshtein, M. V.: 1959, Configurational Statistics of Polymeric Chains, Acad. Sci. USSR, Moscow.
(283) Flory, P. J.: 1969, Statistical Mechanics of Chain Molecules, Wiley-Interscience, New York.
(284) Flory, P. J.: 1974, Macromolecules *7*, pp. 381—392.
(285) Gō, N., and Scheraga, H. A.: 1969, J. Chem. Phys. *51*, pp. 4751—4767.
(286) Lewis, P. N., Momany, F. A., and Scheraga, H. A.: 1973, Isr. J. Chem. *11*, pp. 121—152.
(287) Mark, J. E.: 1977, J. Chem. Phys. *67*, pp. 3300—3302.
(288) Mattice, W. L.: 1977, J. Am. Chem. Soc. *99*, pp. 2324—2330.
(289) Sisido, M., and Shimada, K.: 1977, J. Am. Chem. Soc. *99*, pp. 7785—7792.
(290) Vaughan, W. E.: 1978, Advan. Mol. Relax. Interact. Proces. *13*, pp. 171—187.
(291) Hagler, A. T., Stern, P. S., Sharon, R., Becker, J. M., and Naider, F.: 1979, J. Am. Chem. Soc. *101*, pp. 6842—6852.
(292) Scheraga, H. A.: 1971, Chem. Rev. *71*, pp. 195—217.
(293) Hopfinger, A. J.: 1973, Conformational Properties of Macromolecules, Academic Press, New York.
(294) Ikegami, A.: 1981, Advan. Chem. Phys. *46*, pp. 363—413.
(295) Lielmezs, J., and Bondi, A.: 1965, Chem. Eng. Sci. *20*, pp. 706—709.
(296) Frankiss, S. G.: 1974, J. Chem. Soc., Faraday Trans. *II 70*, pp. 1516—1521.
(297) Chen, S. S., Wilhoit, R. C., and Zwolinski, B. J.: 1975, J. Phys. Chem. Ref. Data *4*, pp. 859—869.
(298) Hald, N. C. P., and Rasmussen, K.: 1978, Acta Chem. Scand. *A 32*, pp. 879—886.
(299) Durig, J. R., and Compton, D. A. C.: 1980, J. Phys. Chem. *84*, pp. 773—781.
(300) Durig, J. R., Guirgis, G. A., and Compton, D. A. C.: 1980, J. Phys. Chem. *84*, pp. 3547 to 3554.
(301) Durig, J. R., Gerson, D. J., and Compton, D. A. C.: 1980, J. Phys. Chem. *84*, pp. 3554 to 3561.
(302) Compton, D. A. C., Montero, S., and Murphy, W. F.: 1980, J. Phys. Chem. *84*, pp. 3587—3591.
(303) Durig, J. R., and Gerson, D. J.: 1981, J. Phys. Chem. *85*, pp. 426—434.
(304) Compton, D. A. C., and Murphy, W. F.: 1981, J. Phys. Chem. *85*, pp. 482—488.
(305) Compton, D. A. C.: 1981, in Vibrational Spectra and Structure, Vol. 9, Ed. J. R. Durig, Elsevier, Amsterdam, pp. 255—404.
(306) Durig, J. R., Brletic, P. A., and Church, J. S.: 1982, J. Chem. Phys. *76*, pp. 1723—1734.

(307) HOPFINGER, A. J.: 1971, Macromolecules *4*, pp. 731—737.
(308) GODLESKI, S. A., SCHLEYER, P. v. R., ŌSAWA, E., and WIPKE, W. T.: 1981, Progr. Phys. Org. Chem. *13*, pp. 63—117.
(309) SLANINA, Z.: 1979, Int. J. Quantum Chem. *16*, pp. 79—86.
(310) SLANINA, Z.: 1977, Collect. Czech. Chem. Commun. *42*, pp. 1914—1921.
(311) GREEN, S., and HERBST, E.: 1979, Astrophys. J. *229*, pp. 121—131.
(312) HERZBERG, G.: 1980, Highlights of Astronomy *5*, pp. 3—26.
(313) ISHIDA, K., MOROKUMA, K., and KOMORNICKI, A.: 1977, J. Chem. Phys. *66*, pp. 2153—2156.
(314) REDMON, L. T., PURVIS III, G. D., and BARTLETT, R. J.: 1980, J. Chem. Phys. *72*, pp. 986 to 991.
(315) HERNDON, W. C., FEUER, J., and HALL, L. H.: 1968, Theor. Chim. Acta 11, pp. 178—181.
(316) COWLEY, A. H., WHITE, W. D., and DAMASCO, M. C.: 1969, J. Am. Chem. Soc. *91*, pp. 1922 to 1928.
(317) FAVINI, G., RUBINO, C., and TODESCHINI, R.: 1977, J. Mol. Struct. *41*, pp. 305—313.
(318) CETINA, R., RUBIO, M., SALMÓN, M., and BERNAL, J.: 1978, Aust. J. Chem. *31*, pp. 1911 to 1915.
(319) ZIELINSKI, T. J., and REIN, R.: 1978, Int. J. Quantum Chem. *14*, pp. 851—860.
(320) CZERMIŃSKI, R., LESYNG, B., and POHORILLE, A.: 1979, Int. J. Quantum Chem. *16*, pp. 1141—1148.
(321) DOUGHERTY, D. A., and MISLOW, K.: 1979, J. Am. Chem. Soc. *101*, pp. 1401—1405.
(322) NIEMEYER, H. M.: 1979, J. Mol. Struct. *57*, pp. 241—244.
(323) LOOS, D., and LEŠKA, J.: 1980, Collect. Czech. Chem. Commun. *45*, pp. 187—200.
(324) SCHARFENBERG, P., and SAUER, J.: 1980, Int. J. Quantum Chem. *18*, pp. 1309—1337.
(325) PEARSON, P. K., SCHAEFER III, H. F., RICHARDSON, J. H., STEPHENSON, L. M., and BRAUMAN, J. I.: 1974, J. Am. Chem. Soc. *96*, pp. 6778—6779.
(326) TALATY, E. R., SCHWARTZ, A. K., and SIMONS, G.: 1975, J. Am. Chem. Soc. *97*, pp. 972 to 978.
(327) KARLSTRÖM, G., JÖNSSON, B., ROOS, B., and WENNERSTRÖM, H.: 1976, J. Am. Chem. Soc. *98*, pp. 6851—6854.
(328) KAO, J.: 1977, Inorg. Chem. *16*, pp. 3347—3349.
(329) NIEMEYER, H. M.: 1978, J. Mol. Struct. *50*, pp. 123—126.
(330) REDMON, L. T., PURVIS, G. D., and BARTLETT, R. J.: 1978, J. Chem. Phys. *69*, pp. 5386 to 5392.
(331) BARTHELAT, J.-C., TRINQUIER, G., and BERTRAND, G.: 1979, J. Am. Chem. Soc. *101*, pp. 3785—3789.
(332) MOFFAT, J. B.: 1979, J. Mol. Struct. *52*, pp. 275—280.
(333) NOACK, W.-E.: 1979, Theor. Chim. Acta *53*, pp. 101—119.
(334) VAN DEN BERG, A. B. A., and DEN BOER, D. H. W.: 1979, Recl. Trav. Chim. Pays-Bas *98*, pp. 432—437.
(335) WHITESIDE, R. A., KRISHNAN, R., DEFREES, D. J., POPLE, J. A., and SCHLEYER, P. v. R.: 1981, Chem. Phys. Lett. *78*, pp. 538—540.
(336) WHITESIDE, R. A., KRISHNAN, R., FRISCH, M. J., POPLE, J. A., and SCHLEYER, P. v. R.: 1981, Chem. Phys. Lett. *80*, pp. 547—551.
(337) ANRADE, J. G., CHANDRASEKHAR, J., and SCHLEYER, P. v. R.: 1981, J. Comput. Chem. *2*, pp. 207—211.
(338) WRIGHT, J. S.: 1973, Can. J. Chem. *51*, pp. 139—146.
(339) LUCCHESE, R. R., and SCHAEFER III, H. F.: 1977, J. Chem. Phys. *67*, pp. 848—849.
(340) HARDING, L. B., and GODDARD III, W. A.: 1977, J. Chem. Phys. *67*, pp. 2377—2379.
(341) BURTON, P. G.: 1977, Int. J. Quantum Chem., Quantum Chem. Symp. *11*, pp. 207—213.

(342) KARLSTRÖM, G., ENGSTRÖM, S., and JÖNSSON, B.: 1978, Chem. Phys. Lett. *57*, pp. 390 to 394.
(343) BOUMA, W. J., POPPINGER, D., and RADOM, L.: 1977, J. Am. Chem. Soc. *99*, pp. 6443—6444.
(344) Bouma, W. J., and RADOM, L.: 1978, Aust. J. Chem. *31*, pp. 1649—1660.
(345) RADOM, L., and POPLE, J. A.: 1970, J. Am. Chem. Soc. *92*, pp. 4786—4795.
(346) POPLE, J. A.: 1974, Tetrahedron *30*, pp. 1605—1615.
(347) MOFFAT, J. B.: 1976, J. Mol. Spectrosc. *61*, pp. 211—215.
(348) EPIOTIS, N. D., CHERRY, W. R., SHAIK, S., YATES, R. L., and BERNARDI, R.: 1977, Fortschr. Chem. Forsch. *70*, pp. 1—242.
(349) BARALDI, I., BRUNI, M. C., MOMICCHIOLI, F., LANGLET, J., and MALRIEU, J. P.: 1977, Chem. Phys. Lett. *51*, pp. 493—500.
(350) BINKLEY, J. S., and POPLE, J. A.: 1977, Chem. Phys. Lett. *45*, pp. 197—200.
(351) RADOM, L., STILES, P. J., and VINCENT, M. A.: 1978, J. Mol. Struct. *48*, pp. 259—270.
(352) OLSEN, J. F., and HOWELL, J. M.: 1979, J. Mol. Struct. *51*, pp. 257—266.
(353) OLSEN, J. F.: 1979, J. Mol. Struct. *57*, pp. 245—250.
(354) OLSEN, J. F.: 1979, J. Mol. Struct. *57*, pp. 251—258.
(355) LABARRE, J.-F.: 1978, Structure and Bonding *35*, pp. 1—35.
(356) DILL, J. D., SCHLEYER, P. v. R., and POPLE, J. .A: 1975, J. Am. Chem. Soc. *97*, pp. 3402 to 3409.
(357) APELOIG, Y., SCHLEYER, P. v. R., BINKLEY, J. S., and POPLE, J. A.: 1976, J. Am. Chem. Soc. *98*, pp. 4332—4334.
(358) APELOIG, Y., SCHLEYER, P. v. R., BINKLEY, J. S., POPLE, J. A., and JORGENSEN, W. L.: 1976, Tetrahedron Lett. *43*, pp. 3923—3926.
(359) JEMMIS, E. D., POPPINGER, D., SCHLEYER, P. v. R., and POPLE, J. A.: 1977, J. Am. Chem. Soc. *99*, pp. 5796—5798.
(360) JEMMIS, E. D., ALEXANDRATOS, S., SCHLEYER, P. v. R., STREITWIESER JR., A., and SCHAEFER III, H. F.: 1978, J. Am. Chem. Soc. *100*, pp. 5695—5700.
(361) JEMMIS, E. D., CHANDRASEKHAR, J., and SCHLEYER, P. v. R.: 1979, J. Am. Chem. Soc. *101*, pp. 527—533.
(362) JEMMIS, E. D., CHANDRASEKHAR, J., and SCHLEYER, P. v. R.: 1979, J. Am. Chem. Soc. *101*, pp. 2848—2856.
(363) KROGH-JESPERSEN, K., CREMER, D., POPPINGER, D., POPLE, J. A., SCHLEYER, P. v. R., and CHANDRASEKHAR, J.: 1979, J. Am. Chem. Soc. *101*, pp. 4843—4851.
(364) KOS, A., POPPINGER, D., SCHLEYER, P. v. R., and THIEL, W.: 1980, Tetrahedron Lett. *21*, pp. 2151—2154.
(365) BOWEN, R. D., and WILLIAMS, D. H.: 1977, J. Am. Chem. Soc. *99*, pp. 6822—6828.
(366) DEWAR, M. J. S., and RZEPA, H. S.: 1977, J. Am. Chem. Soc. *99*, pp. 7432—7439.
(367) DITS, H., NIBBERING, N. M. M., and VERHOEVEN, J. W.: 1977, Chem. Phys. Lett. *51*, pp. 95—98.
(368) GOETZ, D. W., SCHLEGEL, H. B., and ALLEN, L. C.: 1977, J. Am. Chem. Soc. *99*, pp. 8118 to 8120.
(369) HEHRE, W. J.: 1977, in Applications of Electronic Structure Theory, Ed. H. F. Schaefer III, Plenum Press, New York, pp. 277—331.
(370) BOUMA, W. J., MACLEOD, J. K., and RADOM, L.: 1979, J. Am. Chem. Soc. *101*, pp. 5540 to 5545.
(371) WENKE, G., and LENOIR, D.: 1979, Tetrahedron *35*, pp. 489—498.
(372) LIPKOWITZ, K. B., LARTER, R. M., and BOYD, D. B.: 1980, J. Am. Chem. Soc. *102*, pp. 85—92.
(373) BRUNA, P. J., and MARIAN, C. M.: 1979, Chem. Phys. *37*, pp. 425—444.

(374) Hehre, W. J., Pople, J. A., Lathan, W. A., Radom, L., Wasserman, E., and Wasserman, Z. R.: 1976, J. Am. Chem. Soc. *98*, pp. 4378—4383.
(375) Dykstra, C. E., Lucchese, R. R., and Schaefer III, H. F.: 1977, J. Chem. Phys. *67*, pp. 2422—2426.
(376) Lucchese, R. R., and Schaefer III, H. F.: 1977, J. Am. Chem. Soc. *99*, pp. 6765—6766.
(377) Roos, B. O., and Siegbahn, P. M.: 1977, J. Am. Chem. Soc. *99*, pp. 7716—7718.
(378) Gustav, K., Sühnel, J., and Wild, V. P.: 1978, Helv. Chim. Acta *61*, pp. 2100—2107.
(379) Hood, D. M., and Schaefer III, H. F.: 1978, J. Chem. Phys. *68*, pp. 2985—2986.
(380) Kapur, A., Steer, R. P., and Mezey, P. G.: 1978, J. Chem. Phys. *69*, pp. 968—975.
(381) Laidig, W. D., and Schaefer III, H. F.: 1978, J. Am. Chem. Soc. *100*, pp. 5972—5973.
(382) van Lenthe, J. H., and Ruttink, P. J. A.: 1978, Chem. Phys. Lett. *56*, pp. 20—24.
(383) Wetmore, R. W., and Schaefer III, H. F.: 1978, J. Chem. Phys. *69*, pp. 1648—1654.
(384) Herzberg, G.: 1966, Molecular Spectra and Molecular Structure. III. Electronic Spectra and Electronic Structure of Polyatomic Molecules, Van Nostrand Reinhold Comp., New York.
(385) Scheraga, H. A., 1968, Advan. Phys. Org. Chem. *6*, pp. 103—184.
(386) Allinger, N. L.: 1976, Advan. Phys. Org. Chem. *13*, pp. 1—82.
(387) Niketić, S. R., and Rasmussen, K.: 1977, The Consistent Force Field: A Documentation, Springer-Verlag, Berlin.
(388) White, D. N. J., and Morrow, C.: 1979, Comput. Chem. *3*, pp. 33—48.
(389) Baas, J. M. A., van de Graaf, B., Tavernier, D., and Vanhee, P.: 1981, J. Am. Chem. Soc. *103*, pp. 5014—5021.
(390) Maier, W. F., and Schleyer, P. v. R.: 1981, J. Am. Chem. Soc. *103*, pp. 1891—1900.
(391) Lide Jr., D. R., and Mann, D. E.: 1959, J. Chem. Phys. *31*, pp. 1129—1130.
(392) Colburn, C B.., Johnson, F. A., and Haney, C.: 1965, J. Chem. Phys. *43*, pp. 4526 to 4527.
(393) Slanina, Z.: 1975, J. Fluor. Chem. *6*, pp. 465—475.
(394) Slanina, Z.: 1977, Chem. Phys. Lett. *52*, pp. 117—120.
(395) Curtin, D. Y.: 1954, Rec. Chem. Progr. *15*, pp. 111—128.
(396) Hammett, L. P.: 1976, Physikalische Organische Chemie, Akademie-Verlag, Berlin, p. 121.
(397) Gordon, M. S., and Fischer, H.: 1968, J. Am. Chem. Soc. *90*, pp. 2471—2476.
(398) Valko, L., and Kovařík, P.: 1976, J. Phys. Chem. *80*, pp. 19—25.
(399) Howell, J. M., and Kirschenbaum, L. J.: 1976, J. Am. Chem. Soc. *98*, pp. 877—885.
(400) Seeger, R., Krishnan, R., Pople, J. A., and Schleyer, P. v. R.: 1977, J. Am. Chem. Soc. *99*, pp. 7103—7105.
(401) Favini, G., and Todeschini, R.: 1978, J. Mol. Struct. *50*, pp. 191—193.
(402) Penkovskii, V. V.: 1978, Dokl. Akad. Nauk SSSR *243*, pp. 375—377.
(403) Cimiraglia, R., Persico, M., and Tomasi, J.: 1978, Theor. Chim. Acta *49*, pp. 13—23.
(404) Peterson, M. R., and Csizmadia, I. G.: 1979, J. Am. Chem. Soc. *101*, pp. 1076—1079.
(405) Schleker, W., and Fleischhauer, J.: 1979, Z. Naturforsch. *34a*, pp. 488—492.
(406) Chen, M. M. L., Wetmore, R. W., and Schaefer III, H. F.: 1981, J. Chem. Phys. *74*, pp. 2938—2944.
(407) Favini, G., and Nava, A.: 1973, Theor. Chim. Acta *31*, pp. 261—268.
(408) Figuera, J. M., Shevlin, P. B., and Worley, S. D.: 1976, J. Am. Chem. Soc. *98*, pp. 3820—3825.
(409) Jug, K.: 1980, Theor. Chim. Acta *54*, pp. 263—300.
(410) Slanina, Z.: 1983, Collect. Czech. Chem. Commun. *48*, pp. 3027—3032.
(411) Slanina, Z.: 1982, Z. Phys. Chem., Neue Folge *132*, pp. 41—54.
(412) Slanina, Z.: 1984, React. Kinet. Catal. Lett. *25*, pp. 293—296.
(413) Slanina, Z.: 1983, Int. J. Quantum Chem. *23*, pp. 1553—1561.

(414) Slanina, Z.: 1983, Chem. Phys. Lett. *96*, pp. 579—583.
(415) 1973, Organic Reactive Intermediates, Ed. S. P. McManus, Academic Press, New York.
(416) Brown, H. C.: 1979, Top Curr. Chem. *80*, pp. 1—18.
(417) 1981, Reactive Intermediates, Vol. 2, Ed. M. Jones Jr., and R. A. Moss, Wiley-Interscience, New York.
(418) Laidler, K. J.: 1965, Chemical Kinetics, McGraw-Hill, London.
(419) Gillespie, P. D., and Ugi, I.: 1971, Angew. Chem., Int. Ed. Engl. *10*, pp. 503—506.
(420) Halgren, T. A., Pepperberg, I. M., and Lipscomb, W. N.: 1975, J. Am. Chem. Soc. *97*, pp. 1248—1250.
(421) Komornicki, A., and McIver Jr., J. W.: 1976, J. Am. Chem. Soc. *98*, pp. 4553—4561.
(422) Cone, C., Dewar, M. J. S., and Landman, D.: 1977, J. Am. Chem. Soc. *99*, pp. 372—376.
(423) Dewar, M. J. S., Ford, G. P., McKee, M. L., Rzepa, H. S., and Wade, L. E.: 1977, J. Am. Chem. Soc. *99*, pp. 5069—5073.
(424) Niemeyer, H. M.: 1977, Tetrahedron *33*, pp. 2267—2270.
(425) Sana, M., Leroy, G., Nguyen, M.-T., and Elguero, J.: 1979, Nouv. J. Chim. *3*, pp. 607 to 621.
(426) Morokuma, K., Kato, S., and Hirao, K.: 1980, J. Chem. Phys. *72*, pp. 6800—6802.
(427) Tasaka, M., Ogata, M., and Ichikawa, H.: 1981, J. Am. Chem. Soc. *103*, pp. 1885—1891.
(428) Cremer, D.: 1981, J. Am. Chem. Soc. *103*, pp. 3627—3633.
(429) Doering, W. v. E., Toscano, V. G., and Beasley, G. H.: 1971, Tetrahedron *27*, pp. 5299 to 5306.
(430) Benzon, M. S.: 1972, Ph. D. Thesis, Cornell University, Ithaca.
(431) Goldstein, M. J., and Benzon, M. S.: 1972, J. Am. Chem. Soc. *94*, pp. 7147—7149.
(432) Lowry, T. M., and John, W. T.: 1910, J. Chem. Soc. *97*, pp. 2634—2645.
(433) Kollmar, H., Carrion, F., Dewar, M. J. S., and Bingham, R. C.: 1981, J. Am. Chem. Soc. *103*, pp. 5292—5303.
(434) Schuster, P., Jakubetz, W., and Marius, W.: 1975, Top. Curr. Chem. *60*, pp. 1—107.
(435) Schuster, P.: 1976, in The Hydrogen Bond, Vol. I. Theory, Ed. P. Schuster, G. Zundel, and C. Sandorfy, North-Holland Publ. Comp., Amsterdam, pp. 25—163.
(436) Hobza, P., and Zahradník, R.: 1979, Weak Intermolecular Interactions in Chemistry and Biology, Academia—Elsevier, Prague—Amsterdam.
(437) Schuster, P.: 1981, Angew. Chem., Int. Ed. Engl. *20*, pp. 546—568.
(438) Prissette, J., and Kochanski, E.: 1978, J. Am. Chem. Soc. *99*, pp. 7352—7353.
(439) Prissette, J., and Kochanski, E.: 1978, J. Am. Chem. Soc. *100*, pp. 6609—6613.
(440) Kochanski, E.: 1982, J. Chem. Phys. *77*, pp. 2691—2692.
(441) Skaarup, S., Skancke, P. N., and Boggs, J. E.: 1976, J. Am. Chem. Soc. *98*, pp. 6106 to 6109.
(442) Skancke, P. N., and Wislöff-Nilsen, E.: Int. J. Quantum Chem., in press.
(443) Brigot, N., Odiot, S., Walmsley, S. H., and Whitten, J. L.: 1977, Chem. Phys. Lett. *49*, pp. 157—159.
(444) Brigot, N., Odiot, S., and Walmsley, S. H.: 1982, Chem. Phys. Lett. *88*, pp. 543—546.
(445) Slanina, Z.: 1983, J. Mol. Struct. *94*, pp. 401—405.
(446) Frurip, D. J., Curtiss, L. A., and Blander, M.: 1978, J. Phys. Chem. *82*, pp. 2555—2561.
(447) Curtiss, L. A.: 1977, J. Chem. Phys. *67*, pp. 1144—1149.
(448) Brink, G., and Glasser, L.: 1981, J. Comput. Chem. *2*, pp. 14—19.
(449) Curtiss, L. A.: 1977, Int. J. Quantum Chem., Quantum Chem. Symp. *11*, pp. 459—467.
(450) Curtiss, L. A., Frurip, D. J., and Blander, M.: 1978, J. Am. Chem. Soc. *100*, pp. 79—86.
(451) Brink, G., and Glasser, L.: 1981, J. Comput. Chem. *2*, pp. 177—181.
(452) Jönsson, B., Karlström, G., and Wennerström, H.: 1975, Chem. Phys. Lett. *30*, pp. 58—59.

(453) PRISSETTE, J., SEGER, G., and KOCHANSKI, E.: 1978, J. Am. Chem. Soc. *100*, pp. 6941 to 6947.
(454) HOBZA, P., SZCZĘŚNIAK, M. M., and LATAJKA, Z.: 1980, Chem. Phys. Lett. *74*, pp. 248—251.
(455) SLANINA, Z.: 1981, Chem. Phys. Lett. *82*, pp. 33—35.
(456) HOBZA, P., SZCZĘŚNIAK, M. M., and LATAJKA, Z.: 1981, Chem. Phys. Lett. *82*, pp. 469—472.
(457) SLANINA, Z.: 1981, Chem. Phys. Lett. *83*, pp. 418—422.
(458) EVERETT, D. H., HAYNES, J. M., and MCELROY, P. J.: 1971, Sci. Prog. Oxf. *59*, pp. 279—308.
(459) KISTENMACHER, H., LIE, G. C., POPKIE, H., and CLEMENTI, E.: 1974, J. Chem. Phys. *61*, pp. 546—561.
(460) MATSUOKA, O., CLEMENTI, E., and YOSHIMINE, M.: 1976, J. Chem. Phys. *64*, pp. 1351—1361.
(461) THIEL, W.: 1978, Theor. Chim. Acta *48*, pp. 357—359.
(462) SLANINA, Z.: Proc. 13th Symp. Rarefied Gas Dynamics, in press.
(463) EVETT, A. A., and MARGENAU, H.: 1953, Phys. Rev. *90*, pp. 1021—1023.
(464) GALLUP, G. A.: 1977, J. Chem. Phys. *66*, pp. 2252—2256.
(465) LLOYD-EVANS, D. J. R.: 1966, Mol. Phys. *10*, pp. 377—380.
(466) HOLLENSTEIN, H., BAUDER, A., and GÜNTHARD, Hs. H.: 1980, Chem. Phys. *47*, pp. 269—285.
(467) SLANINA, Z.: 1982, Collect. Czech. Chem. Commun. *47*, pp. 3206—3220.
(468) NOVICK, S. E., JANDA, K. C., and KLEMPERER, W.: 1976, J. Chem. Phys. *65*, pp. 5115—5121.
(469) KLEMPERER, W.: 1981, Personal Communication.
(470) LEVY, D. H.: 1980, in Quantum Dynamics of Molecules, Ed. R. G. Woolley, Plenum Press, New York, pp. 115—142.
(471) FELDER, P., and GÜNTHARD, Hs. H.: 1982, Chem. Phys. *71*, pp. 9—25.
(472) MCGINTY, D. J.: 1971, J. Chem. Phys. *55*, pp. 580—588.
(473) HOARE, M. R., and PAL, P.: 1971, Nature Phys. Sci. *230*, pp. 5—8.
(474) HOARE, M. R., and PAL, P.: 1971, Advan. Phys. *20*, pp. 161—196.
(475) HOARE, M. R., and PAL, P., 1972, Nature Phys. Sci. *236*, pp. 35—37.
(476) MCGINTY, D. J.: 1972, Chem. Phys. Lett. *13*, pp. 525—528.
(477) BURTON, J. J.: 1972, Chem. Phys. Lett. *17*, pp. 199—202.
(478) HOARE, M. R., and PAL, P.: 1972, J. Cryst. Growth *17*, pp. 77—96.
(479) BURTON, J. J.: 1973, J. Chem. Soc., Faraday Trans. *II 69*, pp. 540—550.
(480) HOARE, M. R., and MCINNES, J.: 1976, Faraday Discuss. Chem. Soc. *61*, pp. 12—24.
(481) BAUER, S. H., and FRURIP, D. J.: 1977, J. Phys. Chem. *81*, pp. 1015—1024.
(482) HOARE, M. R., PAL, P., and WEGENER, P. P.: 1980, J. Colloid Interface Sci. *75*, pp. 126—137.
(483) MOROCHOV, I. D., PETINOV, V. I., TRUSOV, L. I., and PETRUNIN, V. F.: 1981, Usp. Fiz. Nauk *133*, pp. 653—692.
(484) BERRY, R. S.: 1979, in The Permutation Group in Physics and Chemistry, Ed. J. Hinze, Springer-Verlag, Berlin, pp. 92—120.
(485) BERRY, R. S.: 1980, in Quantum Dynamics of Molecules, Ed. R. G. Woolley, Plenum, New York, pp. 143—195.
(486) DYMOND, J. H., and SMITH, E. B.: 1969, The Virial Coefficients of Gases, Clarendon Press, Oxford.
(487) SLANINA, Z.: 1977, Collect. Czech. Chem. Commun. *42*, pp. 3229—3238.
(488) MASON, E. A., and SPURLING, T. H.: 1969, The Virial Equation of State, Pergamon Press, Oxford.
(489) BROCAS, J.: 1976, in Chemical Applications of Graph Theory, Ed. A. T. Balaban, Academic Press, London, pp. 223—259.
(490) RICE, S. A.: 1975, Top. Curr. Chem. *60*, pp. 109—200.
(491) BEVERIDGE, D. L., and SCHNUELLE, G. W.: 1974, J. Phys. Chem. *78*, pp. 2064—2069.
(492) PULLMAN, A.: 1980, in Quantum Theory of Chemical Reactions, Vol. II, Ed. R. Daudel, A. Pullman, L. Salem, and A. Veillard, D. Reidel Publ. Comp., Dordrecht, pp. 1—24.

(493) Tapia, O.: 1982, Molecular Interactions, Vol. 3, Ed. H. Ratajczak, and W. J. Orville--Thomas, Wiley, New York, pp.47—117.
(494) Némethy, G., and Scheraga, H. A.: 1962, J. Chem. Phys. *36*, pp. 3382—3400.
(495) Némethy, G., and Scheraga, H. A.: 1962, J. Chem. Phys. *36*, pp. 3401—3417.
(496) Hagler, A. T., Scheraga, H. A., and Némethy, G.: 1972, J. Phys. Chem. *76*, pp. 3229 to 3243.
(497) Lentz, B. R., Hagler, A. T., and Scheraga, H. A.: 1974, J. Phys. Chem. *78*, pp. 1531 to 1550.
(498) Frank, H. S., and Wen, W.-Y.: 1957, Discuss. Faraday Soc. *24*, pp. 133—140.
(499) Eley, D. D.: 1939, Trans. Faraday Soc. *35*, pp. 1281—1293.
(500) Pierotti, R. A.: 1963, J. Phys. Chem. *67*, pp. 1840—1845.
(501) Pierotti, R. A.: 1965, J. Phys. Chem. *69*, pp. 281—288.
(502) Zahradník, R., Hobza, P., and Slanina, Z.: 1975, Collect. Czech. Chem. Commun. *40*, pp. 799—808.
(503) Popkie, H., Kistenmacher, H., and Clementi, E.: 1973, J. Chem. Phys. *59*, pp. 1325—1336.
(504) Kistenmacher, H., Popkie, H., Clementi, E., and Watts, R. O.: 1974, J. Chem. Phys. *60*, pp. 4455—4465.
(505) Lie, G. C., Clementi, E., and Yoshimine, M.: 1976, J. Chem. Phys. *64*, pp. 2314—2323.
(506) Clementi, E.: 1976, Determination of Liquid Water Structure, Coordination Numbers for Ions and Solvation for Biological Molecules, Springer-Verlag, Berlin.
(507) Mruzik, M.: 1977, Chem. Phys. Lett. *48*, pp. 171—175.
(508) Owicki, J. C., and Scheraga, H. A.: 1977, J. Am. Chem. Soc. *99*, pp. 7403—7412.
(509) Owicki, J. C., and Scheraga, H. A.: 1977, J. Am. Chem. Soc. *99*, pp. 7413—7418.
(510) Jorgensen, W. L.: 1979, J. Am. Chem. Soc. *101*, pp. 2016—2021.
(511) Clementi, E.: 1980, Computational Aspects for Large Chemical System, Springer-Verlag, Berlin.
(512) Jorgensen, W. L.: 1981, J. Am. Chem. Soc. *103*, pp. 335—340.
(513) Jorgensen, W. L.: 1981, J. Am. Chem. Soc. *103*, pp. 341—345.
(514) Jorgensen, W. L.: 1981, J. Am. Chem. Soc. *103*, pp. 345—350.
(515) Jönsson, B., and Karlström, G.: 1981, J. Chem. Phys. *74*, pp. 2896—2903.
(516) Ree, F. H.: 1971, in Physical Chemistry. An Advanced Treatise, Vol. VIII A, Ed. D. Henderson, Academic Press, New York, pp. 157—266.
(517) Metropolis, N., Rosenbluth, A. W., Rosenbluth, M. N., Teller, A. H., and Teller, E.: 1953, J. Chem. Phys. *21*, pp. 1087—1092.
(518) McGinty, D. J.: 1973, J. Chem. Phys. *58*, pp. 4733—4742.
(519) Abraham, F. F.: 1974, J. Chem. Phys. *61*, pp. 1221—1225.
(520) Briant, C. L., and Burton, J. J.: 1975, J. Chem. Phys. *63*, pp. 2045—2058.
(521) Kaelberer, J. B., Etters, R. D., and Raich, J. C.: 1976, Chem. Phys. Lett. *41*, pp. 580—582.
(522) Kaelberer, J. B., and Etters, R. D.: 1977, J. Chem. Phys. *66*, pp. 3233—3239.
(523) Etters, R. D., Flurchick, K., Pan, R. P., and Chandrasekharan, V.: 1981, J. Chem. Phys. *75*, pp. 929—936.
(524) Baetzold, R. C.: 1976, Advan. Catal. *25*, pp. 1—55.
(525) Dunken, H. H., and Lygin, V. I.: 1978, Quantenchemie der Adsorption an Festkörperoberflächen, VEB Deutscher Verlag für Grundstoffindustrie, Leipzig.
(526) Simonetta, M., and Gavezzotti, A.: 1980, Advan. Quantum Chem. *12*, pp. 103—158.
(527) Haber, J., and Witko, M.: 1981, Accounts Chem. Res. *14*, pp. 1—7.
(528) Bennett, A. J., McCarroll, B., and Messmer, R. P.: 1971, Phys. Rev. *B 3*, pp. 1397—1406.
(529) Hayns, M. R.: 1975, Theor. Chim. Acta *39*, pp. 61—74.
(530) Bauschlicher Jr., C. W., Bender, C. F., and Schaefer III, H. F.: 1976, Chem. Phys. *15*, pp. 227—235.

(531) Gavezzotti, A., and Simonetta, M.: 1977, Chem. Phys. Lett. *48*, pp. 434—438.
(532) Gadiyak, G. V., Karpuhsin, A. A., Morokov, Yu. N., and Repinskii, S. M.: 1980, Zh. Fiz. Khim. *54*, pp. 419—422.
(533) Hoflund, G. B., and Merrill, R. P..: 1981, J. Phys. Chem. *85*, pp. 2037—2041.
(534) Dovesi, R., Pisani, C., and Roetti, C.: 1981, Chem. Phys. Lett. *81*, pp. 498—502.
(535) Tantardini, G. F., and Simonetta, M.: 1981, Surf. Sci. *105*, pp. 517—535.
(536) Post, D., and Baerends, E. J.: 1981, Surf. Sci. *109*, pp. 167—190.
(537) Matsumura, K., Yamabe, S., Yanagisawa, Y., and Huzimura, R.: 1981, Surf. Sci. *109*, pp. 311—319.
(538) Marcusson, P., Opitz, Ch., and Müller, H.: 1981, Surf. Sci. *111*, pp. L 657—L 661.
(539) Anderson, A. B., and Hoffmann, R.: 1974, J. Chem. Phys. *61*, pp. 4545—4549.
(540) Anderson, A. B.: 1977, J. Am. Chem. Soc. *99*, pp. 696—707.
(541) Slanina, Z., Beran, S., and Knížek, P.: 1979, in Heterogeneous Catalysis, Part 2, Ed. D. Shopov, A. Andreev, A. Palazov, and L. Petrov, Publ. House Bulgarian Acad. Sci., Sofia, pp. 69—74.
(542) Slanina, Z.: 1983, Int. J. Quantum Chem. *23*, pp. 1563—1570.
(543) Slanina, Z.: 1982, Theor. Chim. Acta *60*, pp. 589—598.
(544) Beran, S., and Slanina, Z.: 1983, Zh. Fiz. Khim. *57*, pp. 1168—1171.
(545) Slanina, Z.: 1983, Zh. Fiz. Khim. *57*, pp. 1164—1167.
(546) Miller, W. H.: 1976, J. Chem. Phys. *65*, pp. 2216—2223.
(547) Pollak, E., and Pechukas, P.: 1979, J. Chem. Phys. *70*, pp. 325—333.
(548) Chesnavich, W. J., Bass, L., Su, T., and Bowers, M. T.: 1981, J. Chem. Phys. *74*, pp. 2228—2246.
(549) Garrett, B. C., Truhlar, D. G., and Grev, R. S.: 1981, J. Phys. Chem. *85*, pp. 1569—1572.
(550) Garrett, B. C., and Truhlar, D. G.: 1982, J. Chem. Phys. *76*, pp. 1853—1858.
(551) Valko, L., and Šimon, P.: J. Chem. Phys., in press.
(552) Tardy, D. C., and Rabinovitch, B. S.: 1977, Chem. Rev. *77*, pp. 369—408.
(553) van Kampen, N. G.: 1981, Stochastic Processes in Physics and Chemistry, North-Holland, Amsterdam.
(554) Berne, B. J., De Leon, N., and Rosenberg, R. O.: 1982, J. Phys. Chem. *86*, pp. 2166—2177.
(555) Tanaka, H., Nakanishi, K., and Watanabe, N.: 1983, J. Chem. Phys. *78*, pp. 2626—2634.
(556) Balescu, R.: 1975, Equilibrium and Nonequilibrium Statistical Mechanics, Wiley, New York.

7 Résumé

An attempt was made in this work to demonstrate that the phenomenon of chemical isomerism (for which experimental studies have yielded a great deal of interesting information) can be fruitfully described by various theoretical approaches. In addition, in certain situations (extreme observational conditions) theoretical techniques are preferable, or even the only source of information. Theoretical studies further emphasize the fact (widely established on the basis of observations) that isomerism must be considered as a general and universal feature of chemical systems (and also as a general, intrinsic feature of both higher and lower forms of organization of matter). The degree of uniqueness of interrelated isomers can vary widely between complete distinctness and fluxional behaviour.

In the framework of the Born-Oppenheimer approximation, the central concept is the potential energy hypersurface, at present typically characterized in terms of its stationary points. The isomers are then identified as local minima on the hypersurface, while transitions between the minima occur through its saddle points of transition state type. In this connection, techniques for the localization of stationary points on quantum-chemical or phenomenological hypersurfaces have been analyzed. In addition to the development of means for ensuring the completeness of localization of stationary points, interest and progress can be expected in the improvement of the quality of description of rotational-vibrational motions (both individual in the isolated stationary points and collective, interconverting the individual structures). This will, apparently, be closely connected with a gradual transition from representation of the hypersurface in terms of isolated stationary points to analytical description.

Although at present the utilization of quantum-chemical methods is a key practical tool in the theoretical study of isomerism, rigorous quantum-mechanical studies are also developing successfully in close connection with quantum concepts of molecular structure and the tunnelling effect. This approach includes both limiting forms of isomerism – the uniqueness of some isomers and the experimental indistinguishability of others.

An algebraic and, especially, a topological approach to the problems of isomeric chemistry stresses the intrinsic mathematical structure lying behind physical or chemical reality. These primarily involve techniques for the topological reduction and characterization of the potential energy hypersurface, enumeration of isomeric structures and treatment of stationary point symmetry. The algebraic generalization of isomerism related to computer design of syntheses is a more practical aspect.

Combination with the methods of statistical mechanics permits the inclusion of the temperature factor, while utilization of the methods of the theory of rate processes involves the time factor. Generalization of the conventional theory of chemical reactivity, in which each component of a chemical process (either equilibrium or rate) can exhibit isomerism, is discussed; situations where this isomerism was distinguished only during theoretical studies are especially important. It is pointed out that the simulation of experimentally observed multi-isomeric characteristics by theoretical values for (any) individual structure involved cannot be generally considered as acceptable. This is documented by many numeric examples of the weighting of contributions for individual isomeric structures. Here again, future interest will also undoubtedly be increasingly shifted towards the study of isomeric systems using the whole potential energy hypersurface, e.g. in connection with the concepts of the Monte Carlo and molecular dynamics methods. However, the extreme variety of experimental situations will apparently also necessitate the use of the theory of stochastic processes or nonequilibrium thermodynamics.

The phenomenon of isomerism is not limited to chemical species and it would thus be desirable to extend the rigorous quantum mechanical description of isomerism so that a single unified concept would include both chemical isomerism and isomerism at the atomic level, possibly also including biological isomerism. Thus the ideas of the Greek atomists, almost 25 centuries old, would finally find a rigorous basis at a microscopic level....

8 Postscript

As it was mentioned in the Preface, this Postscript or note added in proof (June 1985) is an attempt to summarize the important works related to this study that appeared after completion of this work in the original manuscript (November, 1982), and those not cited earlier. As this note is limited to about four printed pages only, no more than one hundred and fifty additional references could be listed. These references are divided according to the five main chapters of the book connected with basic concepts in chemical isomerism[1–10], its quantum chemical study in the framework of the potential hypersurface concept[11–65], rigorous quantum-mechanical study[66–75], algebraic aspects of isomerism[76–100], and the relationships between isomerism and the theory of chemical reactivity[101–158].

REFERENCES

(1) Mislow, K., and Siegel, J.: 1984, J. Am. Chem. Soc. *106*, pp. 3319—3328.
(2) Fang, H. L., and Swofford, R. L.: 1984, Chem. Phys. Lett. *105*, pp. 5—11.
(3) Glukhovcev, M. N., Simkin, B. Ja., and Minkin, V. I.: 1985, Usp. Khim. *54*, pp. 86—125.
(4) Wulff, G., Kemmerer, R., Vietmeier, J., and Poll, H.-G.: 1982, Nouv. J. Chim. *6*, pp. 681—687.
(5) Sundaralingam, M., and Rao, S. T.: 1983, Int. J. Quantum Chem., Quantum Biol. Symp. *10*, pp. 301—308.
(6) Dileila, D. P., Taylor, K. V., and Moskovits, M.: 1983, J. Phys. Chem. *87*, pp. 524—527.
(7) Schuster, P.: 1984, in Aspects of Chemical Evolution, Ed. G. Nicolis, J. Wiley & Sons, New York, p. 109.
(8) Mason, S. F.: 1983, Int. Rev. Phys. Chem. *3*, pp. 217—241.
(9) Mason, S.: 1983, Nature *305*, pp. 12—13.
(10) Hargittai, I.: 1983, J. Chem. Educ. *60*, p. 94.
(11) Mills, I. M.: 1984, J. Phys. Chem. *88*, pp. 532—536.
(12) Fischer, G.: 1984, Vibronic Coupling: The Interaction between the Electronic and Nuclear Motions, Academic Press, London.
(13) Makarewicz, J., and Wierzbicki, A.: 1984, Chem. Phys. Lett. *108*, pp. 155—160.
(14) Wagner, M.: 1985, J. Chem. Phys. *82*, 3207—3213.
(15) Davidson, E. R., and Borden, W. T.: 1983, J. Phys. Chem. *87*, pp. 4783—4790.
(16) Bersuker, I. B.: 1984, The Jahn-Teller Effect and Vibronic Interactions in Modern Chemistry, Plenum Press, New York.

(17) BERSUKER, I. B.: 1984, The Jahn-Teller Effect. A Bibliographic Review, Plenum Press, New York.
(18) CLARY, D. C., and CONNOR, J. N. L.: 1983, Chem. Phys. Lett. *94*, pp. 81–84.
(19) MOISEYEV, N.: 1984, Chem. Phys. Lett. *106*, pp. 354–355.
(20) MANZ, J., MEYER, R., POLLAK, E., RÖMELT, J., and SCHOR, H. H. R.: 1984, Chem. Phys. *83*, pp. 333–343.
(21) MANZ, J., MEYER, R., and SCHOR, H. H. R.: 1984, J. Chem. Phys. *80*, pp. 1562–1568.
(22) PULAY, P.: 1983, J. Chem. Phys. *78*, pp. 5043–5051.
(23) VINCENT, M. A., SAXE, P., and SCHAEFER III, H. F.: 1983, Chem. Phys. Lett. *94*, pp. 351–354.
(24) JØRGENSEN, P., and SIMONS, J.: 1983, J. Chem. Phys. *79*, pp. 334–357.
(25) PULAY, P.: 1983, Int. J. Quantum Chem., Quantum Chem. Symp. *17*, pp. 257–263.
(26) FOX, D. J., OSAMURA, Y., HOFFMANN, M. R., GAW, J. F., FITZGERALD, G., YAMAGUCHI, Y., and SCHAEFER III, H. F.: 1983, Chem. Phys. Lett. *102*, pp. 17–19.
(27) PULAY, P.: 1983, J. Mol. Struct. *103*, pp. 57–66.
(28) OSAMURA, Y., YAMAGUCHI, Y., SAXE, P., FOX, D. J., VINCENT, M. A., SCHAEFER III, H. F.: 1983, J. Mol. Struct. *103*, pp. 183–196.
(29) HOFFMANN, M. R., FOX, D. J., GAW, J. F., OSAMURA, Y., YAMAGUCGI, Y., GREV, R. S., FITZGERALD, G., SCHAEFER III, H. F., KNOWLES, P. J., and HANDY, N. C.: 1984, J. Chem. Phys. *80*, pp. 2660–2668.
(30) SIMONS, J., JØRGENSEN, P., and HELGAKER, T. U.: (1984), Chem. Phys. *86*, pp. 413–432.
(31) PAGE, M., SAXE, P., ADAMS, G. F., and LENGSFIELD III, B. H.: 1984, J. Chem. Phys. *81*, pp. 434–439.
(32) KERTESZ, M.: 1984, Chem. Phys. Lett. *106*, pp. 443–446.
(33) GAW, J. F., YAMAGUCHI, Y., and SCHAEFER III, H. F.: 1984, J. Chem. Phys. *81*, pp. 6395 to 6396.
(34) SELLERS, H.: 1985, Chem. Phys. Lett. *116*, pp. 153–154.
(35) BÁLINT, I., and BÁN, M. I.: 1983, Theor. Chim. Acta *63*, pp. 255–268.
(36) BELL, S., and CRIGHTON, J. S.: 1984, J. Chem. Phys. *80*, pp. 2464–2475.
(37) BANERJEE, A., ADAMS, N., SIMONS, J., and SHEPARD, R.: 1985, J. Phys. Chem. *89*, pp. 52–57.
(38) BASILEVSKY, M. V.: 1983, Pure Appl. Chem. *55*, pp. 207–212.
(39) PAGE, M., SAXE, P., ADAMS, G. F., and LENGSFIELD III, B. H.: 1984, Chem. Phys. Lett. *104*, pp. 587–590.
(40) SCHLEGEL, H. B.: 1984, Theor. Chim. Acta *66*, pp. 333–340.
(41) DASHEVSKII, V. G., and RABINOVICH, A. L.: 1983, Dokl. Akad. Nauk SSSR *273*, pp. 375 to 379.
(42) BUNKER, P. R.: 1983, Ann. Rev. Phys. Chem. *34*, pp. 59–75.
(43) SMEYERS, Y. G., and BELLIDO, M. N.: 1983, Int. J. Quantum Chem. *22*, pp. 507–516.
(44) SILVERMAN, J. N.: 1984, Int. J. Quantum Chem. *25*, pp. 915–928.
(45) RACKOVSKY, S., and SCHERAGA, H. A.: 1984, Acc. Chem. Res. *17*, pp. 209–214.
(46) MEZEY, P. G.: 1984, Canad. J. Chem. *62*, pp. 1356–1357.
(47) KATAOKA, M., and NAKAJIMA, T.: 1984, Theor. Chim. Acta *66*, pp. 121–131.
(48) ŌSAWA, E., and MUSSO, H.: 1982, Top. Stereochem. *13*, pp. 117–193
(49) ŌSAWA, E., and MUSSO, H.: 1983, Angew. Chem., Int. Ed. Engl. *22*, pp. 1–12.
(50) RASMUSSEN, K.: 1985, Potential Energy Functions in Conformational Analysis, Springer-Verlag, Berlin.
(51) BASILEVSKY, M. V.: 1983, J. Mol. Struct. *103*, pp. 139–152.
(52) YAMASHITA, K., and YAMABE, T.: 1983, Int. J. Quantum Chem., Quantum Chem. Symp. *17*, pp. 177–189.

(53) McKelvey, J. M., and Hamilton Jr., J. F.: 1984, J. Chem. Phys. *80*, pp. 579—580.
(54) Simons, J., Jørgensen, P., Taylor, H., and Ozment, J.: 1983, J. Phys. Chem. *87*, pp. 2745—2753.
(55) Miller, W. H.: 1983, J. Phys. Chem. *87*, pp. 3811—3819.
(56) Quapp, W., and Heidrich, D.: 1984, Theor. Chim. Acta *66*, pp. 245—260.
(57) Davidson, E. R., and McMurchie, L. E.: 1982, in Excited States, Ed. E. C. Lim, Academic Press, New York, pp. 1—39.
(58) Beyer, A., Karpfen, A., and Schuster, P.: 1984, Top. Curr. Chem. *120*, pp. 1—40.
(59) Camp, R. N., King, H. F., McIver Jr., J. W., and Mullaly, D.: 1983, J. Chem. Phys. *79*, pp. 1088—1089.
(60) Nauts, A., and Chapuisat, X.: 1983, Chem. Phys. *76*, pp. 349—366.
(61) Strauss, H. L.: 1983, Ann. Rev. Phys. Chem. *34*, pp. 301—328.
(62) Fredkin, D. R., Komornicki, A., White, S. R., and Wilson, K. R.: 1983, J. Chem. Phys. *78*, pp. 7077—7092.
(63) Botschwina, P.: 1983, Chem. Phys. *81*, pp. 73—85.
(64) Simons, J., and Jørgensen, P.: 1984, Int. J. Quantum Chem. *25*, pp. 1135—1150.
(65) Murrell, J. N., Carter, S., Farantos, S. C., Huxley, P., and Varandas, A. J. C.: 1984, Molecular Potential Energy Functions, J. Wiley & Sons, New York.
(66) Pfeifer, P.: 1983, in Energy Storage and Redistribution in Molecules, Plenum Press, New York, pp. 315—326.
(67) Woolley, R. G.: 1982, Structure and Bonding *52*, pp. 1—35.
(68) Haase, D.: 1984, Theor. Chim. Acta *64*, pp. 421—430.
(69) Farantos, S. C., and Tennyson, J.: 1985, J. Chem. Phys. *82*, pp. 800—809.
(70) 1985, Chaotic Behaviour in Quantum Systems. Theory and Applications, Ed. G. Casati, Plenum Press, New York.
(71) Mason, S. F., and Tranter, G. E.: 1983, J. Chem. Soc., Chem. Commun., pp. 117—119.
(72) Harris, R. A., and Silbey, R.: 1983, J. Chem. Phys. *78*, pp. 7330—7333.
(73) Tennakone, K.: 1984, Chem. Phys. Lett. *105*, pp. 444—446.
(74) Hameka, H. F., and De La Vega, J. R.: 1984, J. Am. Chem. Soc. *106*, pp. 7703—7705.
(75) Tranter, G. E.: 1985, Chem. Phys. Lett. *115*, pp. 286—290.
(76) Kvasnička, V., Kratochvíl, M., and Koča, J.: 1983, Collect. Czech. Chem. Commun. *48*, pp. 2284—2304.
(77) McLarnan, T. J.: 1983, Theor. Chim. Acta *63*, pp. 195—207.
(78) Hässelbarth, W.: 1984, Theor. Chim. Acta *66*, pp. 91—110.
(79) Dias, J. R.: 1984, J. Chem. Info. Comput. Sci. *24*, pp. 124—135.
(80) Flurry Jr., R. L.: 1984, J. Chem. Educ. *61*, pp. 663—665.
(81) Ruch, E., and Klein, D. J.: 1983, Theor. Chim. Acta *63*, pp. 447—472.
(82) Mezey, P. G.: 1983, Theor. Chim. Acta *63*, pp. 9—33.
(83) Mezey, P. G.: 1983, J. Mol. Struct. *103*, pp. 81—99.
(84) Mezey, P. G.: 1983, Int. J. Quantum Chem., Quantum Biol. Symp. *10*, pp. 153—160.
(85) Shevchenko, S. M.: 1983, Teor. Eksp. Khim. *19*, pp. 672—678.
(86) Sinanoğlu, O.: 1984, Chem. Phys. Lett. *103*, pp. 315—327.
(87) King, R. B.: 1983, Theor. Chim. Acta *63*, pp. 323—338.
(88) King, R. B.: 1984, Theor. Chim. Acta *64*, pp. 439—452.
(89) Bonchev, D., and Mekenyan, O.: 1984, J. Chem. Soc., Faraday Trans. 2 *80*, pp. 695—712.
(90) Hass, E. C., and Plath, P. J.: 1985, J. Mol. Catal. *29*, pp. 181—200.
(91) Randić, M.: 1983, J. Comput. Chem. *4*, pp. 73—83.
(92) Randić, M., and Davis, M. I.: 1984, Int. J. Quantum Chem. *26*, pp. 69—89.
(93) Serre, J.: 1984, Int. J. Quantum Chem. *26*, pp. 593—605.
(94) Balasubramanian, K.: 1984, Theor. Chim. Acta *65*, pp. 49—58.

(95) 1983, Symmetries and Properties of Non-rigid Molecules. A Comprehensive Survey, Ed. J. Maruani and J. Serre, Elsevier, Amsterdam.

(96) TRINAJSTIĆ, N.: 1983, Chemical Graph Theory, Vols I and II, CRC Press, Boca Raton.

(97) 1983, Chemical Applications of Topology and Graph Theory, Ed. R. B. King, Elsevier, Amsterdam.

(98) KNOP, J. V., MÜLLER, W. R., SZYMANSKI, K., and TRINAJSTIĆ, N.: 1985, Computer Generation of Certain Classes of Molecules, Union of Chemists and Technologists of Croatia, Zagreb.

(99) Comp. & Maths. with Appls. Symmetry Special Issue, Ed. I. Hargittai, in press.

(100) Applications of Mathematical Concepts to Chemistry, Ed. N. Trinajstić, E. Horwood Ltd., Chichester, in press.

(101) ISAACSON, A. D., and TRUHLAR, D. G.: 1984, J. Chem. Phys. *80*, pp. 2888—2896.

(102) BERNSTEIN, L. S., and WORMHOUDT, J.: 1984, J. Chem. Phys. *80*, pp. 4630—4639.

(103) MEZEY, P. G.: 1984, Int. J. Quantum Chem. *25*, pp. 853—861.

(104) TRUHLAR, D. G., HASE, W. L., and HYNES, J. T.: 1983, J. Phys. Chem. *87*, pp. 2664—2682.

(105) PELIKÁN, P., and LIŠKA, M.: 1984, Collect. Czech. Chem. Commun. *49*, pp. 2837—2856.

(106) COLWELL, S. M.: 1984, Mol. Phys. *51*, pp. 1217—1233.

(107) PERIĆ, M., MLADENOVIĆ, M., PEYERIMHOFF, S. D., and BUENKER, R. J.: 1984, Chem. Phys. *86*, pp. 85—103.

(108) URBAN, J., KLIMO, V., and TIŇO, J.: 1984, Collect. Czech. Chem. Commun. *49*, pp. 1440 to 1447.

(109) LAGANÀ, A., HERNANDEZ, M. L., and ALVARIÑO, J. M.: 1984, Chem. Phys. Lett. *106*, pp. 41—47.

(110) ZAWADZKI, A. G., and HYNES, J. T.: 1985, Chem. Phys. Lett. *113*, pp. 476—482.

(111) SAID, M., and MALRIEU, J.-P.: 1983, Chem. Phys. Lett. *102*, pp. 312—316.

(112) LADANYI, B. M., and EVANS, G. T.: 1983, J. Chem. Phys. *79*, pp. 944—952.

(113) LEROY, G.: 1985, Advan. Quantum Chem. *17*, pp. 1—95.

(114) JORGENSEN, W. L.: 1983, J. Phys. Chem. *87*, pp. 5304—5314.

(115) BLANDAMER, M. J., BURGESS, J., ROBERTSON, R. E., and SCOTT, J. M. W.: 1982, Chem. Rev. *82*, pp. 259—286.

(116) MORITA, A.: 1984, J. Phys. Chem. *88*, pp. 1678—1680.

(117) ŌKI, M.: 1984, Acc. Chem. Res. *17*, pp. 154—159.

(118) BÖHM, S., and KUTHAN, J.: 1982, Collect. Czech. Chem. Commun. *47*, pp. 2735—2745.

(119) BRUNO, A. E., STEER, R. P., and MEZEY, P. G.: 1983, J. Comput. Chem. *4*, 104—109.

(120) MURTO, J., RÄSÄNEN, M., ASPIALA, A., and HOMANEN, L.: 1983, J. Mol. Struct. *92*, pp. 45—56.

(121) SCHLEYER, P. v. R.: 1983, Pure Appl. Chem. *55*, pp. 355—362.

(122) PONEC, R.: 1983, Z. phys. Chem. (Leipzig) *264*, pp. 964—968.

(123) CHANDRASEKHAR, J., SCHLEYER, P. v. R., BAUMGÄRTNER, R. O. W., and REETZ, M. T.: 1983, J. Org. Chem. *48*, pp. 3453—3457.

(124) NEMUKHIN, A. V., DEMENTEV, A. I., KOLESNIKOV, A. I., STEPANOV, N. F., and POLYAKOV, V. I.: 1983, Teor. Eksp. Khim. *19*, pp. 715—719.

(125) ROOS, S. C., and BUNKER, P. R.: 1983, J. Mol. Spectrosc. *101*, pp. 199—211.

(126) HAMADA, Y., HASHIGUCHI, K., HIRAKAWA, A. Y., TSUBOI, M., NAKATA, M., TASUMI, M., KATO, S., and MOROKUMA, K.: 1983, J. Mol. Spectrosc. *102*, pp. 123—147.

(127) SEN, K. D., OHTA, K., and MOROKUMA, K.: 1984, J. Mol. Struct. *109*, pp. 287—292.

(128) NAKAMURA, S., and DEDIEU, A.: 1984, Theor. Chim. Acta *64*, pp. 461—467.

(129) KAO, J., and SEEMAN, J. I.: 1984, J. Comput. Chem. *5*, pp. 200—206.

(130) GUTMAN, I., GRAOVAC, A., and POLANSKY, O. E.: 1985, Chem. Phys. Lett. *116*, pp. 206 to 209.

(131) Nachbar Jr., R. B., Hounshell, W. D., Naman, V. A., Wennerström, O., Guenzi, A., and Mislow, K.: 1983, J. Org. Chem. *48*, pp. 1227—1232.
(132) Ivanov, P. M., and Ōsawa, E.: 1984, J. Comput. Chem. *5*, pp. 307—313.
(133) Favini, G., Moro, G., Todeschini, R., and Simonetta, M.: 1984, J. Comput. Chem. *5*, pp. 343—348.
(134) Frisch, M. J., Raghavachari, K., Pople, J. A., Bouma, W. J., and Radom, L.: 1983, Chem. Phys. *75*, pp. 323—329.
(135) Huang, M. B., Goscinski, O., Jonsäll, G., and Ahlberg, P.: 1984, J. Chem. Soc., Perkin Trans. II, pp. 1327—1330.
(136) Radom, L., Bouma, W. J., Nobes, R. H., and Yates, B. F.: 1984, Pure Appl. Chem. *56*, pp. 1831—1842.
(137) Ha, T.-K., and Nguyen, M. T.: 1984, J. Phys. Chem. *88*, pp. 4295—4298.
(138) Koch, W., and Schwarz, H.: 1985, Chem. Phys. Lett. *113*, pp. 145—150.
(139) Natanson, G., Amar, F., and Berry, R. S.: 1983, J. Chem. Phys. *78*, pp. 399—408.
(140) Doxtader, M. M., Gulis, I. M., Schwartz, S. A., and Topp, M. R.: 1984, Chem. Phys. Lett. *112*, pp. 483—490.
(141) Van de Wall, B. W.: 1983, J. Chem. Phys. *79*, pp. 3948—3961.
(142) Carrington Jr., T., Hubbard, L. M., Schaefer III, H. F., and Miller, W. H.: 1984, J. Chem. Phys. *80*, pp. 4347—4354.
(143) Kellö, V., Urban, M., and Boldyrev, A. I.: 1984, Chem. Phys. Lett. *106*, pp. 455—459.
(144) Osamura, Y., Kato, S., Morokuma, K., Feller, D., Davidson, E. R., and Borden, W. T.: 1984, J. Am. Chem. Soc. *106*, pp. 3362—3363.
(145) Slanina, Z.: 1984, Chem. Phys. Lett. *105*, pp. 531—534.
(146) Slanina, Z.: 1984, Thermochim. Acta *78*, pp. 47—54.
(147) Slanina, Z.: 1984, Kinam, Rev. Fis. *6*, pp. 115—137.
(148) Minkin, V. I., Olekhovich, L. P., and Zhdanov, Yu. A.: 1977, Molecular Design of Tautomeric Systems, Publishing House of Rostov University, Rostov on Don, (in Russian).
(149) Nabo, G. Ya., Roganov, G. N., Frenkal, M. L.: Thermodynamics and Equilibria of Isomers, Publishing House of Belorussian University, Minsk, in press, (in Russian).
(150) Minkin, V. I., Simkin, V. Ya., and Minyaev, R. M.: Quantum Chemistry of Organic Compounds. Reaction Mechanisms, Khimiya, Moscow, in press, (in Russian).
(151) Ugi, I., Dugundji, J., Kopp, R., and Marquarding, D.: 1984, Perspectives in Theoretical Stereochemistry, Springer-Verlag, Berlin.
(152) Natanson, G. A.: 1985, Advan. Chem. Phys. *58*, pp. 55—126.
(153) Gribov, L. A., and Prokofeva, N. I.: 1985, Zh. Strukt. Khim. *26*, pp. 49—54.
(154) Alberty, R. A.: 1983, J. Phys. Chem. *87*, pp. 4999—5002.
(155) Alberty, R. A.: 1985, J. Phys. Chem. *89*, pp. 880—883.
(156) Andrews, L., and Johnson, G. L.: 1983, J. Chem. Phys. *79*, pp. 3670—3677.
(157) Mark, T. D., and Castleman, A. W.: 1985, Advan. Atom. Mol. Phys. *20*, pp. 65—172.
(158) Curtiss, L. A., Pochatko, D. J., Reed, A. E., and Weinhold, F.: 1985, J. Chem. Phys. *82*, pp. 2679—2687.

SUBJECT INDEX

The size of this index was *a priori* limited to about 5 typed pages; consequently only important terms have been included.

A

Adiabatic approximation 30—35, 63
Adjacency matrix 117
Adsorption-complex isomerism 224—227
AHMOS 117
Algebraic structures 2
Alloisomerism 14
Ammonium cyanate 8, 9
Anharmonicity 40, 42, 75, 76, 165—167
Anionotropy 10, 11
Antipodes 14, 16
Artificial minima 53
Asymmetrical carbon atom 9
Asymmetry
— left-handed 3
— of biomolecules 4
— right-handed 3
Asymptotic enumeration formulae 127, 132
Atomic hypothesis 111
Atomic integrals 46, 48, 75
Atropisomers 15, 16
Attractors 96
Autoisomerization 36, 106, 219, 220
Automerizations 111
Axiomatic methods 2

B

Barriers
— flexion 19
— tension 19
— torsion 19
Basins 97
BEBO method 64
Be-matrices 114, 115
Berry mechanism 118
Binary relation 112
Bioisomerism 3
Biomolecules 4, 41, 132
Born-Oppenheimer approximation 29—35, 41, 60, 63, 70, 76, 77, 91, 94, 95, 99, 100, 108, 113, 140, 159, 243
Bullvalene 21
Burnside's lemma 148

C

Cartesian product 112
Catchment region 137
Cc-matrix 114
CFF method 41, 46, 47
Charge distribution 95, 96
Chemical distance 115
Chemical hysteresis 61
Chemical metric 115, 116
Chirality 13, 16, 17, 21, 23
— functions 21
Chiral physical field 4
CICLOPS 117
Clamped nuclei 31
Clusters 35, 37, 221—224
— isomerism of 213—221
CNPE 117
Combinatorial entropy 130, 131
Computer-assisted synthesis design 111, 116—118
Configuration 10, 13, 22
Configuration interaction 45, 48, 49, 53, 70, 73
— nuclear 75
Conformation 10, 13, 44
Conformers 14, 29, 44
Coordinate domains 137
— excluded 137
Coordinates
— Cartesian 70—73
— internal 31, 70—72
— mass weighted 58, 60, 70
— redundant 71

Coordinates symmetry 70, 71
Cope rearrangement 204–212
Correlation diagram 37, 38, 141
Coulomb interactions 29, 33
Critical points 46, 96, 97, 138
Curvature 36
Cycle index 119, 121, 123, 124, 129
Cyclohexane, conformational isomers of 14, 15, 20, 129, 202

D

Desmotropy 10
Density matrix method 108
Diastereoisomers 14, 15
DIM method 64
Dissociation energies 33
Distinguishability of isomers 22–24
Double coset formalism 148
Double minimum potential 102–108, 166

E

Electron correlation 46
Elementary rate processes 34
Ellipsoid
– elongated rotational 3
– flattened rotational 3
Enantiomers 14, 15, 127
Enantiomorphism
– in microorganisms 3
– in tobacco 3, 4
Energy barrier 14, 19, 20, 23, 29, 36, 59, 60, 175, 176, 198, 203
Energy gradient 47–52
Equilibrium hypothesis 159, 172, 198
Equilibrium processes 5, 177–181
– with isomerism of components 181–184, 192–195
Equivalence class 112, 119
Equivalence relation 20, 112, 113
EROS 117
Euclidian norm of gradient 51
Expectation value of position 105
External fields 29

F

Family of isomeric ensembles of molecules 113, 116
Finite difference method 52
Fixed nuclei 30
Fluxional molecules 21–23, 36, 40, 41, 121
Force constant matrix 47, 52, 56, 71–74
Force constants 29, 47, 50, 52, 64, 75
Force field 44
Framework group 143
Fulminic acid 8

G

Gaussian orbitals 45, 48
Gear-shaped molecules 17
General equilibrium problem 184–190
– microscopic formulation 190, 191
Generalized isomerism 112–116, 117, 148, 244
Generating functions 118–124
Generator coordinate method 33, 98
Global minimum 50, 53
Gradient methods 47
Graph-like state 129–132

H

Hamiltonian walks 131
Harmonic oscillator 38, 70
Hartree-Fock limit 45, 49
Heats of formation 46
Hellmann-Feynman theorem 32, 49, 72
Hessian matrix 47, 97, 137, 138
Hilbert space 95, 101
Homosynthetic compounds 8
Homotopy 121

I

Indeterminacy of energy 106
– of position 105
Intensities of vibrational spectra 74, 75
Internal rotation 14, 44, 166, 167
– free 131, 168–170
Intersection of hypersurfaces 34
Inversion equilibrium problem 195–196
Inversion motion 39, 168
Isomer enumeration 6, 111, 118–129, 132, 243
– iterative 125–127, 132
– of non-rigid molecules 127–129
Isomeric ensembles of molecules 113–116
Isomerism
– atomic nucleus 3, 4
– biological 3, 4
– bond 11
– chain 10, 19
– *cis-trans* 14, 15, 127, 199–201
– classification of 10–21
– constitutional 10

Isomerism coordination 11, 13
— definition of 9
— distortion 18
— dynamic 10
— functional group 10, 19
— geometrical 13, 14, 15, 19
— hydrate 11, 13
— inversion 14, 19, 20
— ionic 11
— ionization 13
— ligand 11, 13
— linkage 11, 13
— nuclear spin 19, 20
— optical 13, 14, 19
— orientation 19, 20
— permutational 20, 22
— phase 17
— place 10, 19
— positional 121
— rotational 14, 19, 20
— skeletal 10
— spatial 13
— structural 10, 11, 13, 19, 127, 128
— substituent 10
— *syn-anti* 15
— topological 16, 17
— torsional 14
— valence 5, 11

J

Jahn-Teller effect 18, 34, 35, 144, 147

K

Kinematix matrix 70
Kryptomerism 10

L

Lagrangian equations 35, 56, 57, 70
Lagrangian multipliers 48, 192
Large vibrational amplitude 3, 21
Legendre polynomials 67
LEP method 135
LEPS method 63
LHASA 117
Linear internal coordinate path 52
Liquid state, cluster concept of 222—224
Localized Hamiltonians 101
Logical structure 2, 111, 116
Longuet-Higgins groups 128, 141—143, 147

M

Mapping parameter 61
Mass polarization 31
Mesomers 9, 20
Metamerism 11
Metastable states 3
Microclusters 220—221
Minimal basis set 45, 46
Molecular dynamics methods 224, 244
Molecular energy 44—46, 53
Molecular propellers 16
Molecular structure 23, 29, 31, 35—44, 53, 108, 138
— criticism of 91—100
Molecular wave function 101—108
Monte Carlo simulation 67, 223, 224, 244
Murrell-Laidler theorem 144—147

N

Newton-Raphson method 50
Non-adiabatic approximation 31—35, 98
Non-adiabatic reactions 137
Nonclassical ions 18, 40
Non-crossing rule 38
Non-rigidity parameter 37—39
Non-rigid molecules 21—23, 36, 40, 41, 43, 70, 128, 129
— symmetry groups of 141—144
Normal coordinates 58
Nuclear motion 31
Nuclear spin isomers 19, 20
Nucleus deformation parameter 3

O

Open shells 48
Optical activity
— inorganic 14
— organic 14
— oscillations in 108
Optical isomers 4, 13, 14, 19, 107, 127, 173, 174
Optically active compounds, reactions of 5, 173, 174
Ordered sets 2
Orthonormal set 30, 48

P

Pairwise harmonic forces 37
Parallel activated-complex isomerism 196—202
Partial geometry optimization 54
Partition functions 160—172

Partition RRHO approximation of 163—172, 216
Path of steepest descent 55, 57, 60, 61, 146
Permutation groups 118—124, 127, 129
Perturbation expansion 33
Plasticity 18
Point-by-point definition 54
Pólya's theorem 118—125, 126—128, 132, 148
Polymerism 8
Polymorphism 19, 20
Polynomial roots 68
Polytopic molecules 41
Potential energy hypersurface 5, 29—35, 39, 44, 46, 62—69, 132, 133, 136—138, 140, 144, 243
— asymptotic behaviour of 63, 75
Potential field 31
Principle of least motion 61
Probability distributions 93, 94, 99, 100
Property hypersurfaces 69
Prototropy 10, 11
Pseudo-Jahn-Teller effect 18, 147, 168
Pseudomerism 10
Pseudomonotropy 20
Pseudorotation 14, 21, 167, 168

Q

Quantum topology 95—97
Quasi-free particles 37

R

Rate characteristics 172—177
— quantum-chemical evaluation of 177—181
Rate processes 5, 172—177
Reaction coordinates 54—62
— intrinsic 55—62, 73
— meta-intrinsic 59
— topological 115
Reaction ergodography 59
Reaction matrix 114, 115
Reaction networks 116
Reaction operator 115
Reaction paths 51, 54—62, 148
— minimum energy 60, 61, 137
— — stability of 61
Reaction pathway, shortest 116
Reflexivity 112
Renner effect 34
Residual isomers 23
Resonance 10, 11
Rigid rotator 38
RKR method 65
Rotamers 14, 44, 194, 195
Rotational invariance 46

S

SECS 117
Secular equations 36
Separability of isomers 23, 24, 36, 44
Separation of motions 29, 31, 33—35, 38, 71, 161
Sequential activated-complex isomerism 202—213
Set mapping 2
Set theory 2
Shoulder 47
Silver cyanate 8, 9
Silver fulminate 8, 9
Simplex method 47, 61
Site-symmetry groups 143
Skeleton 10, 20
Slater orbitals 45, 46, 48
Spherical symmetry 65
Stationary points 46, 111, 132—134, 136, 137
— bounds for the number of 138—140
— symmetry properties of 140—148
Stationary state functions 102
Steepest descent method 50
Stepwise structuration 98
Stereoisomerism 4, 10, 13, 19, 20, 114, 127
Stoichiomers 13
Structural fluctuations 41
Structure-activity relationships 111
Structure, average 42—44
Structure, equilibrium 37, 42
Structure, substitution 43
Symmetricity 112
Symmetry numbers 173—175
SYNCHEM 117
Synchronous transit paths 61
Synchronous transit technique 51

T

Target molecule 116
Tautomerism 10—13, 19, 21
— dyadic 10, 11
— ring-chain 10, 11
— triadic 10, 11
— valence 11
Theory of absolute reaction rates 62, 172—177

Thermodynamic characteristics 159—172
— quantum-chemical evaluation of 177—181
— summary and partial 184—190
— — limiting properties of 191, 192
Thermodynamic equilibrium 113
Time-averaged values 22, 61
Time factor 1, 22, 23
Time scales 22, 23
Topological analyses of hypersurfaces 133—140
Topological codes 117
Topological entropy 130
Topological indices 111, 118
Topological spaces 2
Totally symmetric representation 50
Trace algebra 5
Trajectory, classical 36, 54, 56, 62, 176, 177, 197, 198
Transition-state symmetry 144—148
Transition vector 146
Transitivity 112
Translational continuum 31
Tree type graph 130, 133, 134, 137
Trial and error method 47
Tunnelling 36, 40, 102—108, 175, 176, 200, 201, 243
— times 102
Turnstile mechanism 118

U

Uncertainty principle 113
Unified statistical theory 212, 227
Unitary group 38
Urea 8, 9

V

van der Waals molecules 35, 41, 65—67, 171
— isomerism of 213—215
Variable metric methods 49—51
Vibrational analysis 6, 70—75
Vibronic coupling 34
Virial coefficient, second 65—67, 221, 222
— third 65, 221

W

Walsh diagrams 40
Water dimer 66, 219, 220
Weak interactions 4, 107, 108
Weighted averages 39
Weight factors 186—188, 190, 191
Woodward-Hoffmann rules 5, 11, 141, 148
Wreath product method, generalized 128, 129

X

X method 51

Y

Yields of reactions 117

Z

Zero-point motion 91, 172